Graduate Texts in Contemporary Physics

Springer
New York
Berlin
Heidelberg
Barcelona
Hong Kong
London
Milan
Paris
Singapore
Tokyo

Graduate Texts in Contemporary Physics

S.T. Ali, J.P. Antoine, and J.P. Gazeau: **Coherent States, Wavelets and Their Generalizations**

A. Auerbach: **Interacting Electrons and Quantum Magnetism**

B. Felsager: **Geometry, Particles, and Fields**

P. Di Francesco, P. Mathieu, and D. Sénéchal: **Conformal Field Theories**

A. Gonis and W.H. Butler: **Multiple Scattering in Solids**

J.H. Hinken: **Superconductor Electronics: Fundamentals and Microwave Applications**

J. Hladik: **Spinors in Physics**

Yu.M. Ivanchenko and A.A. Lisyansky: **Physics of Critical Fluctuations**

M. Kaku: **Introduction to Superstrings and M-Theory, 2nd Edition**

M. Kaku: **Strings, Conformal Fields, and M-Theory, 2nd Edition**

H.V. Klapdor (ed.): **Neutrinos**

J.W. Lynn (ed.): **High-Temperature Superconductivity**

H.J. Metcalf and P. van der Straten: **Laser Cooling and Trapping**

R.N. Mohapatra: **Unification and Supersymmetry: The Frontiers of Quark-Lepton Physics, 2nd Edition**

H. Oberhummer: **Nuclei in the Cosmos**

G.D.J. Phillies: **Elementary Lectures in Statistical Mechanics**

R.E. Prange and S.M. Girvin (eds.): **The Quantum Hall Effect**

B.M. Smirnov: **Clusters and Small Particles: In Gases and Plasmas**

M. Stone: **The Physics of Quantum Fields**

(continued following index)

Antonios Gonis
William H. Butler

Multiple Scattering in Solids

With 44 Illustrations

Springer

Antonios Gonis
Chemistry and Materials Science Group
Lawrence Livermore Laboratory
Livermore, CA 94551
USA

William H. Butler
Metal and Ceramics Division
Oak Ridge National Laboratory
Oak Ridge, TN 37831
USA

Series Editors
R. Stephen Berry
Department of Chemistry
University of Chicago
Chicago, IL 60637
USA

Joseph L. Birman
Department of Physics
City College of CUNY
New York, NY 10031
USA

Jeffrey W. Lynn
Department of Physics
University of Maryland
College Park, MD 20742
USA

Mark P. Silverman
Department of Physics
Trinity College
Hartford, CT 06106
USA

H. Eugene Stanley
Center for Polymer Studies
Physics Department
Boston University
Boston, MA 02215
USA

Mikhail Voloshin
Theoretical Physics Institute
Tate Laboratory of Physics
University of Minnesota
Minneapolis, MN 55455
USA

Library of Congress Cataloging-in-Publication Data
Gonis, Antonios, 1945–
Multiple scattering in solids / Antonios Gonis, William H. Butler.
p. cm. — (Graduate texts in contemporary physics)
Includes bibliographical references and index.
ISBN 0-387-98853-X (alk. paper)
1. Multiple scattering (Physics) 2. Energy-band theory of solids.
I. Butler, W.H. (William H.), 1943– . II. Title. III. Series.
QC173.4.M85G66 1999
530.4′16—dc21 99-14736

Printed on acid-free paper.

Production managed by Steven Pisano; manufacturing supervised by Joe Quatela.
Photocomposed pages prepared from the authors' LaTeX files.
Printed and bound by R.R. Donnelley and Sons, Harrisonburg, VA.
Printed in the United States of America.

9 8 7 6 5 4 3 2 1

ISBN 0-387-98853-X Springer-Verlag New York Berlin Heidelberg SPIN 10726373

Preface

The origins of multiple scattering theory (MST) can be traced back to Lord Rayleigh's publication of a paper treating the electrical resistivity of an array of spheres, which appeared more than a century ago. At its most basic, MST provides a technique for solving a linear partial differential equation defined over a region of space by dividing space into nonoverlapping subregions, solving the differential equation for each of these subregions separately and then assembling these partial solutions into a global physical solution that is smooth and continuous over the entire region. This approach has given rise to a large and growing list of applications both in classical and quantum physics. Presently, the method is being applied to the study of membranes and colloids, to acoustics, to electromagnetics, and to the solution of the quantum-mechanical wave equation. It is with this latter application, in particular, with the solution of the Schrödinger and the Dirac equations, that this book is primarily concerned. We will also demonstrate that it provides a convenient technique for solving the Poisson equation in solid materials. These differential equations are important in modern calculations of the electronic structure of solids.

The application of MST to calculate the electronic structure of solid materials, which originated with Korringa's famous paper of 1947, provided an efficient technique for solving the one-electron Schrödinger equation. Later, Kohn and Rostoker derived the same equations from a different approach and the MST technique, when applied to periodic solids, came to be known as the Korringa-Kohn-Rostoker, or KKR, method of band theory. Originally it was thought that MST/KKR theory was limited to potentials of the muffin-tin (MT) form, that is, potentials that are spherically symmetric and confined within nonoverlapping spheres. However, after many years of uncertainty, it has become clear that this restriction is not necessary and that MST can be applied more generally. Because the resolution of this issue has proved difficult and controversial, it will be discussed at some length.

There are several unique features of MST that make it particularly attractive for electronic structure calculations. First, it is extremely flexible. No other technique can be applied so easily to ordered and disordered solids, to surfaces, and to molecules and clusters. Second, it offers the capability for direct calculation of the Green function. Knowledge of the Green function is critical for treating perturbations such as impurities or applied fields, and for improving on the one-electron approximation. Finally, its efficiency in terms of basis set size helps to make calculations on large and complicated systems tractable.

This book is an attempt both to collect in a single volume and to present in a coherent manner the fundamental underpinnings of multiple scattering theory. Our intention is to make the theory accessible both to those embarking on a career in the field of electronic structure, as well as to more seasoned scientists who may wish to gain a working knowledge of the method. Thus, most of the formalism is developed within the concrete and computationally expedient angular momentum representation. Explicit formulae are derived for the secular equation, the wave function, and the Green function of a solid in restricted cases, such as that of spherical potentials, and for the general case of space-filling potential cells.

We have also aimed to make this volume a useful reference. Occasionally, important arguments and equations are repeated in order to relieve the reader from searching for them through previous pages. Material judged to be too technical and possibly not of immediate interest to a reader has been designated by an asterisk and can be skipped in a first reading.

The process of performing a MST calculation is the same for the case of general nonoverlapping potentials as for muffin-tins. It involves the following three stages:

1. Solve for the scattering characteristics, i.e., the scattering matrix or phase functions of individual scatterers. These solutions are discussed in Chapter 3.
2. Construct the structure constants of the underlying lattice.
3. Solve the secular equation of MST whose solutions yield the band structure, the wavefunctions, and the single-particle Green function. MST for both MT and space-filling potentials is discussed extensively in Chapters 4–7. For the sake of easy reference, the most important equations of MST are collected together in the final appendix.

In composing the material for a book on MST, we relied heavily on expositions of scattering theory itself and on research papers on the subject. Most importantly, we wish to acknowledge our indebtedness to many colleagues and collaborators who have contributed substantially to the development of MST, and in whose company our own thoughts were shaped and sharpened. Although a complete list would be too long to provide in this space, special thanks for many discussions and insights are due to J. S. Faulkner, B. L. Györffy, D. M. Nicholson, James M. MacLaren, X.-G. Zhang, G. M. Stocks, and Y. Wang. At the same time, it goes without saying that the responsibility for any remaining errors or omissions lies entirely with the authors. The staff at Springer-Verlag has worked hard to eliminate technical and stylistic discrepancies from the final product. Our thanks to all who directly or indirectly have contributed to the completion of this project. Finally, we express our thanks to Jamie and Dena for their patience during the months that were required to write this book.

Antonios Gonis
William H. Butler

Contents

1

Introduction

1.1 Basic Characteristics of MST

Multiple scattering theory was initiated in 1892 with the publication by Lord Rayleigh [1] of a paper on the propagation of heat or electricity through inhomogeneous media. At the time of its inception, nothing foretold its rise to a robust formalism that encompasses large parts of both classical and quantum physics. As an example of the power of the method, consider the following property of multiple scattering theory: The solution of the single-particle Schrödinger equation for any number of nonoverlapping potentials V_1, V_2, ...,V_n, can be obtained entirely in terms of the solutions for each of the potentials separately. This is an important strength of multiple scattering theory, because it allows one to treat a large, complicated system by treating smaller parts of it separately and then assembling the solutions for the pieces into an overall solution.

Since its inception, multiple scattering theory has evolved within the contexts of both classical and quantum physics. Within the former, MST has been applied to the study of problems ranging from the scattering of sound and electromagnetic waves [2, 3, 4], to the dielectric and elastic properties of composite materials [5]. Within a quantum-mechanical context, MST has been used in the study of a number of physical phenomena including low-energy-electron diffraction (LEED) spectra [6]; point defects and substitutionally disordered alloys [7]; surfaces, interfaces, grain boundaries and other two-dimensional extended defects in ordered or disordered materials [8, 9]; transport phenomena [10]; and photoemission and other

Equation	$\mathcal{D}\psi = [\phi]$
Schrödinger	$(-\nabla^2 + V - E)\psi = 0$
Dirac	$(-i\alpha \cdot \nabla + \beta + V - W)\psi = 0$
Laplace	$\nabla \cdot \epsilon \nabla \psi = 0$
Poisson	$\nabla^2 \psi = -4\pi\rho$
Debye-Huckel	$(-\nabla^2 + \lambda^2)\psi = 0$
Vector Wave	$\nabla \times [\nabla \times \mathbf{E}] - \kappa^2 \epsilon(\mathbf{r})\mathbf{E} = 0$

TABLE 1.1. Multiple scattering theory can be applied to most of the linear partial differential equations of mathematical physics.

spectroscopies [11, 12]. In many of these applications, the fundamental idea of MST, namely, the calculation of the scattering properties of a complex system in terms of the corresponding properties of its constituent parts turns out to be both computationally efficient and conceptually advantageous. In fact, MST allows a unified treatment of many of the linear partial differential equations (PDEs) of modern physics and engineering. Table 1.1 summarizes some prominent partial differential equations that can be treated within MST.

Of course, MST is not the only, nor necessarily the most efficient, method for solving these problems. Finite element techniques are probably the most commonly used general approach for solving linear PDEs [13]. Within the context of electronic structure theory, methods such as that of augmented plane waves (APW) [14] and certain of its more recent refinements [15, 16], of linear muffin-tin orbitals [LMTO] [17], of orthogonalized plane waves (OPW) [18, 19], and of pseudopotentials [20], to mention just a handful, have been used successfully to calculate the electronic structure of materials and related properties. However, MST is the approach that is applicable to the widest spectrum of physical studies within essentially the same mathematical framework, providing, in principle, an exact treatment of the single-particle Schrödinger (and the Dirac) equation for a single particle in an external field. Furthermore, many of the other methods for the calculation of the electronic structure can be shown to be closely related to MST [21, 22].

Multiple scattering theory possesses particular advantages for applications to disordered systems because it leads directly to the single-particle Green function. The Green function is usually required for discussing disordered systems because the statistical average of the Green functions of a statistical ensemble can be used to calculate the average physical properties of the system, whereas the ensemble average of the wave function cannot easily be related to physical observables.

Finally, as a theory that can be developed on the basis of a scattering formalism, MST is closer to experimental reality than other techniques when the physical properties under consideration can be viewed in terms of a scattering experiment. Applications such as LEED, photoemission, and XAFS come to mind in this regard.

1.2 Electronic Structure Calculations

Our primary concern in this book is the formulation of multiple scattering theory in a manner that facilitates its application to the calculation of the electronic structure of solid materials. Therefore, much emphasis is placed on presenting MST within an angular-momentum representation. Indeed, MST has been used for this purpose and in this form ever since the pioneering paper by Korringa [23] that showed that MST can be used to derive a secular equation that determines the eigenvalues and eigenvectors associated with the electronic states of a translationally invariant system.

Seven years later, Kohn and Rostoker [24] derived the same equation within a variational formalism. Subsequently, Ham and Segal [25] derived expressions for the so-called structure constants of a lattice and provided formulae for their calculation. The method that has evolved out of these works has come to be known as the Korringa-Kohn-Rostoker (KKR), or the Green-function method because it leads in a natural way to the Green function of the material under consideration. A characteristic feature of the KKR method, as originally presented and applied, is that it affords a complete separation of the potential aspects of a material, embodied in the cell scattering matrices (transition matrices or t-matrices), from the structural aspects, represented by the structure constants of the underlying lattice, which describe electron propagation between lattice points. In addition to being attractive in a conceptual sense, this separation leads to a very efficient computational technique for the calculation of the band structure and the eigenfunctions (Bloch functions) of a translationally invariant solid.

Applications of the KKR method were first made within the so-called muffin-tin (MT) approximation for the potential. In this approximation, it is assumed that the potential is confined within nonoverlapping spheres and that, in addition, it is spherically symmetric. Although spherical symmetry is not formally necessary, the restriction that the scattering cells be far enough apart so that their bounding spheres do not overlap was thought to be necessary for the validity of the theory. In this regard, it is of some interest to note that the original paper by Korringa [23] is formulated in a manner that makes clear the author's intention to apply the theory to space-filling potential cells. However, most, if not all, applications of the KKR method were carried out within the MT approximation. In such applications, one spherically averages the potential inside a sphere inscribed within the Wigner-Seitz (WS) cell of a lattice (the muffin-tin sphere), while setting the potential outside all MT spheres, in the *interstitial region*, to a constant value, often chosen to be equal to zero (the MT zero). Even though the approximate nature of the MT construction has been fairly well appreciated from the beginning, the calculations of the electronic structure of metals and many of their alloys that were carried out within MST in this approximation were so successful [26] that the use of the theory in this mode has continued until the present. It is only when the desire and necessity arose to treat systems with low rotational symmetry, e.g., covalently bonded materials, surfaces, grain boundaries, and various kinds of impurities, or to calculate quantities whose determination demanded both the use of a Green-function formalism and of a precision higher than had been hitherto necessary, e.g., the calculation of lattice relaxation effects and the forces around and between impurities in metals [7] that the need for a theory able to account properly for the potential throughout the WS cell became pointedly evident. Chapter 6 is devoted to an extended exposition of recent formal advances that have allowed the application of MST to space filling potential cells.

1.3 The Aim of This Book

The goal of this book is to present a self-contained exposition of MST within the computationally convenient angular-momentum representation. The formalism is geared toward the calculation of the electronic structure of solid materials, and is written at a level that should make it accessible to a graduate student embarking on a career in this field. At the same time, some of the results are relatively new (e.g., that MST is valid for space-filling cells, and may be of interest to more seasoned scientists). Although many of these results are in the scientific literature, the authors hope that a coherent and holistic exposition of them will increase their accessibility and allow a better appreciation of their significance.

The book has been arranged in a way that should facilitate its use by a diverse readership. Those who wish to use it mostly as a manual of the most relevant results of MST, and their primary use, can bypass all sections and chapters marked with an asterisk. On the other hand, those who wish to delve deeper into the properties of MST are encouraged to read all of the sections.

References

[1] Lord Rayleigh, Philos. Mag. **34**, 481 (1892); *The Theory of Sound*, 2nd Ed. vol. 2, p. 328 (1892)(reprinted by Dover Publications, New York (1976)).

[2] V. K. Varadan, *Acoustic, Electromagnetic, and Elastic Wave Scattering-Focus on the T-Matrix Approach*, edited by V. K. Varadan and V. V. Varadan (Pergamon, New York, 1980), pp. 103-134.

[3] V. V. Varadan and V. K. Varadan, Phys. Rev. D **21**, 388 (1980).

[4] R. Bruce Thompson, James H. Rose, and S. Ahmed, *Applications of Multiple Scattering Theory to Materials Science*, edited by W. H. Butler, P. H. Dederichs, A. Gonis, and R. L. Weaver, Materials Science Society, symposium proceedings, Vol. 253 (1992), p. 135.

[5] Pedro Villasenor-Gonzalez, Cecilia Noguez, and Ruben G. Barrera, *Applications of Multiple Scattering Theory to Materials Science*, edited by W. H. Butler, P. H. Dederichs, A. Gonis, and R. L. Weaver, Materials Research Society, symposium proceedings, Vol. 253 (1992), p. 123.

[6] J. B. Pendry, *Low Energy Electron Diffraction* (Academic, New York, 1974).

[7] P. H. Dederichs, B. Drittler, and R. Zeller, *Applications of Multiple Scattering Theory to Materials Science*, edited by W. H. Butler, P. H. Dederichs, A. Gonis, and R. L. Weaver, Materials Science Society, symposium proceedings, Vol. 253 (1992), p. 185.

[8] J. S. Faulkner, in *Progress in Materials Science*, edited by J. W. Christian, P. Hassen, and T. B. Massalski (Pergamon, New York, 1982), Vols. 1 and 2, and references therein.

[9] See Part VI in *Applications of Multiple Scattering Theory to Materials Science*, edited by W. H. Butler, P. H. Dederichs, A. Gonis, and R. L. Weaver, Materials Science Society, symposium proceedings, Vol. 253 (1992), p. 355.

[10] W. H. Butler, Phys. Rev. B**31**, 3260 (1985).

[11] See Part VIII, in *Applications of Multiple Scattering Theory to Materials Science*, edited by W. H. Butler, P. H. Dederichs, A. Gonis, and R. L. Weaver, Materials Science Society, symposium proceedings, Vol. 253 (1992), p. 459.

[12] P. J. Durham J. Phys. F: Metal Phys. **11**, 2475 (1981).

[13] E. B. Becker, G. F. Carey, J. Tinsley Oden *Finite Elements*, Prentice-Hall, Englewood Cliffs, (1981).

[14] Terry Loucks, *Augmented Plane Wave Method* (Benjamin, New York, 1967).

[15] H. Krakauer, M. Posternak, and A. J. Freeman, Phys. Rev. B**19**, 1706 (1979).

[16] E. Wimmer, H. Krakauer, M. Weinert, and A. J. Freeman, Phys. Rev. B**24**, 864 (1981).

[17] H. L. Skriver, *The LMTO Method* (Springer-Verlag, Berlin 1984).

[18] C. Herring, Phys. Rev. **57**, 1169 (1940).

[19] Joseph Callaway, *Quantum Theory of the Solid State* (Academic Press, New York, 1974).

[20] W. A. Harrison, *Pseudopotentials in the Theory of Metals* (Benjamin, New York, 1966).

[21] G. J. Morgan, Proc. Phys. Soc. **89**, 365 (1966).

[22] The Green function embedding method proposed by J. Inglesfield is reviewed by F. Garciá-Moliner and V. R. Velasco, Phys. Rep. **200**, 85 (1991). See also the original articles, J. Inglesfield, J. Phys. C**10**, 4067 (1977); C**14**, 3795; Surf. Sci. **76**, 379 (1978).

[23] J. Korringa, Physica **13**, 392 (1947).

[24] W. Kohn and N. Rostoker, Phys. Rev. **94**, 1111 (1954).

[25] F. S. Ham and B. Segal, Phys. Rev. **124**, 1786 (1961).

[26] V. L. Morruzi, J. F. Janak, and A. R. W. Williams, *Calculated Electronic Properties of Metals* (Pergamom, New York, 1978).

2
Intuitive Approach to MST

2.1 Huygens' Principle and MST

The fundamental construction upon which multiple scattering theory is based can be understood through an application of the principle originally proposed by Huygens in 1678 [1] as governing the propagation of light waves. Given that Huygens had no hint of the electromagnetic nature of light – Maxwell's theory of electromagnetism appeared only after the lapse of a century – and did not even know whether light was a transverse or longitudinal wave, the principle he postulated has proved to be of great didactic as well as practical value. Huygens' theory of light propagation is based on a geometric construction – *Huygens' principle* – which allows one to determine the position of an advancing wavefront at time $t + \Delta t$, from its position at a previous time, t. This principle can be stated as follows: *Each point on an advancing wavefront acts as an independent source of (spherical) wavelets. The surfaces of tangency to these secondary wavelets determine successive positions of the wavefront at later times.* For isotropic media the wavelets are spherical, but they can assume more complex shapes in general, anisotropic media, i.e., when the index of refraction depends on direction. This apparently simple description and the geometric construction that derives from it are sufficient to explain such commonly occurring phenomena as refraction and diffraction of light.

Huygens' principle finds applications not only in the realm of classical optics, but throughout quantum physics as well. It is somewhat ironic that Huygens' principle is mathematically *wrong* for the case of classical optics but it is *correct* for the case of quantum mechanics [2]. The reason for this is that the classical wave equation involves the second derivative with respect to time, thus requiring the specification of *both* the amplitude and its time derivative before the subsequent position of the wavefront can be determined. However, the quantum wave equation is of the first order in time and Huygens' principle holds intact. The physical picture is slightly different from the optical case because the velocity of an electron wave depends on its frequency. This means that for electrons the wave fronts will not remain sharp but will disperse with the progress of time.

In this chapter, we are concerned with the quantum-mechanical applications of Huygens' principle, and in particular with its role in providing an understanding of the multiple scattering processes of electrons in matter. Our objective is to describe the quantum-mechanical analog of Huygens' principle and to use it to introduce the reader to the basic notions of multiple scattering theory. For the sake of clarity, we begin our discussion with a time-dependent description of a scattering process, whereas in our subsequent exposition of multiple scattering theory we follow a time-independent approach. The time-dependent picture seems to provide a more satisfactory framework for an informal development since it is fairly easy to "visualize" waves being scattered and propagating in time from point to point. The time-independent picture, on the other hand, is more convenient for developing the formalism of multiple scattering theory, particularly in connection with time-independent potentials such as those usually used to characterize materials. These pictures are, of course, related by a simple Fourier transformation which replaces the time variable by its conjugate one, the energy. More importantly, the various conclusions reached about multiple scattering processes are equally valid in both frameworks.

2.1.1 Informal Discussion: Point Scatterers

We begin with the time-dependent form of the single-particle Schrödinger equation,

$$\mathrm{i}\hbar\frac{\partial\Psi}{\partial t} = H\Psi, \tag{2.1}$$

which describes the wave function for a single particle moving in a perturbing potential, $V(\mathbf{r}, t)$, which depends on position and time. In the coordinate representation, the Hamiltonian operator, H, can be written in the form,

$$H = -\frac{\hbar^2}{2m}\nabla^2 + V(\mathbf{r}, t). \tag{2.2}$$

In the following we shall use atomic units in which lengths are measured in units of the Bohr radius ($a_0 = \frac{\hbar^2}{me^2}$) and energy in Rydbergs ($\frac{e^4 m}{2\hbar^2}$). This allows us to set $\hbar = 1$, and $2m = 1$ so that these quantities do not appear in the equations. In the absence of a perturbing potential, the Hamiltonian operator becomes

$$H_0 = -\nabla^2, \tag{2.3}$$

and the Schrödinger equation can be solved exactly in terms of the free-particle propagator or Green function, $G_0^+(\mathbf{r}-\mathbf{r}', t)$, which satisfies the equation

$$\left[\mathrm{i}\frac{\partial}{\partial t} - H_0\right] G_0^+(\mathbf{r}, t; \mathbf{r}', t') = \delta(\mathbf{r}-\mathbf{r}')\delta(t-t'). \tag{2.4}$$

Explicit expressions for $G_0^+(\mathbf{r}, t; \mathbf{r}', t')$ are given in later sections. For the moment, it suffices to note that the free-particle Green function connects the values of a wave function at the space-time point $(\mathbf{r}', t')$ to its value at $(\mathbf{r}, t)$. Hence the wave function at time t is determined by the wave function at time $t = 0$ according to

$$\Psi(\mathbf{r}, t) = \int \mathrm{d}\mathbf{r}' G_0^+(\mathbf{r}-\mathbf{r}', t)\Psi(\mathbf{r}', 0). \tag{2.5}$$

Note that every point in space where the wave function is nonzero at $t = 0$ serves as a source for producing the wave function at time t. This is quite analogous to Huygens' description of wave propagation in terms of wavelets emanating from points on an advancing wave. In fact, the Green function $G_0^+(\mathbf{r}, t)$ can be thought of as the wave function amplitude at time t if the wave function amplitude is a delta function at the origin at $t = 0$. However, we note that the Green function for the time-dependent Schrödinger equation differs from that which describes light propagation primarily because the velocity of an electron wave (described by the Schrödinger equation) depends strongly upon its energy whereas light waves in vacuum have a velocity that is energy independent. This lack of dispersion for light waves yields the familiar spherical wavelets in Huygens' principle, whereas the strong dispersion of particles described by the Schrödinger equation yields a more complicated Green function. Components of the wave that have different energies propagate with different speeds, and the Green function must reflect that fact.

As it stands, Eq.(2.5) describes the propagation of a wave in free space between two different points in space-time. Consider now how the time development of the wave function will be modified by the presence of a perturbing potential. For simplicity we assume that the perturbing potential is confined to a small region of space and that it can be approximated as a point-scatterer, whose strength will be denoted by $\hat{t}_1$. We also assume that the potential is independent of time. If there is only one such point-scatterer located at the position $\mathbf{r} = \mathbf{r}_1$, the time development of the wave function will be given by

$$\Psi(\mathbf{r}, t) = \int \mathrm{d}\mathbf{r}' G_0^+(\mathbf{r}-\mathbf{r}', t)\Psi(\mathbf{r}', 0)$$

$$+\int_0^t dt_1 \int d\mathbf{r}' G_0^+(\mathbf{r}-\mathbf{r}_1, t-t_1)\hat{t}_1 \times G_0^+(\mathbf{r}_1-\mathbf{r}', t_1)\Psi(\mathbf{r}', 0). \tag{2.6}$$

We can think, by analogy with Huygens' principle, of waves propagating either directly (first integral in the equation above) or by being scattered from the single point-scatterer.[1] If there are two point-scatterers located at $\mathbf{r}_1$ and $\mathbf{r}_2$, the waves can propagate freely, they can scatter from either scatterer, they can scatter from both scatterers (in either order) or they can bounce back and forth between the scatterers any number of times. Thus, the wave function for two scatterers will develop in time according to the expression

$$\begin{aligned}
\Psi(\mathbf{r},t) = & \int d\mathbf{r}' G_0^+(\mathbf{r}-\mathbf{r}', t)\Psi(\mathbf{r}', 0) \\
& + \int_0^t dt_1 \int d\mathbf{r}' G_0^+(\mathbf{r}-\mathbf{r}_1, t-t_1) \times \hat{t}_1 G_0^+(\mathbf{r}_1-\mathbf{r}', t_1)\Psi(\mathbf{r}', 0) \\
& + \int_0^t dt_1 \int d\mathbf{r}' G_0^+(\mathbf{r}-\mathbf{r}_1, t-t_1)\hat{t}_2 G_0^+(\mathbf{r}_1-\mathbf{r}', t_1)\Psi(\mathbf{r}', 0) \\
& + \int_0^t dt_2 \int_0^{t_2} dt_1 \int d\mathbf{r}' G_0^+(\mathbf{r}-\mathbf{r}_2, t-t_2) \\
& \times \hat{t}_2 G_0^+(\mathbf{r}_2-\mathbf{r}_1, t_2-t_1)\hat{t}_1 G_0^+(\mathbf{r}_1-\mathbf{r}', t_1)\Psi(\mathbf{r}', 0) \\
& + \int_0^t dt_1 \int_0^{t_1} dt_2 \int d\mathbf{r}' G_0^+(\mathbf{r}-\mathbf{r}_1, t-t_1) \\
& \times \hat{t}_1 G_0^+(\mathbf{r}_1-\mathbf{r}_2, t_1-t_2)\hat{t}_2 G_0^+(\mathbf{r}_2-\mathbf{r}', t_2)\Psi(\mathbf{r}', 0) \\
& + \int_0^t dt_1' \int_0^{t_1'} dt_2 \int_0^{t_2} dt_1 \int d\mathbf{r}' G_0^+(\mathbf{r}-\mathbf{r}_1, t-t_1')\hat{t}_1 \\
& \times G_0^+(\mathbf{r}_1-\mathbf{r}_2, t_1'-t_2)\hat{t}_2 G_0^+(\mathbf{r}_2-\mathbf{r}_1, t_2-t_1)\hat{t}_1 \\
& \times G_0^+(\mathbf{r}_1-\mathbf{r}', t_1)\Psi(\mathbf{r}', 0) + \cdots .
\end{aligned} \tag{2.7}$$

In the two equations above, $\hat{t}_1$ and $\hat{t}_2$ represent the so-called t-matrices, or scattering matrices, of the point scatterers. The relation between the potential and the t-matrix will be developed in the next section. For the present case of point scatterers, $\hat{t}_1$ and $\hat{t}_2$ are simply complex numbers that relate the scattered amplitude to the amplitude of the incident wave function. We emphasize the structure of the equations that describe the time development of the wave function in the presence of one or more scatterers. The final scattering amplitude is given by the sum of the amplitudes for all possible ways that the particle can get from time $t = 0$ to time t. For

[1] As will become clear in the following discussion, $\hat{t}_1$ describes the complete effect of scattering from a spatially bounded potential region, containing an infinite number of repeated scatterings of the wave by the potential.

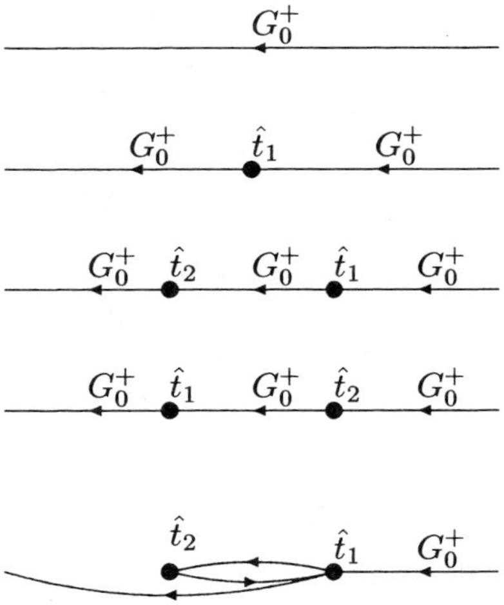

FIGURE 2.1. Some of the scattering processes that can occur with two scatterers.

two or more scatterers, the wave can scatter an arbitrary number of times from each scatterer (represented by a scattering matrix, $\hat{t}$), with only the restriction that it cannot scatter twice in succession from the same scatterer. This last restriction follows from the fact that a t-matrix describes fully the scattering from the corresponding potential. Figure 2.1 gives a pictorial representation of some of the processes contributing to Eq.(2.7).

It now follows that we can write the system wave function at time t in the presence of any number of scatterers in the form

$$\Psi(\mathbf{r}, t) = \int d\mathbf{r}' G^+(\mathbf{r} - \mathbf{r}', t)\Psi(\mathbf{r}', 0), \tag{2.8}$$

where $G^+(\mathbf{r} - \mathbf{r}', t)$ is the total Green function or propagator that describes propagation in the presence of the scatterers, and can be written in the form

$$\begin{aligned} G^+(\mathbf{r}, \mathbf{r}', t - t') &= G_0^+(\mathbf{r} - \mathbf{r}', t - t') \\ &+ \int_0^t dt_1 \int_0^{t_1} dt_2 \int d\mathbf{r}_1 \int d\mathbf{r}_2 G_0^+(\mathbf{r} - \mathbf{r}_1, t - t_1) \\ &\times T(\mathbf{r}_1, \mathbf{r}_2; t_1, t_2) G_0^+(\mathbf{r}_2 - \mathbf{r}', t_2 - t'). \end{aligned} \tag{2.9}$$

Here the total system t-matrix, T, is given by the expression

$$T = \sum_{i,j} T^{ij}, \tag{2.10}$$

where T^{ij} is the sum of all scattering sequences that start at scatterer i and end at scatterer j and which is given (in a notation that suppresses the integrations) by the *multiple scattering series*,

$$T^{ij} = \hat{t}^i \delta_{ij} + \hat{t}^i G_0^+ \hat{t}^j (1 - \delta_{ij}) + \sum_{k \neq j,i} \hat{t}^i G_0^+ \hat{t}^k G_0^+ \hat{t}^j + \cdots . \tag{2.11}$$

In the equation above, t^i represents one of the point scatterers (that we are using for didactic purposes), but it can equally well represent any *combination* of these scatterers. Thus, for example, we could collect together elementary point scatterers labeled 1, 7, 3, and 12 and call that combination the "blue" scatterer or scatterer "1." Equation (2.11) is equally valid for these combination scatterers. This is a useful point to keep in mind when we discuss multiple scattering theory for nonspherical full-cell potentials in subsequent chapters. Usually, the individual t-matrices will represent the scattering that arises from a definite, contiguous region of space, e.g. a sphere or a cell that surrounds an atom in a solid.

Equation (2.11) can be written in the form,

$$T^{ij} = \hat{t}^i \delta_{ij} + \hat{t}^i \sum_{k \neq i} G_0^+ T^{kj}, \tag{2.12}$$

as can easily be established by iteration. In interpreting the last equation, it should be remembered that the time and position arguments of the scattering matrices and propagators have been suppressed or, alternatively, the equation can be viewed as expressed in terms of abstract operators. In a coordinate representation, Eq.(2.12) becomes

$$\begin{aligned} T^{ij}(\mathbf{r}, \mathbf{r}'; t, t') &= \hat{t}^i(\mathbf{r}, \mathbf{r}'; t, t')\delta_{ij} \\ &+ \sum_{k \neq i} \int_{t'}^{t} \mathrm{d}t_1 \int_{t'}^{t_1} \mathrm{d}t_2 \int \mathrm{d}\mathbf{r}_1 \int \mathrm{d}\mathbf{r}_2 \hat{t}^i(\mathbf{r}, \mathbf{r}_1; t, t_1) \\ &\times G_0^+(\mathbf{r}_1 - \mathbf{r}_2; t_1 - t_2) T^{kj}(\mathbf{r}_2, \mathbf{r}'; t_2, t'). \end{aligned} \tag{2.13}$$

Much of this book is devoted to demonstrating that Eqs.(2.12) and (2.13) can be solved by first using a Fourier transform with respect to time to eliminate the temporal integrations and, second, by using angular-momentum expansions to convert the spatial integrals into sums over angular-momentum indices. This, of course, amounts to converting integral equations into matrix equations, which are much easier to handle computationally. The final result will be that the total system t-matrix, T, can be written formally as the inverse of a matrix, M, which can be put in the form

$$M^{ij} = (t^i)^{-1} \delta_{ij} - G_0^+ (1 - \delta_{ij}), \tag{2.14}$$

or in forms algebraically equivalent to it. Because the poles of the system t-matrix are poles of the system Green function and correspond to stationary states of the system, these states may be found by searching for the conditions under which M is singular, i.e., $\det M = 0$.

2.1.2 *Formal Presentation*

In this section, we provide a somewhat more detailed derivation of the fundamental equations of multiple scattering theory, in particular Eq.(2.12). In the absence of a perturbing potential, the Hamiltonian operator is given simply by Eq.(2.3),

$$H_0 = -\nabla^2, \tag{2.15}$$

and the stationary solutions of Eq.(2.1)

$$\left(\mathrm{i}\frac{\partial}{\partial t} - H_0\right)\chi(\mathbf{r},t) = 0, \tag{2.16}$$

corresponding to energy E, are given by the expression,

$$\chi(\mathbf{r},t) = (2\pi)^{-3/2}\exp(\mathrm{i}\mathbf{k}\cdot\mathbf{r})\exp(-\mathrm{i}Et), \tag{2.17}$$

where we have employed the usual δ-function normalization of plane waves.

We now return to the full time-dependent Schrödinger equation, Eq.(2.1),

$$\left[\mathrm{i}\frac{\partial}{\partial t} - H_0\right]\Psi(\mathbf{r},t) = V(\mathbf{r},t)\Psi(\mathbf{r},t). \tag{2.18}$$

Formally, the solution of this equation can be written in the form,

$$\Psi(\mathbf{r},t) = \chi(\mathbf{r},t) + \int G_0(\mathbf{r},t;\mathbf{r}',t')V(\mathbf{r}',t')\Psi(\mathbf{r}',t')\mathrm{d}^3r'\mathrm{d}t', \tag{2.19}$$

where the *time-dependent Green function* or *propagator*, $G_0(\mathbf{r},t;\mathbf{r}',t')$, satisfies Eq.(2.4), and $\chi(\mathbf{r},t)$ is a solution of the homogeneous, free-space ($V=0$), Schrödinger equation, Eq.(2.16). Equation (2.19) is often referred to as the Lippmann-Schwinger equation. As shown in Appendix A, in the coordinate representation G_0 takes the forms

$$G_0^+(\mathbf{r},t;\mathbf{r}',t') = -\mathrm{i}\left[\frac{1}{4\pi\mathrm{i}(t-t')^{3/2}}\right]\exp\left(\frac{\mathrm{i}|\mathbf{r}-\mathbf{r}'|^2}{4(t-t')}\right)\Theta(t-t'), \tag{2.20}$$

and

$$G_0^-(\mathbf{r},t;\mathbf{r}',t') = \mathrm{i}\left[\frac{1}{4\pi\mathrm{i}(t-t')^{3/2}}\right]\exp\left(\frac{\mathrm{i}|\mathbf{r}-\mathbf{r}'|^2}{4(t-t')}\right)\Theta(t'-t). \tag{2.21}$$

We note that G_0^+ vanishes for past times, $t < t'$, whereas G_0^- vanishes into the future, $t > t'$. The distinction between these two parts of the free-particle Green function is based on the physical concept of *causality.* In order to ensure *causal propagation*, i.e., that $\Psi(\mathbf{r},t)$ should be determined by values of $\Psi(\mathbf{r},t')$ at times $t' < t$ lying in the past, we must use the *causal* or *retarded* Green function, G_0^+. The other Green function, G_0^-, is called the *advanced* propagator. Even though it appears to be rather nonphysical, the advanced Green function plays an important role in quantum mechanics in general, and in scattering theory in particular. It is fairly straightforward to show by direct application that if G_0 satisfies Eq.(2.4), then the function $\Psi(\mathbf{r},t)$ given by Eq.(2.19) is a solution of Eq.(2.18).

Iteration of Eq.(2.19) leads to the expression,

$$\Psi(\mathbf{r},t) = \chi(\mathbf{r},t) + \int G_0^+(\mathbf{r},t;\mathbf{r}',t')T(\mathbf{r}',t';\mathbf{r}'',t'')\chi(\mathbf{r}'',t'')\mathrm{d}^3r'\mathrm{d}t'\mathrm{d}^3r''\mathrm{d}t'', \tag{2.22}$$

where the system t-matrix, $T(\mathbf{r}',t';\mathbf{r}'',t'')$, may be written (in a condensed notation in which the variables $\mathbf{r}_n, t_n$ are denoted simply by n) as

$$\begin{aligned} T(4,1) &= V(4)\delta(4-1) + V(4)G_0^+(4,1)V(1) \\ &\quad + \int V(4)G_0^+(4,5)V(5)G_0^+(5,1)V(1)\mathrm{d}\tau_5 + \cdots \\ &= V(4)\delta(4-1) + \int V(4)G_0^+(4,5)T(5,1)\mathrm{d}\tau_5, \end{aligned} \tag{2.23}$$

where $\mathrm{d}\tau_n = \mathrm{d}^3r_n\mathrm{d}t_n$.

In Eq.(2.23), the *transition*, *t-matrix*, or *scattering* matrix,[2] T, represents the effect of the entire potential viewed as a single unit. The multiple scattering expressions arise when we consider V to be a sum of potentials,

$$V = \sum_i V^i. \tag{2.24}$$

Upon substituting Eq.(2.24) into Eq.(2.23), using the definition,

$$t^i(\alpha,\beta) = V^i(\alpha)\delta(\alpha-\beta) + \int V^i(\alpha)G_0^+(\alpha,\gamma)t^i(\gamma,\beta)\mathrm{d}\tau_\gamma, \tag{2.25}$$

with a corresponding power-series like expansion as in Eq.(2.23) but in terms of individual potentials, V^i, and collecting terms, we can write,

$$\begin{aligned} T(\alpha,\beta) &= \sum_i t^i(\alpha,\beta) + \sum_i\sum_{j\neq i}\int\int t^i(\alpha,\gamma)G_0^+(\gamma,\delta) \\ &\quad \times t^j(\delta,\beta)\mathrm{d}\tau_\gamma\mathrm{d}\tau_\delta \\ &\quad + \sum_i\sum_{j\neq i}\sum_{k\neq j}\int\int\int\int t^i(\alpha,\gamma)G_0^+(\gamma,\delta)t^j(\delta,\epsilon) \\ &\quad \times G_0^+(\epsilon,\zeta)t^i(\zeta,\beta)\mathrm{d}\tau_\gamma\mathrm{d}\tau_\delta\mathrm{d}\tau_\epsilon\tau_\zeta + \cdots, \end{aligned} \tag{2.26}$$

where we have assumed that sums and integrals are interchangeable. In purely abstract form, we have

$$T = \sum_{i,j} T^{ij}$$

[2]The term "scattering matrix" is somewhat inappropriate when applied to this quantity because it is also commonly used for a slightly different quantity in scattering theory. However, we will follow long-standing practice by using the term to refer to t or T.

$$
\begin{aligned}
&= \sum_i t^i + \sum_i \sum_{j \neq i} t^i G_0^+ t^j \\
&+ \sum_i \sum_{j \neq i} \sum_{k \neq j} t^i G_0^+ t^j G_0^+ t^k + \cdots .
\end{aligned} \tag{2.27}
$$

The individual cell t-matrices, t^i, defined in Eq.(2.25) correspond to the "point" scatterers $\hat{t}$ that were introduced in the more intuitive discussion of the previous section. Note that the definition of t^i, Eq.(2.25), shows explicitly that it describes the complete scattering from the ith potential, V^i, since it includes an infinite number of scattering events from V^i. Therefore, no two successive scattering events can involve the same t-matrix, a condition that is expressed through the exclusions in the sums in the last equation. It is important to emphasize that Eq.(2.27) describes the total t-matrix as the sum of all possible scattering paths (events) that connect two points in space-time.

In addition to being explicitly time-dependent, Eq.(2.27) contains a further generalization when compared to Eq.(2.12). Because no restriction was placed on the spatial extent of the potentials V^i, the last multiple scattering expression, Eq.(2.27), is applicable to scattering potentials of arbitrary extent and shape. As long as the potentials do not overlap, the application of Huygens' principle made above leads to the powerful result that multiple scattering theory is valid in *all* cases in which a scattering region consists of a collection of arbitrarily shaped scattering centers. (If the potentials overlap, then it may no longer be legitimate to use the free-particle propagator to connect scattering events. Although it is possible to formulate a multiple scattering theory in a medium other than free space [3], we will concern ourselves primarily with the separable case.) This property is a simple manifestation of the fact that the various terms contributing to the final t-matrix of multiple scattering theory can be summed in any convenient order. Thus, it is possible to collect together all contributions arising from the interior of a given region or potential, V^i, to form the corresponding t-matrix, t^i. This allows one to express the t-matrix of an assembly of scatterers in terms of the individual t-matrices, rather than the potentials of the cells. This last feature is one of the most useful results of multiple scattering theory.

As a simple illustration of the freedom of combining scatterers, we note that T in Eq.(2.27), which corresponds to a collection of potential cells, can always be written in the form,

$$
T = \sum_{i,j} T^{ij}, \tag{2.28}
$$

where T^{ij} is the sum of all scattering sequences that start at scatterer i and end at scatterer j, irrespective of the manner in which individual cells are grouped into units. It is easy to show that the quantities T^{ij} satisfy the *equation of motion,*[3]

$$T^{ij} = t^i\delta_{ij} + t^i \sum_{k\neq i} G_0 T^{kj}, \tag{2.29}$$

which is formally identical to Eq.(2.12) that was derived for the case of point scatterers. The formal validity of this expression can be established upon iteration and replacement into Eq.(2.28), which immediately yields the original Eq.(2.27). Now, it follows from Eq.(2.29) that the T^{ij} are the matrix elements of the inverse of a matrix, $\underline{M}$, with matrix elements,

$$M^{ij} = m^i\delta_{ij} - G_0^{ij}(1-\delta_{ij}), \tag{2.30}$$

where $m^i = (t^i)^{-1}$, and the quantities G_0^{ij} are the off-diagonal elements of the free-particle propagator connecting the scattering centers i and j. We can now summarize our discussion up to this point as follows: *The total scattering matrix of an assembly of point scatterers, or of extended but nonoverlapping potentials is given by Eq.(2.28), with T^{ij} being the elements of the inverse of the matrix $\underline{M}$ defined in Eq.(2.30).*

2.2 Time-Independent Green Functions

Although the time-dependent picture of scattering is very convenient for developing an intuitive understanding of Huygens' principle and multiple scattering theory, it is unnecessarily complicated for the treatment of time-independent potentials, such as those commonly entering the determination of the electronic structure of materials. In this section, we show how, for time-independent potentials, the explicit presence of the time can be eliminated from the equations defining the Green function in favor of the variable conjugate to time, namely the energy. The results obtained here will be used in later chapters to show that the equations of multiple scattering theory take identical forms in the time-dependent and time-independent pictures (with appropriate interpretations of the various parameters).

To develop the time-independent formalism, we begin with Eq.(2.19). Using the explicit expression for the retarded free-particle Green function derived in Appendix A, Eq.(A.18), we obtain the expression,

$$\begin{aligned}\Psi(\mathbf{r},t) = \chi(\mathbf{r},t) - \mathrm{i}(2\pi)^{-3}\int \mathrm{d}^3r' \int_{-\infty}^{t} \mathrm{d}t'\mathrm{d}^3k' \exp\{\mathrm{i}\mathbf{k}'\cdot(\mathbf{r}-\mathbf{r}')\} \\ \times \exp\{-\mathrm{i}k'^2(t-t')\}\exp\{-\epsilon(t-t')\}V(\mathbf{r}',t')\Psi(\mathbf{r}',t'). \end{aligned}\tag{2.31}$$

[3] Here, and in the following we will often suppress the superscript "+" on the Green function in order to simplify the notation. Unless explicitly indicated otherwise, a Green function without a + or − superscript will be retarded.

We now restrict the potential to being time-independent, and look for stationary solutions of the form,

$$\Psi(\mathbf{r},t) = \psi_{\mathbf{k}_i}(\mathbf{r})\exp(-\mathrm{i}E_i t), \tag{2.32}$$

where $\mathbf{k}_i$ is the incident wave vector, and $E_i = k_i^2$. The free-particle solutions, $\chi(\mathbf{r},t)$, have the form,

$$\chi(\mathbf{r},t) = (2\pi)^{-3/2}\exp(\mathrm{i}\mathbf{k}_i\cdot\mathbf{r})\exp\{-\mathrm{i}E_i t\}, \tag{2.33}$$

so that Eq.(2.31) becomes

$$\begin{aligned}\psi_{\mathbf{k}_i}(\mathbf{r})\exp(-\mathrm{i}E_i t) &= (2\pi)^{-3/2}\exp(\mathrm{i}\mathbf{k}_i\cdot\mathbf{r})\exp\{-\mathrm{i}E_i t\}\\ &\quad - \mathrm{i}(2\pi)^{-3}\int \mathrm{d}^3r'\int_{-\infty}^{t}\mathrm{d}t'\mathrm{d}^3k'\exp\{\mathrm{i}\mathbf{k}'\cdot(\mathbf{r}-\mathbf{r}')\}\\ &\quad \times\exp\{-\mathrm{i}k'^2(t-t')\}\exp\{-\epsilon(t-t')\}\\ &\quad \times V(\mathbf{r})\psi_{\mathbf{k}_i}(\mathbf{r})\exp(-\mathrm{i}E_i t).\end{aligned} \tag{2.34}$$

The integral over t' can be evaluated using the result,[4]

$$\int_{-\infty}^{t}\exp\{\mathrm{i}(k'^2 - k_i{}^2 - \mathrm{i}\epsilon)t'\}\mathrm{d}t' = \frac{\exp\{\mathrm{i}(k'^2 - k_i{}^2 - \mathrm{i}\epsilon)t\}}{\mathrm{i}(k'^2 - k_i{}^2 - \mathrm{i}\epsilon)}, \tag{2.35}$$

and, after canceling the common factor $\exp(-\mathrm{i}E_i t)$, we can cast Eq.(2.34) into the form,

$$\begin{aligned}\psi_{\mathbf{k}_i}(\mathbf{r}) &= (2\pi)^{-3/2}\exp(\mathrm{i}\mathbf{k}_i\cdot\mathbf{r})\\ &\quad -(2\pi)^{-3}\lim_{\epsilon\to 0+}\int \mathrm{d}^3r'\int \mathrm{d}^3k'\frac{\exp[\mathrm{i}\mathbf{k}'\cdot(\mathbf{r}-\mathbf{r}')]}{k'^2 - k_i{}^2 - \mathrm{i}\epsilon}\\ &\quad \times V(\mathbf{r})\psi_{\mathbf{k}_i}(\mathbf{r}')\mathrm{d}^3r'.\end{aligned} \tag{2.36}$$

This last equation can also be rewritten as follows,

$$\psi_{\mathbf{k}_i}(\mathbf{r}) = (2\pi)^{-3/2}\exp(\mathrm{i}\mathbf{k}_i\cdot\mathbf{r}) + \int G_0^+(\mathbf{r},\mathbf{r}';E_i)V(\mathbf{r}')\psi_{\mathbf{k}_i}(\mathbf{r}')\mathrm{d}^3r', \tag{2.37}$$

where the time-independent but energy-dependent retarded propagator is defined by the expression

$$G_0^+(\mathbf{r},\mathbf{r}';E_i) = (2\pi)^{-3}\lim_{\epsilon\to 0}\int \mathrm{d}^3k'\frac{\exp[\mathrm{i}\mathbf{k}'\cdot(\mathbf{r}-\mathbf{r}')]}{k'^2 - E_i - \mathrm{i}\epsilon}. \tag{2.38}$$

As is shown in Appendix B, in the coordinate representation $G_0^+(\mathbf{r}-\mathbf{r}';E_i)$ takes the form

$$G_0^+(\mathbf{r}-\mathbf{r}';E_i) \equiv G_0^+(\mathbf{r}'-\mathbf{r};E_i) = -\frac{1}{4\pi}\frac{\mathrm{e}^{\mathrm{i}\sqrt{E_i}|\mathbf{r}-\mathbf{r}'|}}{|\mathbf{r}-\mathbf{r}'|}. \tag{2.39}$$

---

[4]There is no contribution from the lower limit because of the increasingly rapid oscillations as the limit is approached.

In this representation, the Green function satisfies the equation,

$$(\nabla^2 + E)G_0^+(\mathbf{r} - \mathbf{r}'; E) = \delta(\mathbf{r} - \mathbf{r}'). \tag{2.40}$$

Equation (2.37) is the well-known *Lippmann-Schwinger* equation of scattering theory. This equation will be made the formal basis for the development of multiple scattering theory in the following chapters. In much of our discussion, the explicit dependence of various quantities such as the Green function on energy will often be suppressed, but is to be understood throughout.

There are no fundamental difficulties connected with the application of multiple scattering theory, Eqs.(2.27)- (2.30), to arbitrarily shaped spatially extended scattering regions. This is an important point to keep in mind when we formulate multiple scattering theory in the angular-momentum representation. Controversies surrounding attempts to generalize multiple scattering theory have arisen invariably in that representation. In particular, the question arises as to whether the t-matrix suffices to represent the wave function inside the moon region of a cell, i.e., outside the cell but inside a sphere circumscribing the cell. Huygens' principle suggests that the t-matrix can determine the wave function *everywhere* outside a cell, including points in the moon region. Of course, there are technical obstacles to be overcome when one uses a spherical-wave basis to represent the scattering from nonspherical cells. However, these obstacles are of a purely *geometrical* rather than physical nature. In the following chapters, we show explicitly how these geometric difficulties can be resolved within the angular-momentum representation, leading to the conclusion that multiple scattering theory holds intact in all cases of nonoverlapping scattering centers, even in the case of contiguous space-filling cells.

References

[1] Christian Huygens published a treatise on light in 1690, but most of the material had been written as early as 1678. Similar ideas had been propounded by Robert Hooke as early as 1664. (See *A Short Account of the History of Mathematics*, W. W. Rouse Ball, 1908, republished by Dover Publications, New York, 1960.)

[2] R. P. Feynman, Rev. Mod. Phys. **20**, 267 (1948).

[3] P. J. Braspenning, *A Multiple Scattering Treatment of Dilute Metal Alloys*, Thesis, University of Amsterdam (1982), unpublished.

3
Single-Potential Scattering

In this chapter we present a partial-wave analysis of single-potential scattering, both for the case of spherically symmetric potentials and for potentials of arbitrary shape. For simplicity we consider only potentials that vanish identically outside the interior of a spatially bounded region. Thus, we assume the existence of a sphere bounding the cell, the *bounding sphere*, beyond which the potential is strictly equal to zero. Our objective in this chapter is to describe the regular solutions of the Schrödinger equation associated with such cells within a partial-wave formalism. For spherically symmetric potentials, such a description is well known from elementary treatments of quantum mechanics [1]. It is relatively straightforward to extend this treatment to nonspherical potentials. The corresponding solutions are important for the generalization of multiple scattering theory to arbitrarily shaped cells in subsequent chapters.

We are primarily concerned with the implications of using a basis of spherical waves to expand the solutions of the Schrödinger equation associated with nonspherical potentials. For the present, we shall assume, therefore, that these solutions exist and can be constructed in some representation. Questions of applicability of angular-momentum expansions to single-cell scattering, as well as to MST, arise even with respect to potentials that are constant throughout the cell. It is questions of this sort, concerning the effect of nonspherical terms in the potential, e.g., cell boundaries, in angular-momentum expansions, that are of interest to us in this chapter and in much of this book.

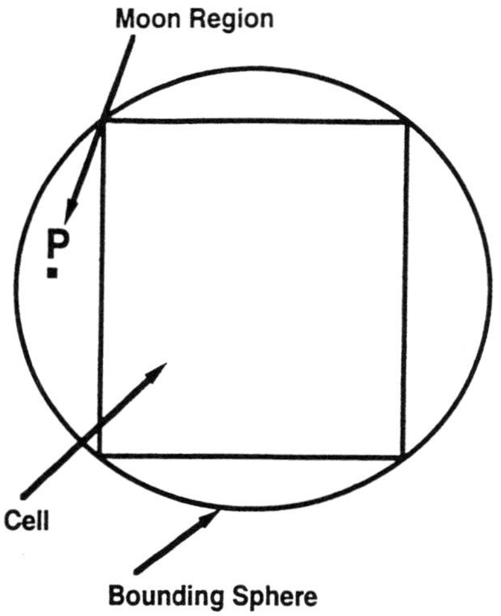

FIGURE 3.1. A cell with a circumscribing sphere and a point, P, in the moon region.

3.1 Partial-Wave Analysis of Single Potential Scattering

We begin with a summary of the partial-wave analysis of scattering from a spherically symmetric potential. In this part of the discussion we follow closely the exposition in Joachain [2]. Similar treatments can be found in a number of elementary texts on quantum mechanics [1, 3, 4]. Then, we proceed with a partial-wave analysis of the wave function for generally shaped, spatially bounded potentials, $V(\mathbf{r})$. We discuss the differential and integral equations satisfied by the wave function, and the form taken by the wave function in various regions of space, such as in the so-called "moon" regions, that is to say, the parts of free space that may lie inside a sphere circumscribing the cell (Figure.3.1) Also, we point out the many similarities that exist between the cases of central and noncentral force fields. With a proper definition of parameters, the formal generalization to nonspherical potentials can be said to require only the generalization from diagonal to nondiagonal matrices. At the same time, very important considerations arise in connection with expansions in angular-momentum eigenfunctions, and questions of convergence that such expansions ultimately entail. We shall attempt to address these as they are brought forth in the discussion, and illustrate them as much as possible by the results of numerical calculations.

3.2 General Considerations

We begin by considering the scattering of a nonrelativistic, spinless particle of mass m and momentum $\mathbf{p}_i$ interacting with a time-independent central potential, V($\mathbf{r}$). In order to simplify our notation, we describe the particle in terms of its wave vector, $\mathbf{k}_i = \mathbf{p}_i/\hbar$, and choose units such that $\hbar = 1$, and $2m = 1$. In this form, the equations also describe the motion of a wavefront of incident wave vector $\mathbf{k}_i$.

From Eq.(2.1), we know that the motion of the particle (wave) is governed by the time-dependent Schrödinger equation, Eq.(2.1), which in the present case we write in the form,

$$[-\nabla^2 + V(\mathbf{r})]\Psi(\mathbf{r},t) = \mathrm{i}\frac{\partial}{\partial t}\Psi(\mathbf{r},t). \tag{3.1}$$

We are interested in the stationary solutions of this equation of the type exhibited in Eq.(2.32), (with $\hbar^2/2m = 1$),

$$\Psi(\mathbf{r},t) = \psi_{\mathbf{k}_i}(\mathbf{r})\exp(-\mathrm{i}E_i t), \tag{3.2}$$

where $E = k_i{}^2$. Substitution of this expression into Eq.(3.1) yields the time-independent Schrödinger equation

$$[-\nabla^2 + V(\mathbf{r})]\psi_{\mathbf{k}_i}(\mathbf{r}) = E_i\psi_{\mathbf{k}_i}(\mathbf{r}). \tag{3.3}$$

For potentials of the type under consideration here (in general, for potentials that vanish faster than r^{-1} as $r \to \infty$), the wave function at regions outside the field of force, i.e., in the limit $r \to \infty$, can be written in the form [2],

$$\psi^{(+)}_{\mathbf{k}_i}(\mathbf{r}) \to A\left(\exp(\mathrm{i}\mathbf{k}_i\cdot\mathbf{r}) + f(\theta,\phi)\frac{\exp(\mathrm{i}kr)}{r}\right), \quad as \quad r \to \infty. \tag{3.4}$$

At large r, the wave function consists of the incident wave (free-particle solution), represented by the first term inside the parentheses on the right-hand side, and a scattered contribution (wave). Note that the solution of Eq.(3.1) involves the causal Green function, (Eq.(B.3)), and, correspondingly, we use a (+) superscript in the wave function. In the last equation, the constant A is independent of r, and the angles θ and ϕ of the outgoing (scattered) wave vector, $\mathbf{k}_s$, are measured in a coordinate system in which the z-axis coincides with the direction of the incident wave vector, $\mathbf{k}_i$. It is important to note that in general the particular solution given by Eq.(3.4) satisfies Eq.(3.3) only asymptotically. There will in general be terms of higher order in $1/r$ in the region in which the potential can be neglected.

The quantity $f(\theta,\phi) = f(\mathbf{k}_i,\mathbf{k}_s)$ in Eq.(3.4) is the so-called "scattering amplitude," a quantity of central importance in scattering theory in general, and in multiple scattering theory in particular. We now proceed with an evaluation of the explicit forms of the wave functions and the scattering amplitudes associated with spherically symmetric potentials. The generalization to nonspherical potentials is given in a subsequent section.

3.3 Spherically Symmetric Potentials

The solution of the Schrödinger equation, Eq.(3.3), is considerably simplified when the potential, $V(\mathbf{r})$, is spherically symmetric. We choose a right-handed coordinate system with the z-axis along the wave vector $\mathbf{k}_i$, and with the origin coinciding with that of the vector $\mathbf{r}$. In spherical polar coordinates, the Hamiltonian operator, Eq.(2.2), reads

$$H = -\left[\frac{1}{r^2}\frac{\partial}{\partial r}\left(r^2\frac{\partial}{\partial r}\right) + \frac{1}{r^2\sin\theta}\frac{\partial}{\partial\theta}\left(\sin\theta\frac{\partial}{\partial\theta}\right) + \frac{1}{r^2\sin^2\theta}\frac{\partial^2}{\partial\phi^2}\right] + V(r), \tag{3.5}$$

where the dependence of V only on the magnitude of $\mathbf{r}$ is explicitly indicated. Now, the Schrödinger equation, Eq.(3.1), for the stationary scattering wave function, $\psi^{(+)}_{\mathbf{k}_i}(\mathbf{r})$, can be written in the form,

$$\begin{aligned} -\left[\frac{1}{r^2}\frac{\partial}{\partial r}\left(r^2\frac{\partial}{\partial r}\right) + \frac{1}{r^2\sin\theta}\frac{\partial}{\partial\theta}\left(\sin\theta\frac{\partial}{\partial\theta}\right) + \frac{1}{r^2\sin^2\theta}\frac{\partial^2}{\partial\phi^2}\right]\psi^{(+)}_{\mathbf{k}_i}(\mathbf{r}) \\ + V(r)\psi^{(+)}_{\mathbf{k}_i}(\mathbf{r}) = E_i\psi^{(+)}_{\mathbf{k}_i}(\mathbf{r}). \end{aligned} \tag{3.6}$$

To solve Eq.(3.6) we make use of the observation that the square of the angular momentum,

$$\mathbf{L} = \mathbf{r} \times \mathbf{p}, \tag{3.7}$$

of a system with spherical symmetry, and its projection, L_z, along the z-axis are constants of the motion. This follows immediately from the fact that they satisfy the relations,

$$[H, L^2] = [H, L_z] = 0, \tag{3.8}$$

where $[A, B] = AB - BA$ denotes the commutator of two operators, A and B. Using the correspondence $\mathbf{p} \to -\mathrm{i}\nabla_{\mathbf{r}}$ in the definition of angular momentum, Eq.(3.7), and the expressions of the Cartesian coordinates, (x, y, z), in terms of polar coordinates, (r, θ, ϕ), one can show that

$$L^2 = -\left[\frac{1}{\sin\theta}\frac{\partial}{\partial\theta}\left(\sin\theta\frac{\partial}{\partial\theta}\right) + \frac{1}{\sin^2\theta}\frac{\partial^2}{\partial\phi^2}\right]. \tag{3.9}$$

Because L^2 commutes with its components, $[L^2, L_i] = 0, \quad i = x, y, z$, it can be diagonalized simultaneously with any one of these components by the same set of eigenfunctions. Choosing that component to lie along the z-axis, the corresponding eigenfunctions are the spherical harmonics [5],

$Y_{lm}(\theta, \phi)$, such that,

$$L^2 Y_{lm}(\theta, \phi) = l(l+1) Y_{lm}(\theta, \phi), \tag{3.10}$$

and

$$L_z Y_{lm}(\theta, \phi) = m Y_{lm}(\theta, \phi). \tag{3.11}$$

Here, l and m are called, respectively, the *orbital angular momentum quantum number*, and the *magnetic or azimuthal quantum number*. A brief description of the spherical harmonics and some of their properties is given in Appendix C.

Now, it follows from Eq.(3.9) that the Hamiltonian operator, Eq.(3.5), can be expressed in the form,

$$H = -\left[\frac{1}{r^2}\frac{\partial}{\partial r}\left(r^2 \frac{\partial}{\partial r}\right) - \frac{L^2}{r^2}\right] + V(r). \tag{3.12}$$

Because of the commutation relations, Eq.(3.8), we look for eigenfunctions that are common to H, L^2, and L_z. Thus, we expand the scattering wave function, $\psi^{(+)}_{\mathbf{k}_i}(\mathbf{r})$, in *partial waves* corresponding to given values of the quantum numbers l and m, in the form,

$$\begin{aligned} \psi^{(+)}(\mathbf{k}_i, \mathbf{r}) &= \sum_{l=0}^{\infty} \sum_{m=-l}^{l} c_{lm}(k) R_{lm}(k, r) Y_{lm}(\theta, \phi) \\ &= \sum_{L} c_L R_L(k, r) Y_L(\hat{\mathbf{r}}), \end{aligned} \tag{3.13}$$

where $k = \sqrt{E}$, and we have introduced the combined index, $L = (l, m)$. It is easily verified directly that because of azimuthal symmetry, the *radial wave functions*, $R_{lm}(k, r)$, are independent of the magnetic quantum number, m, and that each satisfies the *radial* Schrödinger equation,

$$-\left[\frac{1}{r^2}\frac{\mathrm{d}}{\mathrm{d}r}\left(r^2 \frac{\mathrm{d}}{\mathrm{d}r}\right) - \frac{l(l+1)}{r^2}\right] R_l(k, r) + V(r) R_l(k, r) = E R_l(k, r). \tag{3.14}$$

It is clear that expansions of the form given by Eq.(3.13) will satisfy the Schrödinger equation, Eq.(3.1), but not yet clear that they can be made to satisfy the boundary conditions that the wave function be regular at the origin and that it have the form of Eq.(3.4) at large r. That this is indeed the case is proved explicitly below.

It is convenient to introduce the functions,

$$u_l(k, r) = r R_l(k, r), \tag{3.15}$$

which can easily be shown [2] to satisfy the equation,

$$\left[\frac{\mathrm{d}^2}{\mathrm{d}r^2} + k^2 - \frac{l(l+1)}{r^2} - V(r)\right] u_l(k, r) = 0. \tag{3.16}$$

Two features of Eqs.(3.14) and (3.16) are worth emphasizing. First, they are both ordinary differential equations and, second, the functions R_l and u_l can be chosen to be real. This follows because the real and imaginary parts of the complex functions would separately satisfy the differential equations.

Equation (3.16) is particularly convenient for study. However, before considering the solutions of this equation, it is instructive to determine the solutions of the corresponding free-particle, i.e., $V = 0$, equation.

3.3.1 Free-Particle Solutions

Because the spherical harmonics are known functions, the determination of the scattering wave function in Eq.(3.13) requires the knowledge of the radial functions, R_l, and the coefficients, c_{lm}. In this section, we obtain these quantities for the case in which the potential, $V(r)$, vanishes identically. In this case, we obtain from Eq.(3.16) the radial equation for a free particle,

$$\left[\frac{\mathrm{d}^2}{\mathrm{d}r^2} + k^2 - \frac{l(l+1)}{r^2}\right] y_l(k,r) = 0. \tag{3.17}$$

Making a further change of variables to $\rho = kr$ and defining the function

$$f_l(\rho) = y_l/\rho, \tag{3.18}$$

we can rewrite Eq.(3.17) in the form,

$$\left[\frac{\mathrm{d}^2}{\mathrm{d}\rho^2} + \frac{2}{\rho}\frac{\mathrm{d}}{\mathrm{d}\rho^2} + \left(1 - \frac{l(l+1)}{\rho^2}\right)\right] f_l(\rho) = 0. \tag{3.19}$$

This is the well-known *spherical Bessel differential equation.* The independent solutions of this equation that are often used in scattering theory are the spherical Bessel functions, $j_l(\rho)$, the spherical Neumann functions, $n_l(\rho)$, and the spherical Hankel functions, $h_l^{(\pm)}(\rho)$, of the first $(+)$ and the second $(-)$ kind, respectively. The Hankel functions are given in terms of the Bessel and Neumann functions by the relation,

$$h_l^{(\pm)}(\rho) = j_l(\rho) \pm \mathrm{i}n_l(\rho). \tag{3.20}$$

(The definitions and some important properties of these functions are discussed in Appendix C.) As each of the pairs (j_l, n_l) and $(h_l^{(+)}, h_l^{(-)})$ contains two linearly independent solutions of Eq.(3.17), the general solution of that equation can be written in a number of forms, such as,

$$y_l(kr) = kr\left[C_l^{(1)}(k)j_l(kr) + C_l^{(2)}(k)n_l(kr)\right] \tag{3.21}$$

or

$$y_l(kr) = kr\left[D_l^{(1)}(k)h_l^{(+)}(kr) + D_l^{(2)}(k)h_l^{(-)}(kr)\right], \tag{3.22}$$

where the integration constants $(C_l^{(1)}(k)$,$C_l^{(2)}(k))$ and $(D_l^{(1)}(k)$,$D_l^{(2)}(k))$ may still depend on the energy, k^2.

As is shown in Appendix C, the Bessel function, j_l, vanishes as r^l as $r \to 0$ and it is said to be *regular* at the origin. All other functions, $n_l, h_l^{(\pm)}$, fail to vanish at the origin, in fact, they diverge there as $r^{-(l+1)}$, and are said to be *irregular* solutions of the spherical Bessel equation. It now follows that the regular solution, y_l, of Eq.(3.17) vanishes at the origin,

$$y_l(k,0) = krj_l(0) = 0, \tag{3.23}$$

behaving there as

$$y_l(k,r) \to r^{l+1}, \qquad \text{as } r \to 0. \tag{3.24}$$

We are now ready to obtain the coefficients c_{lm} that enter Eq.(3.13) for the case of free motion. This is done by comparing the partial-wave decomposition in Eq.(3.21) or Eq.(3.22) with that of the solutions of the corresponding wave equation in coordinate space. It can readily be verified that the Schrödinger equation for free motion,

$$-\nabla^2\psi(\mathbf{r}) = E\psi(\mathbf{r}), \tag{3.25}$$

has the plane-wave solutions, $\exp(\mathrm{i}\mathbf{k}_i \cdot \mathbf{r})$, which are also the eigenfunctions of the linear momentum operator. Because the eigenfunctions $j_l(kr)Y_{lm}(\hat{r})$ form a complete set, we may expand a plane wave in terms of them (Bauer's identity) to obtain

$$\exp(\mathrm{i}\mathbf{k}_i \cdot \mathbf{r}) = 4\pi \sum_{l=0}^{\infty} \sum_{m=-l}^{l} \mathrm{i}^l j_l(kr) Y_{lm}^*(\hat{\mathbf{k}}_i) Y_{lm}(\hat{\mathbf{r}}). \tag{3.26}$$

Clearly, in the case of free motion the radial wave functions, R_l, are the spherical Bessel functions, j_l. Further comparison of Eqs.(3.13) and (3.26) allows us to identify the coefficients c_{lm} for the case in which the wave vector $\mathbf{k}_i$ points in the direction of the $\hat{z}$ axis. We denote the coefficient for this special case by c_{lm}^0,

$$c_{lm}^0 = \mathrm{i}^l \left[4\pi(2l+1)\right]^{1/2} \delta_{m,0}. \tag{3.27}$$

Here, we have used the relations, $Y_{lm}(0,\phi) = 0$ if $m \neq 0$ and $Y_{l0}(0,\phi) = \sqrt{\frac{2l+1}{4\pi}}$.

This completes our discussion of the partial-wave analysis of free motion. We now turn to the solution of the radial equation, Eq.(3.14) or Eq.(3.16), in the presence of a potential.

3.3.2 The Radial Equation for Central Potentials

To obtain the solution of the radial equation in the presence of a potential, we examine the behavior of the wave function[1] in regions far removed

[1]We will use the term "wave function" generically to include the so-called basis functions, defined below. The difference between these terms is related to

from the range of force. (In the case of spatially bounded potentials this asymptotic behavior is achieved outside a sphere bounding the potential region.) Outside the range of the potential, the particle (or wave) executes essentially free motion and, according to our discussion above, we may write

$$u_l(kr) = kr\left[C_l^{(1)}(k)j_l(kr) + C_l^{(2)}(k)n_l(kr)\right], \tag{3.28}$$

for $r >> a$, where a provides a measure of the extent of the potential region (e.g., the radius of a bounding sphere.) That this representation of the wave function is reasonable can be justified on physical grounds, as argued above, or by means of rigorous arguments based on the method of variation of parameters for solving inhomogeneous second-order differential equations, as is done in Section 3.4 for the general case of nonspherical potentials.

Now, from the fact that for large x, (see Appendix C),

$$j_l(x) \to \frac{1}{x}\sin\left(x - \frac{l\pi}{2}\right) \tag{3.29}$$

$$n_l(x) \to -\frac{1}{x}\cos\left(x - \frac{l\pi}{2}\right) \tag{3.30}$$

$$h_l^{(+)}(x) \to -\mathrm{i}\frac{\exp\{\mathrm{i}(x - \frac{l\pi}{2})\}}{x}, \tag{3.31}$$

and

$$h_l^{(-)}(x) \to \mathrm{i}\frac{\exp\{-\mathrm{i}(x - \frac{l\pi}{2})\}}{x}, \tag{3.32}$$

we can write

$$u_l(k,r) \to A_l(k)\sin\left[kr - \frac{l\pi}{2} + \delta_l(k)\right], \tag{3.33}$$

with

$$A_l(k) = \left\{\left[C_l^{(1)}(k)\right]^2 + \left[C_l^{(2)}(k)\right]^2\right\}^{1/2}, \tag{3.34}$$

and

$$\tan\delta_l(k) = -\frac{C_l^{(2)}(k)}{C_l^{(1)}(k)}. \tag{3.35}$$

From Eq.(3.15) we may also write

$$R_l(k,r) \overset{r\to\infty}{\longrightarrow} A_l'(k)\left[j_l(kr) - \tan\delta_l n_l(kr)\right], \tag{3.36}$$

---

the boundary conditions that they satisfy. We shall use the term "basis functions" to describe solutions to the Schrödinger equation that may not satisfy all of the boundary conditions necessary for them to be physical wave functions.

where the constant $A'_l(k)$ is independent of r.

The quantity $\delta_l(k)$ introduced in Eq.(3.33) is called the *phase shift* for the lth partial wave and contains the effect of the interaction on the scattered wave. Note that in the absence of the interaction, the phase shifts vanish and the radial functions reduce to the spherical Bessel functions. The trigonometric functions $\cos\delta_l(k)$ and $\sin\delta_l(k)$ are called the *phase* functions, and they can be made the basis of a complete analysis of scattering by central fields of force [6]. These functions can be continued inside the range of the potential where they become r-dependent corresponding to the potential being abruptly truncated at a given value of r. We will encounter this use of the generalized phase functions in our discussion of nonspherical potentials in the next section.

Now, the boundary condition at large r can also be expressed in terms of radially incoming [$\exp(-ikr)$] and outgoing [$\exp(ikr)$] waves. Thus, equation (3.36) can also be written in the form,

$$u_l(k,r) \stackrel{r\to\infty}{\longrightarrow} A''_l(k)\left[-(-)^l \mathrm{e}^{-\mathrm{i}kr} + S_l \mathrm{e}^{\mathrm{i}kr}\right], \tag{3.37}$$

where

$$A''_l(k) = A_l \mathrm{i}^l \exp\{-\mathrm{i}\delta_l\}(-)^l/2\mathrm{i} \tag{3.38}$$

and

$$S_l(k) = \exp\{2\mathrm{i}\delta_l(k)\}. \tag{3.39}$$

The coefficient of the outgoing wave $S_l(k)$ is commonly referred to as an *S-matrix element.* [2]

3.3.3 The Scattering Amplitude

We are now in a position to derive an expression for the scattering amplitude. This is done by comparing the asymptotic expressions of the partial-wave expansion, Eq.(3.13), and of the scattering wave, Eq.(3.4). First, using Bauer's identity, Eq.(3.26), and Eq.(3.29) we can rewrite Eq.(3.4) in the form,

$$\begin{aligned}\psi^{(+)}_{\mathbf{k}_i}(k,r) \stackrel{r\to\infty}{\longrightarrow} A(k) & \left[\sum_{l=0}^{\infty}(2l+1)\mathrm{i}^l \frac{\sin(kr-\frac{l\pi}{2})}{kr} P_l(\cos\theta)\right. \\ & + \left. f(k,\theta,\phi)\frac{\mathrm{e}^{\mathrm{i}kr}}{r}\right].\end{aligned} \tag{3.40}$$

[2] We note that the matrix, S, defined here is the *scattering matrix* of formal scattering theory, with t, introduced previously being the *transition matrix.* We trust that the context will avoid confusion when we follow long-standing practice and refer to t as the scattering matrix.

Upon using the relationship between Legendre polynomials and spherical harmonics we also have,

$$\psi_{\mathbf{k}_i}^{(+)}(k,r) \stackrel{r\to\infty}{\longrightarrow} A(k)\left[\sum_{l=0}^{\infty}\sum_{m=-l}^{l}[4\pi(2l+1)]^{1/2}\mathrm{i}^l \right.$$
$$\times\ \frac{\exp\{\mathrm{i}(kr-\frac{l\pi}{2})\}-\exp\{-\mathrm{i}(kr-\frac{l\pi}{2})\}}{2\mathrm{i}kr}Y_{lm}(\hat{r})\delta_{m,0}$$
$$\left. +\ f(k,\theta,\phi)\frac{\mathrm{e}^{\mathrm{i}kr}}{r}\right]. \tag{3.41}$$

At the same time, we can consider the asymptotic behavior of the partial-wave expansion, Eq.(3.13), and use the relation in Eq.(3.15) to write

$$\psi_{\mathbf{k}_i}^{(+)}(k,r) \stackrel{r\to\infty}{\longrightarrow} \sum_{l=0}^{\infty}\sum_{m=-l}^{l} c_{lm}(k)A_l(k)\frac{1}{2\mathrm{i}r}\left[\exp\left\{\mathrm{i}\left(kr-\frac{l\pi}{2}+\delta_l\right)\right\}\right.$$
$$\left. -\ \exp\left\{-\mathrm{i}\left(kr-\frac{l\pi}{2}+\delta_l\right)\right\}\right]Y_{lm}(\hat{r}). \tag{3.42}$$

Upon comparing the coefficients of the incoming spherical waves in Eqs. (3.41) and (3.42) we find,

$$c_{lm}(k)=\frac{A(k)}{kA_l(k)}\left[4\pi(2l+1)\right]^{1/2}\mathrm{i}^l\exp\{\mathrm{i}\delta_l\}\delta_{m,0}. \tag{3.43}$$

Therefore, we may rewrite the partial wave expansion, Eq.(3.13), in the forms,

$$\psi_{\mathbf{k}_i}^{(+)}(k,r)=A(k)\sum_{l=0}^{\infty}\frac{(2l+1)}{kA_l(k)}\mathrm{i}^l\exp\{\mathrm{i}\delta_l\}R_l(k,r)P_l(\cos\theta), \tag{3.44}$$

or

$$\psi_{\mathbf{k}_i}^{(+)}(k,r)=A(k)\sum_{l=0}^{\infty}\frac{\sqrt{4\pi(2l+1)}}{kA_l(k)}\mathrm{i}^l\exp\{\mathrm{i}\delta_l\}R_l(k,r)Y_{l,0}(\theta). \tag{3.45}$$

Finally, by matching the coefficients of the outgoing spherical waves and using Eq.(3.43) we obtain the following expression for the scattering amplitude:

$$f(k,\theta)=\frac{1}{2\mathrm{i}k}\sum_{l=0}^{\infty}(2l+1)\left[\exp\{2\mathrm{i}\delta_l(k)-1\}\right]P_l(\cos\theta). \tag{3.46}$$

This can also be written in the form,

$$f(k,\theta)=\sum_{l=0}^{\infty}(2l+1)a_l(k)P_l(\cos\theta), \tag{3.47}$$

where the *partial wave amplitudes* $a_l(k)$ are given by the expression,

$$a_l(k) = \frac{1}{2ik}\left[\exp\{2i\delta_l(k) - 1\}\right] = \frac{1}{2ik}\left[S_l(k) - 1\right]. \tag{3.48}$$

It follows from this expression that for spherically symmetric potentials the scattering amplitudes are independent of the angles θ and ϕ and are determined completely from a knowledge of the phase shifts. In view of Eq.(3.39) the partial amplitudes $a_l(k)$ can also be written in the form,[3]

$$a_l(k) = \frac{1}{k}\exp\{i\delta_l(k)\}\sin\delta_l(k). \tag{3.49}$$

This form of the partial amplitudes is particularly convenient for use in connection with multiple scattering theory. Anticipating some of the following discussion, we note that the t-matrix, t_l, is given simply by the negative of the last two expressions, $t_l = -a_l$, which makes clear its connection to the S-matrix in the present case.

3.3.4 Normalization of the Scattering Wave Function

Up to this point, we have not considered explicitly the role played by the multiplicative constants, $A_l(k)$. Although these constants serve mainly to fix the normalization of the wave function, their specific choice may have important computational ramifications. This is an especially important feature in the case of nonspherical potentials that will be considered in the next section. Some commonly used normalizations and the corresponding forms of the scattering wave function are listed below. The function u_l may be chosen to be

$$\text{(i)}\quad u_l(k,r) \overset{r\to\infty}{\longrightarrow} \frac{1}{k}\sin\left(kr - \frac{l\pi}{2} + \delta_l\right), \tag{3.50}$$

which corresponds to

$$R_l(k,r) \overset{r\to\infty}{\longrightarrow} \frac{1}{kr}\sin\left(kr - \frac{l\pi}{2} + \delta_l\right). \tag{3.51}$$

This may also be written as

$$u_l(k,r) \overset{r\to\infty}{\longrightarrow} \frac{1}{k}\left[\sin\left(kr - \frac{l\pi}{2}\right)\cos\delta_l - \cos\left(kr - \frac{l\pi}{2}\right)\sin\delta_l\right], \tag{3.52}$$

corresponding to

$$R_l(k,r) \overset{r\to\infty}{\longrightarrow} \left[j_l(kr)\cos\delta_l - n_l(kr)\sin\delta_l\right]. \tag{3.53}$$

---

[3] Some authors define the partial scattering amplitudes so that they absorb the factor $1/k$.

The choice,

$$\text{(ii)}\quad u_l(k,r) \overset{r\to\infty}{\longrightarrow} \frac{1}{k}\left[\sin\left(kr - \frac{l\pi}{2}\right) - \cos\left(kr - \frac{l\pi}{2}\right)\tan\delta_l\right] \tag{3.54}$$

corresponds to the expression,

$$R_l(k,r) \overset{r\to\infty}{\longrightarrow} \left[j_l(kr) - n_l(kr)\tan\delta_l\right]. \tag{3.55}$$

Using the relations in Eqs.(3.20) and Eq.(3.49), we can write,

$$R_l(k,r) \overset{r\to\infty}{\longrightarrow} \left[j_l(kr) - \mathrm{i}kt_l(k)h_l(kr)\right], \tag{3.56}$$

where the *transition matrix*, or t-matrix, t_l, is defined by the expression (see Eq.(3.49)),

$$t_l(k) = -a_l(k) = -\frac{1}{k}\mathrm{e}^{\mathrm{i}\delta_l(k)}\sin\delta_l(k). \tag{3.57}$$

As we mentioned in Chapter 2, the t-matrix can be used as a building block of multiple scattering theory.

Finally, the choice

$$\text{(iii)}\quad u_l(k,r) \overset{r\to\infty}{\longrightarrow} \frac{1}{k}\left[\cot\delta_l\sin\left(kr - \frac{l\pi}{2}\right) - \cos\left(kr - \frac{l\pi}{2}\right)\right] \tag{3.58}$$

corresponds to the expression

$$R_l(k,r) \overset{r\to\infty}{\longrightarrow} \left[\cot\delta_l j_l(kr) - n_l(kr)\right], \tag{3.59}$$

for the radial wave function. From Eq.(3.56), we can write

$$R_l(k,r) \overset{r\to\infty}{\longrightarrow} \left[m_l j_l(kr) - \mathrm{i}kh_l(kr)\right], \tag{3.60}$$

where $m_l = t_l^{-1}$ is the inverse of the t-matrix defined in Eq.(3.57). In the language of multiple scattering theory, the transition matrix or t-matrix, t_l, is commonly referred to as the scattering matrix and, when there is no danger of confusion, we shall follow this practice.

3.3.5 Integral Expressions for the Phase Shifts

It can be shown [2], that various trigonometric functions of the phase shifts can be obtained by means of integral expressions involving the radial wave functions, R_l. The specific trigonometric functions given by these relations depend on the choice of normalization of the radial wave functions. For example, with the wave function normalized as in Eq.(3.55), we have

$$\tan\delta_l = -k\int_0^\infty j_l(kr)V(r)R_l(k,r)r^2\mathrm{d}r. \tag{3.61}$$

On the other hand, with the normalization[4] indicated in Eq.(3.56), one obtains an integral expression for the t-matrix,

$$t_l(k) = -\frac{e^{i\delta_l(k)}\sin\delta_l(k)}{k} = \int_0^\infty j_l(kr)V(r)R_l(k,r)r^2\mathrm{d}r. \tag{3.62}$$

3.4 Nonspherical Potentials

In this section, we discuss the solutions of the Schrödinger equation, or of the Lippmann–Schwinger equation, for the case of a single, spatially bounded potential cell. We place no restriction on the shape of the cell other than it be bounded by a circumscribing sphere of some finite radius. The only requirements placed on the form of the potential is that it cannot support bound states at positive energies, and that it have everywhere a power series expansion in the spatial coordinates, (x, y, z), that is locally convergent except perhaps for a single point (which we take to be the origin) at which it may have a singularity of the form r^{-n} where n is less than two. Potentials used in electronic structure calculations generally satisfy these requirements. An illustrative example of the following formalism in terms of a two-dimensional square-well potential is provided in Appendix E.

The solutions of the Schrödinger equation,

$$(\nabla^2 + k^2)\psi(k,\mathbf{r}) = V(\mathbf{r})\psi(k,\mathbf{r}), \tag{3.63}$$

also satisfy the Lippmann–Schwinger equation,

$$\psi(k,\mathbf{r}) = \chi(k,\mathbf{r}) + \int G_0(\mathbf{r}-\mathbf{r}')V(\mathbf{r}')\psi(k,\mathbf{r}')\mathrm{d}^3r'. \tag{3.64}$$

This can be verified by applying Eq.(3.63) to Eq.(3.64) and using the fact that the Green function, $G_0(\mathbf{r}-\mathbf{r}')$ satisfies the relation,

$$(\nabla^2 + k^2)G_0(\mathbf{r}-\mathbf{r}') = \delta(\mathbf{r}-\mathbf{r}'). \tag{3.65}$$

The function $\chi(k,\mathbf{r})$ is assumed to be a solution of the Helmholtz equation, $(\nabla^2 + k^2)\chi(k,\mathbf{r}) = 0$.

The solutions of Eq. (3.64) are not fully determined until the boundary conditions are specified. Boundary conditions enter the Lippmann–Schwinger equation both through the form of the Green function and through the inhomogeneous term, χ. As we shall show, various solutions to Eq.(3.64) corresponding to different boundary conditions can be obtained. We shall first seek solutions that are regular at the origin. In particular, we shall specify that the solutions tend to $J_L(\mathbf{r}) = j_l(kr)Y_L(\hat{r})$ near the origin.

---

[4] It is hoped that no confusion will arise from the use of the same symbol, R_l, to denote the radial wave function corresponding to different normalizations.

Thus, we expect that an arbitrary regular solution can be expanded as

$$\psi(k,\mathbf{r}) = \sum_L a_L \psi_L(k,\mathbf{r}), \tag{3.66}$$

at least in the vicinity of the origin.

We shall use a form of the Green function appropriate to standing-wave boundary conditions,

$$G_0(E,\mathbf{r}-\mathbf{r}') = -\frac{\cos(k|\mathbf{r}-\mathbf{r}'|)}{4\pi|\mathbf{r}-\mathbf{r}'|} = k\sum_L J_L(\mathbf{r}_<)N_L(\mathbf{r}_>). \tag{3.67}$$

Here $k = \sqrt{E}$ and the notation $J(\mathbf{r}_<)$ means $J(\mathbf{r})$ or $J(\mathbf{r}')$, depending on whether $\mathbf{r}$ or $\mathbf{r}'$ is smaller in absolute value. Similarly $N(\mathbf{r}_>)$ is $N(\mathbf{r})$ or $N(\mathbf{r}')$ depending on which argument is larger in absolute value.

Thus we can write

$$\psi_L(\mathbf{r}) = \chi_L(\mathbf{r}) + \int \mathrm{d}^3r' G_0(E;\mathbf{r}-\mathbf{r}')V(\mathbf{r}')\psi_L(\mathbf{r}') \tag{3.68}$$

which can be expanded as

$$\begin{aligned}\psi_L(\mathbf{r}) = &\sum_{L'} \bar{C}_{LL'} J_{L'}(\mathbf{r}) + k\sum_{L'} J_{L'}(\mathbf{r}) \int_{r'>r} \mathrm{d}^3r' N_{L'}(\mathbf{r}')V(\mathbf{r}')\psi_L(\mathbf{r}') \\ &+ k\sum_{L'} N_{L'}(\mathbf{r}) \int_{r'<r} \mathrm{d}^3r' J_{L'}(\mathbf{r}')V(\mathbf{r}')\psi_L(\mathbf{r}').\end{aligned} \tag{3.69}$$

This has the form

$$\psi_L(\mathbf{r}) = \sum_{L'} [c_{LL'}(r)J_{L'}(\mathbf{r}) - s_{LL'}(r)N_{L'}(\mathbf{r})], \tag{3.70}$$

where

$$c_{LL'}(r) = \bar{C}_{LL'} + k\int_{r'>r} \mathrm{d}^3r' N_{L'}(\mathbf{r}')V(\mathbf{r}')\psi_L(\mathbf{r}'), \tag{3.71}$$

$$s_{LL'}(r) = -k\int_{r'<r} \mathrm{d}^3r' J_{L'}(\mathbf{r}')V(\mathbf{r}')\psi_L(\mathbf{r}'). \tag{3.72}$$

Thus, the solution is a linear combination of regular and irregular solutions with coefficients $\underline{c}$ and $\underline{s}$ which depend on the potential, $V(\mathbf{r})$. The requirement that $\psi_L(\mathbf{r}) \to J_L(\mathbf{r})$ as $r \to 0$ implies that

$$\bar{C}_{LL'} = \delta_{LL'} - k\int \mathrm{d}^3r' N_{L'}(\mathbf{r}')V(\mathbf{r}')\psi_L(\mathbf{r}'), \tag{3.73}$$

which implies that $c_{LL'}(r)$ can be written as

$$c_{LL'}(r) = \delta_{LL'} - k\int_{r'<r} \mathrm{d}^3r' N_{L'}(\mathbf{r}')V(\mathbf{r}')\psi_L(\mathbf{r}'). \tag{3.74}$$

This allows one to solve for $c_{LL'}(r)$ and $s_{LL'}(r)$ by integrating the differential equations

$$\frac{\mathrm{d}c_{LL'}(r)}{\mathrm{d}r} = -k\int_r \mathrm{d}\hat{r}' N_{L'}(\mathbf{r}')V(\mathbf{r}')\psi_L(\mathbf{r}'), \tag{3.75}$$

$$\frac{\mathrm{d}s_{LL'}(r)}{\mathrm{d}r} = -k\int_r \mathrm{d}\hat{r}' J_{L'}(\mathbf{r}')V(\mathbf{r}')\psi_L(\mathbf{r}'), \tag{3.76}$$

together with Eq.(3.70) and the initial conditions,

$$c_{LL'}(0) = \delta_{LL'}, \tag{3.77}$$

$$s_{LL'}(0) = 0, \tag{3.78}$$

which specify that the basis function $\psi_L(\mathbf{r})$ is $J_L(\mathbf{r})$ at the origin.

For $\mathbf{r}$ outside the bounding sphere, $c_{LL'}(r)$ and $s_{LL'}(r)$ take on their asymptotic values which we denote by $C_{LL'}$ and $S_{LL'}$ respectively,

$$C_{LL'} = \bar{C}_{LL'} = \delta_{LL'} - k\int \mathrm{d}^3r' N_{L'}(\mathbf{r}')V(\mathbf{r}')\psi_L(\mathbf{r}') \tag{3.79}$$

$$S_{LL'} = -k\int \mathrm{d}^3r' J_{L'}(\mathbf{r}')V(\mathbf{r}')\psi_L(\mathbf{r}'). \tag{3.80}$$

These can be written as surface integrals over any surface that encloses the potential $V(\mathbf{r})$ by using the results that $V(\mathbf{r})\psi_L(\mathbf{r}) = (\nabla^2 + k^2)\psi_L(\mathbf{r})$, $0 = (\nabla^2 + k^2)J_L(\mathbf{r})$, and $\delta_{LL'} = k\int \mathrm{d}\mathbf{r}\psi_L(\mathbf{r})(\nabla^2 + k^2)N_{L'}(\mathbf{r})$,

$$C_{LL'} = -k\int \mathrm{d}\mathbf{s}\cdot[N_{L'}(\mathbf{r})\nabla\psi_L(\mathbf{r}) - \psi_L(\mathbf{r})\nabla N_{L'}(\mathbf{r})] \tag{3.81}$$

$$S_{LL'} = -k\int \mathrm{d}\mathbf{s}\cdot[J_{L'}(\mathbf{r})\nabla\psi_L(\mathbf{r}) - \psi_L(\mathbf{r})\nabla J_{L'}(\mathbf{r})]. \tag{3.82}$$

3.4.1 Alternative Forms of the Solution

The explicit form given above, Eq.(3.70), for the solution to the Schrödinger equation for a single nonspherical cell corresponds to the solution for a spherically symmetric potential which has the asymptotic behavior, Eq.(3.53),

$$\psi_\ell(k,r) = \cos\delta_\ell j_\ell(kr) - \sin\delta_\ell n_\ell(kr). \tag{3.83}$$

Treating $\underline{C}$ as an invertible matrix, we can also write the basis function $|\psi\rangle = \underline{C}|J\rangle - \underline{S}|N\rangle$ in the form,

$$\underline{C}^{-1}|\psi\rangle = |J\rangle - \underline{C}^{-1}\underline{S}|N\rangle\,, \tag{3.84}$$

for points outside the circumscribing sphere. This corresponds to the form,

$$\psi_\ell(k,r) = j_\ell(kr) - \tan\delta_\ell n_\ell(kr), \tag{3.85}$$

for potentials with spherical symmetry, Eq.(3.55). Thus if the wave function in the region outside the cirumscribing sphere is written in the form,

$$\psi_L = \sum_{L'} [\delta_{LL'} J_{L'} - T_{LL'} N_{L'}], \tag{3.86}$$

the generalized tangent matrix $T_{LL'}$ is given by Eq.(3.80) (with T replacing S). Note that ψ_L will no longer behave like J_L near the origin, but like $\underline{C}^{-1} |J\rangle$.

Similarly, if the phase-function matrix $\underline{S}$ can be treated as being invertible, we can write,

$$\underline{S}^{-1} |\psi\rangle = \underline{S}^{-1} \underline{C} |J\rangle - |N\rangle , \tag{3.87}$$

corresponding to the form,

$$\psi_\ell(k,r) = \cot\delta_\ell j_\ell(kr) - n_\ell(kr), \tag{3.88}$$

for spherical potentials, Eq.(3.59). It should be noted, however, that for nonspherical potentials, S is likely to be singular or nearly singular depending on the potential and on the maximum number of angular momentum values used to represent the matrix, $S_{LL'}$.

The discussion just given was based on the choice of the Neumann functions as the irregular solutions of the free-space Schrödinger equation (Helmholtz equation). Analogous results are obtained when the Hankel functions are used instead. Upon using the relation, Eq.(3.20),

$$H_\ell^\pm = J_\ell \pm \mathrm{i} N_\ell, \tag{3.89}$$

we can write the solution $|\psi\rangle = \underline{C}|J\rangle - \underline{S}|N\rangle$ in the form, (with $h \equiv h^+$),

$$|\psi\rangle = [\underline{C} - \mathrm{i}\underline{S}] |J\rangle + \mathrm{i}\underline{S} |H\rangle . \tag{3.90}$$

Operating on the left by $[\underline{C} - i\underline{S}]^{-1}$ yields yet another form of the basis functions,

$$\begin{aligned} [\underline{C} - \mathrm{i}\underline{S}]^{-1} |\psi\rangle &= |J\rangle + \mathrm{i} [\underline{C} - \mathrm{i}\underline{S}]^{-1} \underline{S} |H\rangle \\ &= |J\rangle - \mathrm{i}k\underline{t} |H\rangle , \end{aligned} \tag{3.91}$$

which can be recognized as a generalization of the form,

$$\psi_\ell(k,r) = j_\ell(kr) - \mathrm{i}kt_\ell h_\ell(kr), \tag{3.92}$$

appropriate to spherically symmetric potentials, Eq.(3.56). Thus, if the function $\psi_L(\mathbf{r})$ is expressed as

$$\psi_L(\mathbf{r}) = J_L(\mathbf{r}) - ik \sum_{L'} t_{LL'} H_{L'}, \tag{3.93}$$

the t-matrix, $t_{LL'}$, can again be obtained from

$$t_{LL'} = \int \mathrm{d}^3 r \psi_L(\mathbf{r}) V(\mathbf{r}) J_{L'}(\mathbf{r}). \tag{3.94}$$

It is still necessary, of course, to use the matrices $c_{LL'}(r)$ and $s_{LL'}(r)$ to solve for $\psi_L(\mathbf{r})$. We shall discuss an alternative technique for obtaining $t_{LL'}$ in the next section.

It is also possible to define the generalizations of the phase shifts to nonspherical potentials by means of the relations,

$$\underline{C} = \cos\underline{\Delta} \quad \text{and} \quad \underline{S} = \sin\underline{\Delta}, \tag{3.95}$$

where $\underline{\Delta}$ is a generally nondiagonal matrix in L-space. In these expressions, the cosine and sine functions (and their inverses), along with the exponential function, of a matrix argument are defined in terms of well-known series expansions.

3.4.2 Direct Determination of the t-Matrix(*)

It is often convenient, e.g., in applications of MST to substitutionally disordered alloys, to use expressions involving the t-matrix rather than the functions $\underline{C}$ and $\underline{S}$. Although the t-matrix can be written in the form $i\{\underline{C} - i\underline{S}\}^{-1}\underline{S}$, inverting the matrix $\{\underline{C} - i\underline{S}\}$can sometimes be troublesome. It can be avoided by solving an integral equation directly for the t-matrix.

This equation can be derived along the following lines: We write Eqs.(3.75) and (3.76) in the forms,

$$\frac{de_{LL'}(r)}{dr} = ik \int_r d\hat{r}' \psi_L(\mathbf{r}')V(\mathbf{r}')H_{L'}(\mathbf{r}') \tag{3.96}$$

and

$$\frac{ds_{LL'}(r)}{dr} = -k \int_r d\hat{r}' \psi_L(\mathbf{r}')V(\mathbf{r}')J_{L'}(\mathbf{r}') \tag{3.97}$$

Note that $c_{LL'}$ has been replaced by $e_{LL'} = c_{LL'} - is_{LL'}$ corresponding to the replacement of N by $H = J + iN$.

With these definitions, the t-matrix is given simply by the relation,

$$\underline{t} = -k^{-1}[\underline{C} - i\underline{S}]^{-1}\underline{S} = -k^{-1}\underline{E}^{-1}\underline{S}, \tag{3.98}$$

while the wave function is given by the expression,

$$\psi_L(k, \mathbf{r}) = \sum_{L'} \left[e_{LL'}(r)J_{L'}(k\mathbf{r}) + is_{LL'}(r)H_{L'}(k\mathbf{r})\right], \tag{3.99}$$

which is simply Eq.(3.70) with the Neumann function replaced by the Hankel function.

Defining an r-dependent t-matrix by the relation,[5] $\underline{t}(r) = -k^{-1}\underline{e}(r)^{-1}\underline{s}(r)$, we find

$$Z_L(\mathbf{r}) = \sum_{L'}[e^{-1}]_{LL'}\psi_{L'}(\mathbf{r}) = J_L(\mathbf{r}) - ik\sum_{L'} t_{LL'}H_{L'}(\mathbf{r}) \tag{3.100}$$

We now note the relations,

$$\left[e^{-1}\frac{\mathrm{d}e}{\mathrm{d}r}\right]_{LL'} = ik\int_r \mathrm{d}\hat{r}' Z_L(\mathbf{r}')V(\mathbf{r}')H_{L'}(\mathbf{r}') \tag{3.101}$$

and

$$\left[e^{-1}\frac{\mathrm{d}s}{\mathrm{d}r}\right]_{LL'} = -k\int_r \mathrm{d}\hat{r}' Z_L(\mathbf{r}')V(\mathbf{r}')J_{L'}(\mathbf{r}'), \tag{3.102}$$

so that from the relation

$$-k\,\dot{\underline{t}}(r) = \underline{e}^{-1}(r)\dot{\underline{s}}(r) - \underline{e}^{-1}(r)\dot{\underline{e}}(r)\underline{e}^{-1}(r)\underline{s}(r) \tag{3.103}$$

where a dot over a variable indicates a derivative with respect to r, we obtain,

$$\begin{aligned}\dot{t}_{LL'}(r) = &\int_r \mathrm{d}\hat{r}' Z_L(\mathbf{r}')V(\mathbf{r}')J_{L'}(\mathbf{r}') \\ &- ik\sum_{L''}\int_r \mathrm{d}\hat{r}' Z_{L'}(\mathbf{r}')V(\mathbf{r}')H_{L''}(\mathbf{r}')t_{L''L'}(r).\end{aligned} \tag{3.104}$$

In view of Eq. (3.100), we obtain

$$\dot{t}_{LL'}(r) = \int_r \mathrm{d}\hat{r}' Z_L(\mathbf{r}')V(\mathbf{r}')\tilde{Z}_{L'}(\mathbf{r}'), \tag{3.105}$$

where $\tilde{Z}_L = J_L - ik\sum_{L'} H_{L'}t_{L'L}$. In the next chapter, this equation will be derived more directly within the formalism of two-potential scattering theory.

Compared to the integral equations for the phase functions, the last expression has the advantage of yielding the t-matrix directly as the solution of a single equation. At first glance, this advantage may seem to be of dubious value as the last equation involves a product of wave functions and is, therefore, of the second degree in the t-matrix. However, at positive energies, where one should not have to consider the existence of bound states, Eq.(3.105) leads to computationally convenient and stable procedures.

---

[5]This is the t-matrix corresponding to the potential that vanishes outside a sphere of radius r.

3.5 Wave Function in the Moon Region

In the previous section we were concerned with obtaining solutions to the Schrödinger equation that were valid inside a nonspherical cell. These solutions are also valid outside the bounding sphere which surrounds the cell. In fact, the form of the wave function which involves the t-matrix, $\psi_L(\mathbf{r}) = J_L(\mathbf{r}) - ik\sum_{L'} t_{LL'}H_{L'}(\mathbf{r})$ is just the generalization for nonspherical potentials of the scattered wave function of Eq. 3.4),

$$\psi^{(+)}_{\mathbf{k}_i}(\mathbf{r}) \to A\left(\exp(i\mathbf{k}_i\cdot\mathbf{r}) + f(\theta,\phi)\frac{\exp(ikr)}{r}\right)$$
$$= 4\pi\sum_L i^l Y_L^*(\hat{k})[J_L(\mathbf{r}) - ikt_l H_L(\mathbf{r})]. \tag{3.106}$$

In this section we consider the convergence of the angular-momentum expansion of the wave function in the moon region, i.e., the region between the boundary of the cell and the bounding sphere.

From our intuitive discussion of multiple scattering theory based on Huygens' principle, it would appear that the t-matrix provides a complete description of the wave function everywhere outside a potential cell, regardless of the shape of the cell. This, however, is not so obvious when scattering theory is cast in the angular-momentum representation, because the multipole expansion in terms of the t-matrix, Eq.(3.91), is only valid *outside* a sphere bounding the cell. In fact, if that expression is used at points inside the sphere, the product,

$$\sum_{L'} t_{LL'}H_{L'}(\mathbf{r}), \tag{3.107}$$

may diverge. To understand the origin of this divergence consider that the t-matrix may be written as

$$t_{LL'} = \int d^3r'\psi_L(\mathbf{r}')V(\mathbf{r}')J_{L'}(\mathbf{r}'), \tag{3.108}$$

where $\psi_L(\mathbf{r})$ is a regular solution to the Schrödinger equation normalized so that it equals $J_L(\mathbf{r}) - ik\sum_{L'} t_{LL'}H_{L'}(\mathbf{r})$ outside the bounding sphere. The divergence in the sum in Eq.(3.107) arises from the fact that the integral over the cell will include some points $\mathbf{r}'$ (argument of J_L) near the corners of the cell that may be further from the origin than the point $\mathbf{r}$ which is the argument of H_L in Eq.(3.107). Since the radius of convergence of the expansion $\sum_L J_L(\mathbf{r})H_L(\mathbf{r}_0)$ is $r = r_0$ for given $\mathbf{r}_0$, the sum in Eq.(3.107) is likely to diverge for points in the moon region.

This behavior may lead one to suspect that in the angular-momentum representation the wave function in the moon region surrounding a cell, i.e., outside the cell but inside a bounding sphere, is not described completely in terms of the cell t-matrix. This has led to the further conjecture that the scattering from two adjacent, generally shaped cells whose bounding spheres overlap cannot be described in terms of the cell-scattering matrices. Instead, (it has been conjectured) multiple scattering theory should be adjusted to take account of the near-field corrections (NFCs)[7–10] arising from the presence of a nonvanishing potential inside the moon region of a given cell. Therefore, within the angular-momentum representation, we are led to ask if information other than that contained in the t-matrix is necessary to describe the wave function in the moon region.

It is not difficult to anticipate that the answer to this question is negative. The wave function in the moon region is indeed determined completely in terms of the t-matrix. After all, we may view the cell (to any desired accuracy) as a collection of nonoverlapping spheres. Then the points in the moon region lie outside *all* spherical surfaces, and the wave disturbance there can be written as the superposition of the waves scattered by the potentials inside the spheres. The resulting multiple scattering expression for the t-matrix or for the wave function, in the angular-momentum representation, yields a summation that by definition becomes equal to the t-matrix of the cell as the number of spheres increases, with their radii chosen so that the spheres cover completely the volume of the cell (except, perhaps, for sets of measure zero that do not contribute to the scattering). This intuitive discussion can also be expressed in mathematical terms and can be used to prove [7] that the t-matrix for an assembly of space-filling cells has the same form as in the case of muffin-tin potentials.

In the following sections, we show that the cell t-matrix alone suffices to determine completely the wave function everywhere outside a cell, even inside the moon region. This is true regardless of the shape of the cell, that is, whether it is convex or concave. In fact, there are at least two different ways of showing this based, respectively, on a shifted-center and shifted-cell approach. Both of these formal arguments are discussed in this section. They both emphasize that the problem is associated with geometry alone, being quite independent of the nature of the potential in the cell. The first line of argument, that of the shifted-center, is much simpler than that of the shifted-cell, and intuitively clearer. However, it is less general and less useful for computations. For example, it is only valid in the case of cells with convex, polyhedral shapes. On the other hand, the shifted-cell argument applies to arbitrarily-shaped, even concave cells. In Section 6.6.3, it will be used to derive the secular equation of multiple scattering theory by matching explicitly the wave function and its derivative across cell boundaries.

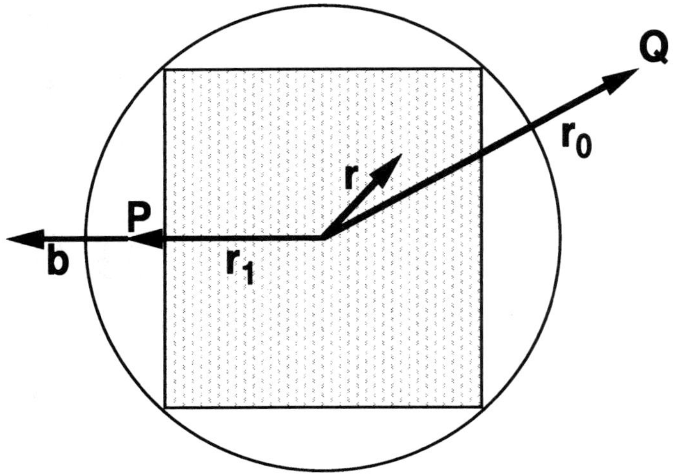

FIGURE 3.2. A cell with a bounding sphere, a cell vector, **r**, an observation vector, $\mathbf{r}_0$, and a vector, **b**, which moves points in the moon region outside the bounding sphere.

3.5.1 Displaced-Center Approach: Convex Cells

We begin with the Lippmann–Schwinger equation written in the form (with energy arguments suppressed),

$$\psi_L(\mathbf{r}_0) = J_L(\mathbf{r}_0) + \int G_0(\mathbf{r} - \mathbf{r}_0)V(\mathbf{r})\psi_L(\mathbf{r})\mathrm{d}^3r, \qquad (3.109)$$

in which the ψ_L is normalized in the manner indicated in Eq.(3.91) or in Eq.(3.86). For points, Q, that lie outside a sphere bounding a cell, (see Fig. 3.2), the vector $\mathbf{r}_0$ is larger than any intracell vector, **r**, and the free-particle propagator can be expanded in the form, Eq.(D.6),

$$G_0(\mathbf{r} - \mathbf{r}_0) = -ik \sum_L H_L(\mathbf{r}_0)J_L(\mathbf{r}). \qquad (3.110)$$

When this expression is substituted into Eq.(3.109) and the integral over **r** is carried out, we obtain the expression,

$$\psi_L(\mathbf{r}_0) = J_L(\mathbf{r}_0) - ik \sum_{L'} H_{L'}(\mathbf{r}_0)t_{L'L}, \qquad (3.111)$$

which is Eq.(3.91) explicitly displayed. (We can also let $\underline{t}$ denote the ratio $\underline{C}^{-1}\underline{S}$, or $\tan\underline{\Delta}$, in a normalization consistent with that displayed in Eq.(3.86).) Evidently, for $\mathbf{r}_0$ inside the bounding sphere, such as at point P in Fig. 3.1, Eq.(3.110) represents a divergent expression because $\mathbf{r}_0$ may be smaller than some cell vector **r**. Consequently Eq.(3.111) may no longer be valid.

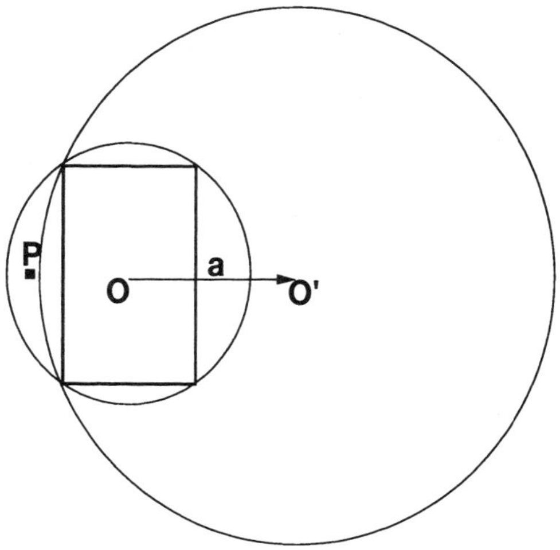

FIGURE 3.3. A point, P, inside the moon region of the sphere centered at point O, lies outside a sphere centered at point O'.

In spite of these difficulties, a convergent expression for the wave function at P in terms of the cell t-matrix can be obtained along the following lines. First, we note that the point P that lies *inside* a sphere centered at O, taken to be the geometric center of the cell shown in Fig. 3.3, lies *outside* a sphere centered at a point O', displaced from O by a vector $\mathbf{a}$. Denoting the cell t-matrix obtained in terms of angular-momentum expansions about O' by $\underline{t}(\mathbf{a})$, we can write the wave function at P in the form,

$$\psi_L(\mathbf{r}_0') = J_L(\mathbf{r}_0' + \mathbf{a}) - ik \sum_{L'} H_{L'}(\mathbf{r}_0') t_{L'L}(\mathbf{a}), \tag{3.112}$$

where $\mathbf{r}_0' = -\mathbf{a} + \mathbf{r}_0$ is the radius vector from the shifted center at O' to P. But, as will be shown in a later section, (Eq.(4.79)) we have,

$$\underline{t}(\mathbf{a}) = \underline{g}(\mathbf{a}) \underline{t} \underline{g}(-\mathbf{a}), \tag{3.113}$$

where $\underline{t}$ is the cell t-matrix evaluated about the original center at O. The last expression illustrates the existence of a similarity transformation, depending only on the amount of shift, which relates the t-matrices expanded about two centers displaced from each other by a vector $\mathbf{a}$. Now, Eq.(3.112) takes the form,

$$\psi_L(\mathbf{r}_0') = J_L(\mathbf{r}_0' + \mathbf{a}) - ik \sum_{L'} H_{L'}(\mathbf{r}_0') \left[\sum_{L_1} \sum_{L_2} g_{L'L_1}(\mathbf{a}) t_{L_1 L_2} g_{L_2 L}(-\mathbf{a}) \right], \tag{3.114}$$

which expresses explicitly the wave function at a point inside the moon region of a cell in terms of the cell t-matrix. Here, the quantities $g_{LL'}(\mathbf{a})$ are the matrix elements of the translation operator in the angular-momentum representation and are defined explicitly in Appendix D. It is important to realize that the sums inside the brackets in Eq.(3.114) must be performed before those outside in order to lead to converged results. In particular, the sum over L_1 must precede that over L'. If the order of these two sums is reversed, the resulting expression reduces formally to Eq.(3.111), which in the present case may diverge.

In the construction just described, we can take the point P to lie arbitrarily close to the cell boundary by allowing the vector $\mathbf{a}$ to become very large (essentially approaching infinity as the point moves right against the face of the cell). Therefore, at least formally, the t-matrix in the angular-momentum representation yields a complete description of the wave function *everywhere* outside a spatially bounded convex potential cell. We point out that the basic feature characterizing the shifted-center construction just described is the replacement of a *single*, possibly divergent sum, that over L' in Eq.(3.111), by a *double*, conditionally convergent sum, that over L_1 and L' in Eq.(3.114). The replacement of single, divergent sums by conditionally convergent double or multiple sums will figure often in our subsequent discussion. Note that this replacement is based *purely* on the underlying geometry, that is, ratios of vectors and the shape of a potential cell, not involving the potential in the cell itself.

It is clear from the discussion so far, that the shifted-center argument can only be used with potential cells of convex shape. Also, its numerical implementation becomes increasingly cumbersome, because of the need to extend the summations to large values of L as the point P approaches the face of a cell, and the length of the vector $\mathbf{a}$ becomes very large. The method described in the next subsection removes completely the first limitation, and also yields an expression that is more convenient for computations.

3.5.2 Displaced-Cell Approach: Convex Cells

Our starting point is again the Lippmann–Schwinger equation in the form of Eq.(3.109). We seek a multipole, convergent expression for the wave function in the moon region, point P in Figs. 3.1 or 3.2, in terms of the cell t-matrix. Such an expression can be obtained as follows. We note that for all points confined to the moon region adjacent to a face of a convex polyhedral cell there exist vectors $\mathbf{b}$ that satisfy the inequalities,

$$|\mathbf{b}| < |\mathbf{r}_0 - \mathbf{r} + \mathbf{b}| , \tag{3.115}$$

and

$$|\mathbf{r}| < |\mathbf{b} + \mathbf{r}_0| , \tag{3.116}$$

for all cell vectors $\mathbf{r}$. Any vector $\mathbf{b}$ that is perpendicular to the face of the cell, directed outward and larger than the largest vertical distance from the bounding sphere to the face satisfies the inequalities above.

We now add and subtract such a vector **b** in the argument of G_0, Eq.(3.109), (see Fig. 3.2) and use the first inequality above to obtain the expression,

$$\psi_L(\mathbf{r}_0) = J_L(\mathbf{r}_0) - ik \sum_{L'} J_{L'}(-\mathbf{b}) \times \left[\int H_{L'}(\mathbf{r} - \mathbf{r}_0 - \mathbf{b}) V(\mathbf{r}) \psi_L(\mathbf{r}) \mathrm{d}^3 r \right] . \tag{3.117}$$

In view of the second inequality, the Hankel function in the last equation can be further expanded to yield the result,

$$\begin{aligned} \psi_L(\mathbf{r}_0) &= J_L(\mathbf{r}_0) - ik \sum_{L'} J_{L'}(-\mathbf{b}) \\ &\quad \times \left[\sum_{L''} G_{L'L''}(-\mathbf{r}_0 - \mathbf{b}) \int J_{L''}(\mathbf{r}) V(\mathbf{r}) \psi_L(\mathbf{r}) \mathrm{d}^3 r \right] , \\ &= J_L(\mathbf{r}_0) - ik \sum_{L'} J_{L'}(-\mathbf{b}) \left[\sum_{L''} G_{L'L''}(-\mathbf{r}_0 - \mathbf{b}) \underline{t}_{L''L} \right] . \end{aligned} \tag{3.118}$$

The quantities $G_{LL'}(\mathbf{R})$ are the so-called real-space structure constants and are the elements of the free-particle propagator in the angular-momentum representation. Explicit expressions for the structure constants are given in Appendix D. Evidently, the last equation yields an expression for the wave function at a point inside the moon region explicitly in terms of the cell t-matrix. It should be kept in mind that the brackets in Eq.(3.118) indicate the proper order in which the sums should be performed in order to obtain a convergent expression, that is, the sum over L'', corresponding to the *second* expansion must be performed before the sum over L' that arose as a result of the *first* expansion.

There are obvious similarities between Eqs.(3.114) and (3.118). In both a divergent sum has been replaced by a conditionally convergent double sum, in a construction that highlights the geometric nature of the problem. However, the expression in Eq.(3.118) holds a number of advantages over that in Eq.(3.114). First, it is of much greater conceptual value since it can be generalized to the case of concave cells, as is shown in the next subsection. Second, it is also of greater computational value, as it only involves sums over values of L that are not excessively large and vectors of finite, and usually small, length in contrast to the large values of L and the infinite vectors that could occur in Eq.(3.114). The results of calculations presented below indicate that converged expressions can be obtained over a broad range of the outside sum, L', when the internal sum, L_1, is taken to values about twice as large as those of the outer sum, L'. Also, the value of b for which optimal convergence rates can be expected was found numerically to be of the order of the radius of the circumscribing sphere. The convergence properties of the double sum in Eq.(3.118) have been discussed at length in connection with the three-dimensional Poisson equation [8] (see also Chapter 9).

It is interesting to compare the physical interpretations of Eqs.(3.118) and Eq.(3.114). The effect of the double expansion in Eq.(3.114) is to shift the position of the cell center. In Eq.(3.118), on the other hand, the points inside the moon region are shifted to positions outside the bounding sphere through the addition of the vector $\mathbf{b}$. Then, the wave function is calculated at these shifted positions and brought back inside the moon region by means of the outer sum involving $J_L(\mathbf{b})$. Alternatively, we may think of the points in the moon region as being kept in place but with the entire cell shifted by $-\mathbf{b}$. In contrast to the shifted-center approach of the previous subsection, during the shifting operations in the present argument the center of the cell remains fixed with respect to the cell boundary, leading to expressions in terms of the cell t-matrix from the beginning.

In summary, we note that Eq.(3.118) provides an alternative but equivalent expression to Eq.(3.112) for the wave function in the asymptotic region. Although Eq.(3.118) is more complicated than Eq.(3.112), it is valid everywhere outside the boundary of a convex cell.

3.5.3 Numerical Example: Convergence for Square Cell

Figure 3.4 illustrates the convergence of the double sum in Eq.(3.118) for the case of a two-dimensional square cell (see related algebra in Section 3.7 below). The figure shows the wave function calculated for a two-dimensional square cell of side equal to π and with a constant potential, $V = -5$ Ry. The thick solid line shows the exact $L = 4$ basis function at the midpoint of

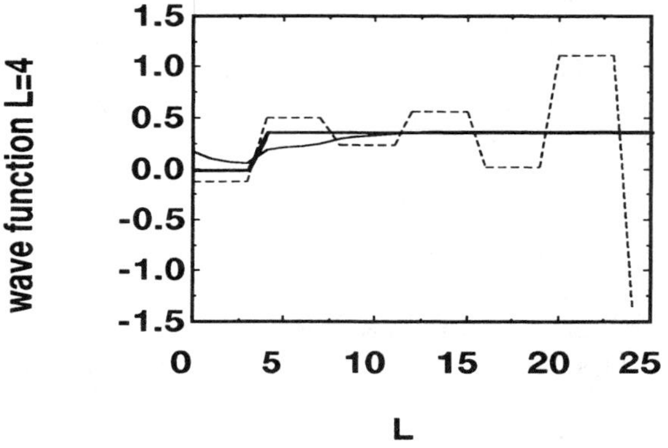

FIGURE 3.4. The wave function inside the moon region of a square cell obtained with the double summation of the shifted-cell construction (thin solid line), compared with the divergent expansion resulting from the ordinary expansion in a single sum (dotted line), which arises when the sum over L' is performed before that over L''. The thick solid line is the exact result, ψ_L, for $L = 4$.

a side of the cell. In this case the exact solution to the Schrödinger equation inside the cell which satisfies the correct boundary condition at the origin[6] is a spherical Bessel function corresponding to a shifted energy, $E' = E+5$. The basis function for $L = 4$ was also calculated using the single expansion in Eq.(3.70), but with the asymptotic values of the phase functions,

$$\psi_L(k, \mathbf{r}) = \sum_L [C_{LL'} J_{L'}(k\mathbf{r}) - S_{LL'} N_{L'}(k\mathbf{r})] \tag{3.119}$$

which is valid *outside* a sphere (here a circle) bounding the cell. The results as a function of the internal summation index, L', are shown by the dashed line in the figure. Of the two terms inside the brackets in Eq.(3.119), the first yields a convergent expression *everywhere* outside the cell. However, the second term diverges giving rise to the dashed curve. The reason for the different behavior of the two terms is discussed in the next section and also in Section 6.5. For the moment, note that the term containing $S_{LL'}$ is the result of an integral, Eq.(3.80), over cell vectors some of which have magnitudes larger than the argument, $\mathbf{r}$, of the Neumann function. Thus, the summation over L' in this term includes an incorrect expansion of the free-particle propagator of the form

$$\sum_L J_L(\mathbf{r}) N_L(\mathbf{r}'), \quad \text{for} \quad r > r', \tag{3.120}$$

leading to the divergent behavior exhibited in Fig. 3.4.

The figure also shows the results obtained when the sum involving $S_{LL'}$ is replaced by a double expansion of the type indicated in Eq.(3.118) so that the basis function in the moon region takes the form

$$\psi_L(k, \mathbf{r}) = \sum_L [C_{LL'} J_{L'}(k\mathbf{r}) + \sum_{L'} J_{L'}(-\mathbf{b}) \left[\sum_{L''} G_{L'L''}(-\mathbf{r} - \mathbf{b}) S_{L''L} \right]. \tag{3.121}$$

Taking $b = 1$ and carrying the sum over L'' to a maximum value of $L''_{\text{max}} = 60$ yields a rapidly converging expansion as L' increases.

As mentioned above, it is necessary to carry out the sum over L'' prior to that over L'. If the order of summations is interchanged, one obtains the divergent expansion of Eq.(3.119). Also, as L' approaches the maximum value of L'' even the double summation in Eq.(3.121) will be seen to diverge, as this equation is again being converted to the divergent expansion in Eq.(3.119). Formally, of course, the internal sum over L'' is to be carried to completion, that is, to infinity, in which case Eq.(3.121) is always convergent with respect to increasing L'. In a practical calculation, it has been found numerically that choosing L''_{max} about twice as large as the largest value of the outer sum (over L'), coupled with a value of b about the size of

[6]This solution is an example of a basis function. The wave function must also satisfy a boundary condition at infinity.

the radius of the bounding sphere usually leads to convergence over a wide range of values of L'. Clearly, both $L''_{\max}$ and the maximum value of L' depend on the value of L that characterizes the basis function. Both must increase sufficiently fast as L increases in order to maintain convergence.

Finally, note that the behavior exhibited by the various sums in Eqs.(3.119) and (3.121) has a purely geometric origin, depending only on the lengths of vectors in the argument of the free-particle propagator, and not on the details of the potential in the cell. These difficulties can be overcome again by a geometric construction, such as the shifted-center or shifted-cell arguments discussed above. In any case, the results shown here illustrate the fact that the asymptotic values of the phase functions, and hence of the t-matrix, suffice to determine the wave function everywhere outside the boundary of a cell, even inside the moon region.

It is because of this property that MST applied to space-filling cells can be expressed in terms of cell t-matrices and structure constants, that is to say in terms of quantities that depend separately on the potential and on the structure (geometry) of the system.

3.5.4 Displaced-Cell Approach: Concave Cells (*)

For parts of the moon region around a concave cell there may be no single vector $\mathbf{b}$ that satisfies the inequalities in (3.115) and (3.116), as is illustrated schematically in Fig. 3.5. For some cell vectors, $\mathbf{r}$, such as the one shown in Fig. 3.5, inequality (3.115) is violated and Eq.(3.117) is no longer applicable.

In order to obtain a convergent expression for the wave function everywhere in the moon region in terms of the cell t-matrix, it is necessary to generalize the approach of the previous subsection. This generalization consists in replacing the single vector $\mathbf{b}$ by a sum of N vectors, $\mathbf{b} = \sum_{\alpha=1}^{N} \mathbf{b}_\alpha$, so chosen that for all α the following inequalities hold:

$$|\mathbf{b}_n| < \left| \sum_{\alpha=1}^{n} \mathbf{b}_\alpha - \mathbf{r} + \mathbf{r}_0 \right| . \tag{3.122}$$

This allows a *series* of consecutive expansions of the type used in Eq.(3.117) by which the point at $\mathbf{r}_0$ is moved outside the bounding sphere, or the cell is displaced in the opposite direction in a number of steps rather than in a single step. A set of vectors $\mathbf{b}_\alpha$ that satisfy the inequality in (3.122) can be determined on the basis of the following process. The vector $\mathbf{b}_{n+1}$ is chosen to have length no larger than the radius of the largest sphere that can be centered at the point $\mathbf{r}_0 + \sum_{\alpha=1}^{n} \mathbf{b}_\alpha$ that does not intersect the cell boundary. Three such vectors, $\mathbf{b}_1$, $\mathbf{b}_2$ and $\mathbf{b}_3$ are shown in Fig. 3.6. As the figure indicates, the vectors $\mathbf{r}_0 - \mathbf{r} + \sum_{\alpha=1}^{n} \mathbf{b}_\alpha$ lie always outside a sphere of radius b_n and are thus larger than b_n. Now, provided that we choose $\mathbf{b}$

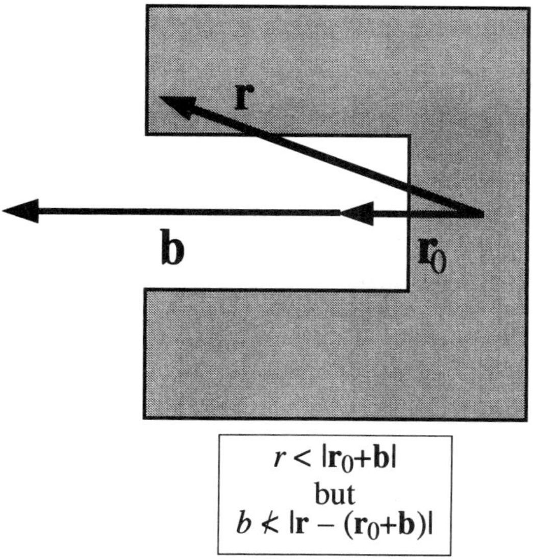

FIGURE 3.5. An illustration that for concave cells a single vector $\mathbf{b}$ may not satisfy all of the conditions for the validity of the shifted-cell construction.

to satisfy the inequality,

$$|\mathbf{r}| < |\mathbf{r}_0 + \mathbf{b}|\,, \tag{3.123}$$

the Lippmann–Schwinger equation can be written in the form,

$$\begin{aligned}\psi_L(\mathbf{r}_0) &= J_L(\mathbf{r}_0) + \sum_{L_1} J_{L_1}(\mathbf{b}_1)\left\{\sum_{L_2} g_{L_1L_2}(\mathbf{b}_2)\left[\cdots\left(\sum_{L_n} g_{L_{n-1}L_n}(\mathbf{b}_n)\right.\right.\right.\\ &\quad\times\left.\left.\left.\int H_{L_n}(\mathbf{r}_0-\mathbf{r}+\mathbf{b})V(\mathbf{r})\psi_L(\mathbf{r})\mathrm{d}^3r\right)\cdots\right]\right\}\\ &= J_L(\mathbf{r}_0) + \sum_{L_1} J_{L_1}(\mathbf{b}_1)\left\{\sum_{L_2} g_{L_1L_2}(\mathbf{b}_2)\right.\\ &\quad\times\left.\left[\cdots\left(\sum_{L_n} g_{L_{n-1}L_n}(\mathbf{b}_n)\ \sum_{L''} G_{L_nL''}(\mathbf{r}_0+\mathbf{b})t_{L''L}\right)\right]\right\}.\end{aligned} \tag{3.124}$$

Here, we have used the inequalities in (3.122) to obtain the set of conditionally convergent sums shown in the first line, and the inequality in (3.123) to obtain the last expression. Note again that the order of summations in this expression must be maintained as indicated by the various sets of brackets. If any of the sums are carried out in order different from the one indicated by the brackets there can be no guarantee that the resulting expressions converge. However, for the correct order of the summations,

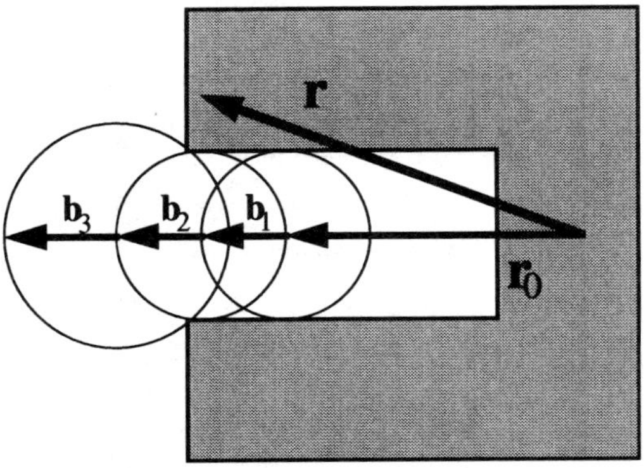

$$r < |\mathbf{r}_0+\mathbf{b}_1+\mathbf{b}_2+\mathbf{b}_3|$$
and
$$b_1 < |\mathbf{r} - (\mathbf{r}_0+\mathbf{b}_1)|$$
$$b_2 < |\mathbf{r} - (\mathbf{r}_0+\mathbf{b}_1+\mathbf{b}_2)|$$
$$b_3 < |\mathbf{r} - (\mathbf{r}_0+\mathbf{b}_1+\mathbf{b}_2+\mathbf{b}_3)|$$

FIGURE 3.6. A series of vectors for expanding the wave function in terms of the t-matrix in the moon region of a concave cell.

Eq.(3.124) provides a valid expression for the solution of the Schrödinger equation inside the moon region of a concave cell in terms of the cell t-matrix.

As it stands, Eq.(3.124) is of little, if any, computational value. The large number of internal summations, each of which must be carried out to values about twice as large as those of the sum that immediately follows it, makes Eq.(3.124) very difficult to use. However, it is of great formal value. As is shown in Section 6.6, it can be used to derive the secular equation of multiple scattering theory for the case of arbitrarily-shaped, space-filling cells through an explicit and direct matching of the wave function across cell boundaries.

Perhaps a point worth emphasizing again at this stage is that the form of the scattered wave function outside a potential region is not uniquely defined in terms of angular-momentum eigenstates. Either the MT form, Eq.(3.112), or any of the forms involving a displaced spherical function, Eqs.(3.114), (3.118), or (3.124) can be used in formal arguments, with the sequence of the various sums being kept in the order prescribed by the particular geometry. The MT form is, of course, the most convenient from a computational point of view, and all other forms reduce to it in the MT case. In the case of space-filling cells, however, or even in the case of MT spheres whose centers have been displaced from their geometric positions, the MT form may lead to divergent results. In such a case any of the other forms can and should be used as this can lead to well-defined and convergent expansions.

The discussion in the last two sections regarding the wave function in the moon region has indicated the types of problems that in the past have beset attempts to generalize multiple scattering theory to space-filling cell potentials. As was pointed out in the discussion, these problems are of purely geometric origin. They arise whenever the various vectors involved in the expansion of the free-particle propagator fail to satisfy the proper conditions for convergence of those expansions. Hence, it is not surprising that the resolution of these difficulties is also of geometric character. In more precise terms, in the angular momentum representation there are multipole expansions which, as expected, diverge when applied outside their region of validity. However, in applications of MST, these divergent summations can be replaced by conditionally convergent double or multiple sums which are completely independent of the near field, that is, the potential in the moon region around a cell. One approach to the generalization of multiple scattering theory to space-filling cells consists of applying properly these conditional summation procedures, carrying out the sums completely as required to obtain the final expressions. When this is done, multiple scattering theory assumes identical forms in the cases of space-filling as well as of muffin-tin potentials. Now, having established the form of the cell wave function in the moon region, we may inquire as to what effect the presence of a potential there has in either a formal or computational sense. This question is addressed in the following section.

3.6 Effect of the Potential in the Moon Region

In this section, we address a question that is closely related to the one posed in the last two sections. Namely, what effect, if any, does the presence of a nonvanishing potential in the moon region have on the cell wave function throughout space? More precisely, given the solution of the Schrödinger equation, for example, Eq.(3.84), corresponding to a nonzero potential in the moon region, is it possible to construct the corresponding solution for the cell potential alone? Somewhat surprisingly, the answer is yes. The reason that one may *not* expect this to be the case is, of course, self-evident. The t-matrix associated only with the potential inside a cell is indeed different from that corresponding to the potential inside a sphere that includes the cell and the moon region which now contains a nonvanishing potential. It is not clear that one can disentangle information from one about the other. In fact, the wave functions for the two cases bear no recognizable relation to each other when written in the form of Eq.(3.91). However, we will show that a judicious choice of basis states or, alternatively, proper choices of the coefficients, $\bar{C}$, in Eq.(3.71) allows the construction of the cell solution (or basis functions) from a knowledge of the solution for the potential inside a sphere circumscribing the cell. In addition, we will see that this feature opens up very useful possibilities of calculating the cell phase functions in ways that may accelerate the convergence of a calculation.

At this point, it is important to emphasize the distinction between *wave functions* and *basis functions.* We shall use the term "basis function" to mean a function that satisfies the Schrödinger equation within a finite region of space containing the origin, and that satisfies a single boundary condition there. An example of basis functions is afforded by what are commonly referred to as the "radial wave functions" in connection with spherically symmetric potentials. A basis function is generally not a meaningful wave function, but a set of basis functions can be used to expand the wave function at a given energy. The coefficients of that expansion can be chosen so that the wave function satisfies a proper boundary condition at infinity. This far from obvious result is demonstrated rigorously in the next chapter. Keeping in mind the distinction between basis functions and wave functions may make some of the results derived below appear more intuitive than otherwise.

Consider the (basis) functions ψ^{Ω} and ψ^{S} that satisfy the Schrödinger equation for the potentials inside a cell, Ω, and a circumscribing sphere, S, respectively. These potentials obviously coincide inside Ω, but may differ essentially in arbitrary fashion in the moon region. Our aim is to show that the functions ψ^{Ω} and ψ^{S} also coincide inside Ω. The discussion below follows that given by Nesbet [9].

The two basis functions have the form shown in Eq.(3.64), which we write symbolically as

$$\psi^{\Omega} = \chi^{\Omega} + \int_{\Omega} G_0 V^{\Omega} \psi^{\Omega} \tag{3.125}$$

and

$$\psi^{S} = \chi^{S} + \int_{S} G_0 V^{S} \psi^{S}. \tag{3.126}$$

The subscripts in the integration signs indicate the region of integration in each case. Let us now impose the requirement that the functions ψ^{Ω} and ψ^{S} coincide inside Ω, and ask what conditions must be fulfilled for this to be the case. We recall at this point that the χ's, the free-particle contributions to the wave functions, are still unspecified. Thus, for points in the cell we subtract Eq.(3.125) from Eq.(3.126) and obtain the relation,

$$0 = \chi^{S} - \chi^{\Omega} + \int_{M} G_0 V^{M} \psi^{S}, \tag{3.127}$$

where the integral now extends only over the moon region, the potential for which is denoted by V^M. Therefore, for a choice such that,

$$\chi^\Omega = \chi^S + \int_M G_0 V^M \psi^S, \tag{3.128}$$

the solutions for the cell and for the sphere coincide in the interior of the cell.

The argument just given can be made more precise [9]. Subtracting Eq.(3.125) from Eq.(3.126), we obtain the general expression,

$$\psi^S - \psi^\Omega = \chi^S - \chi^\Omega + \int_\Omega G_0 V^\Omega \left(\psi^S - \psi^\Omega\right) + \int_M G_0 V^M \psi^S. \tag{3.129}$$

Now, with the vector $\mathbf{r}$ in Ω, we apply the Helmholtz (free-particle) differential operator, $\nabla^2 + k^2$, to both sides of this expression to obtain

$$\left(\nabla^2 + k^2\right)\left(\psi^S - \psi^\Omega\right) = V^\Omega \left(\psi^S - \psi^\Omega\right). \tag{3.130}$$

We have used the fact that the χ's are solutions of the homogeneous ($V = 0$) equation and that the last integral vanishes under this operation because the arguments of G_0 lie in different domains. Now, it follows from Eq.(3.130) that the difference function is a regular solution of the Schrödinger equation inside the cell, Ω. By an appropriate choice of initial conditions, $\mathbf{r} \to 0$, both of the functions, ψ^Ω and ψ^S, can be made to approach the same value and derivative at the origin, so that the difference function is zero with zero derivative in a neighborhood of the origin. Outward integration of this function from the origin produces the zero function throughout the cell, Ω. This matching of solutions near the origin can be directly and easily effected if one chooses the form of the solution indicated in Eqs.(3.79) and (3.80). In this case, both ψ^Ω and ψ^S approach J_L at the origin and it is easy to see that they coincide throughout a sphere inscribed in the cell (the muffin-tin sphere). (Because the Schrödinger equation is local, two integration subroutines beginning at the origin and integrating outward would produce identical numbers in the two cases up to the radius of the inscribed sphere.) The argument given above shows that they are identical *everywhere* inside the cell.

It is clear that the result just established only holds for basis functions, and not, generally, for wave functions. The former are made to satisfy only a single boundary condition at the origin and, consequently, have no knowledge of the potential outside the cell. The wave function, on the other hand, satisfies also a boundary condition at infinity and is affected by the potential throughout space. Thus, in contrast to the basis functions, the wave functions for a sphere and for a cell contained in it are different everywhere if the potential in the moon region is nonzero. On the other hand, the property of the basis functions just discussed can be used to advantage in the computational implementation of MST, as is pointed out in the following section.

3.7 Convergence of Basis Function Expansions (*)

The result just derived is sufficiently counterintuitive as to arouse suspicions concerning its validity. One center of those concerns has been the question of whether or not ψ^{Ω} has a convergent angular-momentum expansion for points in the corners of the cell, for example, points further from the origin than some part of the moon region [10]. In the angular-momentum representation we have,

$$\psi_L^{\Omega}(\mathbf{r}) = \sum_{L'} J_{L'}(\mathbf{r})\bar{C}^{\Omega}_{L'L} + \int_{\Omega} G_0(\mathbf{r}-\mathbf{r}')V(\mathbf{r}')\psi_L^{\Omega}(\mathbf{r}')\mathrm{d}^3r', \qquad (3.131)$$

or, in vector and matrix form,

$$\langle\psi^{\Omega}(\mathbf{r})| = \langle J(\mathbf{r})|\underline{\bar{C}}^{\Omega} + \int_{\Omega} G_0(\mathbf{r}-\mathbf{r}')V^{\Omega}(\mathbf{r}')\,\langle\psi^{\Omega}(\mathbf{r}')|\,\mathrm{d}^3r', \qquad (3.132)$$

and

$$\langle\psi^{S}(\mathbf{r})| = \langle J(\mathbf{r})|\underline{\bar{C}}^{S} + \int_{S} G_0(\mathbf{r}-\mathbf{r}')V^{S}(\mathbf{r}')\,\langle\psi^{S}(\mathbf{r}')|\,\mathrm{d}^3r', \qquad (3.133)$$

where the expression for χ in Eq.(3.69), has been used. Now, Eq.(3.128) yields the following relation between the coefficients, $\underline{C}^{\Omega}$ and $\underline{C}^{S}$,

$$\langle J(\mathbf{r})|\underline{\bar{C}}^{\Omega} = \langle J(\mathbf{r})|\underline{\bar{C}}^{S} + \int_{M} G_0(\mathbf{r}-\mathbf{r}')V^{M}(\mathbf{r}')\,\langle\psi^{S}(\mathbf{r}')|\,\mathrm{d}^3r'. \qquad (3.134)$$

Let us apply this equation to the case in which the cell point $\mathbf{r}$ lies inside a sphere inscribed within the cell. Then the Green function, $G_0(\mathbf{r}-\mathbf{r}')$, can be expanded in the manner indicated in Eq.(3.110), as r' is greater than r, to yield the result,

$$\underline{\bar{C}}^{\Omega} = \underline{\bar{C}}^{S} - k\int_{M} |N(\mathbf{r}')\rangle V^{M}(\mathbf{r}')\langle\psi^{S}(\mathbf{r}')|\,\mathrm{d}^3r'. \qquad (3.135)$$

This equation allows one to obtain directly the coefficients, $\underline{\bar{C}}^{\Omega}$, for the cell from the coefficients, $\underline{\bar{C}}^{S}$, associated with the entire sphere through the subtraction of an integral over the moon region that involves the wave function for the sphere. This relation gives the angular-momentum representation of the general expression in Eq.(3.127). This equation also can provide a convenient approach to the determination of C^{Ω} and S^{Ω}. Here one choses a potential in the in the moon region for which ψ^S can be determined with relative ease and obtains all quantities through Eq. (3.135). For example, for a constant cell potential, the potential can be extended throughout the circumscribing sphere leading to a spherically symmetric wave function, ψ_ℓ independent of m. At the same time, the previous development leads to questions of convergence which must be addressed directly.

If the point $\mathbf{r}$ lies outside the inscribed sphere then the angular-momentum expansion of the Green function in Eq.(3.132),

$$\begin{aligned}\langle J(\mathbf{r})| \underline{\bar{C}}^{\Omega} &= \langle J(\mathbf{r})| \underline{\bar{C}}^{S} + \int_M G_0(\mathbf{r}-\mathbf{r}')V^M(\mathbf{r}')\,\langle \psi^S(\mathbf{r}')|\,\mathrm{d}^3r', \\ &= \langle J(\mathbf{r})| \underline{\bar{C}}^{S} - k\,\langle J(\mathbf{r})| \int_M |N(\mathbf{r}')\rangle\, V^M(\mathbf{r}')\,\langle \psi^S(\mathbf{r}')|\,\mathrm{d}^3r',\end{aligned} \tag{3.136}$$

may not converge because it contains a possibly divergent sum over $J_L(\mathbf{r})$ and $N_L(\mathbf{r}')$. Apparently, then, one does not regain the starting relation, Eq.(3.132).

On close consideration, we may realize that the difficulties associated with the use of Eq.(3.135) are not dissimilar to those arising with the use of the t-matrix to represent the wavefunction inside the circumscribing sphere. In both cases, one attempts to use a multipolar expansion outside its region of validity. In the case of the t-matrix it was possible to use an analytic continuation of the multipolar series, in terms of double or multiple sums, to obtain converged results. A similar process is possible in the present case. In fact, there are at least two different ways of justifying the formal validity of Eq.(3.136).

3.7.1 *First Justification*

We consider explicitly the case of polyhedral cells. We assume that the cell potential is nonvanishing and sufficiently "smooth" so that a regular solution of the Schrödinger equation exists throughout a simply connected neighborhood of the origin. This requirement clearly excludes singular potentials such as delta functions which provide no interior region. We divide the moon region into nonintersecting parts, M_i, each of which is associated with the face, i, of the cell adjacent to it. Now, Eq.(3.135) can be written in the form

$$\underline{\bar{C}}^{\Omega} = \underline{\bar{C}}^{S} - k\sum_i \int_{M_i} |N(\mathbf{r}')\rangle\, V^{M_i}(\mathbf{r}')\,\langle \psi^S(\mathbf{r}')|\,\mathrm{d}^3r'. \tag{3.137}$$

Let us imagine a thin strip of width ϵ and zero potential as separating each part, M_i, of the moon region from its associated face. We will let ϵ go to zero at the end of formal arguments. Now, for a vector $\mathbf{a}_i^1$ perpendicular to face i, pointing outward and away from it, and of magnitude smaller than the radius of the muffin-tin sphere we have,

$$\underline{g}(\mathbf{a}_i^1)\,|N(\mathbf{r})\rangle = |N(\mathbf{r}-\mathbf{a}_i^1)\rangle\,. \tag{3.138}$$

Choose a vector $\mathbf{a}_i^2$ in the same direction as $\mathbf{a}_i^1$ and note that $|\mathbf{a}_i^2| < |\mathbf{r}'-\mathbf{a}_i^1|$, so that we have,

$$\underline{g}(\mathbf{a}_i^2)\,|N(\mathbf{r}-\mathbf{a}_i^1)\rangle = |N(\mathbf{r}-\mathbf{a}_i^1-\mathbf{a}_i^2)\rangle\,. \tag{3.139}$$

Proceeding this way we can shift the argument of the function $|N(\mathbf{r})\rangle$ by an arbitrary amount to obtain $|N(\mathbf{r}-\mathbf{a}_i)\rangle$, by multiplying each partial integral over M_i by a series of matrices $g(\mathbf{a}_i^\alpha)$ with $\sum_\alpha \mathbf{a}_i^\alpha = \mathbf{a}_i$. We can choose $\mathbf{a}_i$ so large that the surface of a sphere bounding the cell and centered at the point $\mathbf{a}_i$ with respect to the cell center passes entirely through the strip of width ϵ. For vectors $\mathbf{r}$ and $\mathbf{r}'$ inside the cell and in the moon region, respectively, we have,

$$|\mathbf{r}-\mathbf{a}_i| < |\mathbf{r}'-\mathbf{a}_i| \tag{3.140}$$

as $\mathbf{r}'-\mathbf{a}_i$ lies outside and $\mathbf{r}-\mathbf{a}_i$ inside the sphere. We then have,

$$\begin{aligned}\langle J(\mathbf{r}-\mathbf{a}_i)| \int_{M_i} |N(\mathbf{r}'-\mathbf{a}_i)\rangle\, V^M(\mathbf{r}')\, \langle\psi^S(\mathbf{r}')|\, \mathrm{d}^3r' \\ = \int_{M_i} G_0(\mathbf{r}-\mathbf{r}')V^M(\mathbf{r}')\, \langle\psi^S(\mathbf{r}')|\, \mathrm{d}^3r'.\end{aligned} \tag{3.141}$$

The construction leading to the last equation for the region M_i can be repeated for all parts of the moon region and the results summed to yield the expression

$$\begin{aligned}\langle J(\mathbf{r})|\, \underline{\bar{C}}^\Omega &= \langle J(\mathbf{r})|\, \underline{\bar{C}}^S - \sum_i \langle J(\mathbf{r}-\mathbf{a}_i)| \\ &\quad \times \int_{M_i} |N(\mathbf{r}'-\mathbf{a}_i)\rangle\, V^M(\mathbf{r}')\, \langle\psi^S(\mathbf{r}')|\, \mathrm{d}^3r' \\ &= \langle J(\mathbf{r})|\, \underline{\bar{C}}^S - \int_M G_0(\mathbf{r}-\mathbf{r}')V^M(\mathbf{r}')\, \langle\psi^S(\mathbf{r}')|\, \mathrm{d}^3r',\end{aligned} \tag{3.142}$$

which clearly has the form of our starting Eq.(3.134). At this point we can let ϵ vanish and obtain Eq.(3.134).

3.7.2 Second Justification

There is an alternative and possibly more elegant way of establishing the convergence of the product on the left-hand side of the last equation. Note that convergence difficulties with this product may conceivably arise when the argument of the Bessel function becomes larger than the argument of the Hankel (or Neumann) function that enters the definition of the quantity $\underline{\bar{C}}^\Omega$, Eq.(3.71). This difficulty, however, can be bypassed. To see this, we observe that the argument of the irregular solution in Eq.(3.71) can be made arbitrarily large without affecting the value of the integral. This is so because the integral can be converted by means of Green's theorem into a surface integral over any surface that includes the potential $V(\mathbf{r})$, with a potential-free region beyond a bounding sphere of radius r. Then, the integrals in Eq.(3.71) or Eq.(3.73) yield the asymptotic value of the coefficients $\underline{C}$ associated with the potential inside a sphere of radius r. That this procedure is allowed becomes clear when one considers the definition of $\underline{C}$ in terms of a surface integral, Eq.(3.81). This asymptotic value is independent of the radius of the sphere enclosing the potential provided that this radius is larger than r itself. Thus, this radius can be chosen infinitely large, leading to an argument of the irregular solution that is always larger than that of the regular solution, and consequently to converged results in Eq.(3.136).

It is also interesting to note that the argument just made does not apply to the product $H(\mathbf{r})\underline{S}$. Here, the argument of the Hankel function is fixed and if the integral defining $\underline{S}$, Eq.(3.82), contains vectors larger than r this product will eventually diverge. Clearly, the same considerations apply to products of the t-matrix with the irregular solution. As we saw in the last section, it is these products that require special treatment in terms of multiple, conditionally convergent summations when applied to regions inside the bounding sphere.

We have now established the following important result. With a proper choice of initial conditions, *the basis function associated with the potential in a cell, Ω, coincides over the interior of the cell with the corresponding basis function associated with a sphere, or any other domain, that includes the cell, independently of any potential that may be present in the moon region, i.e., outside Ω but inside the sphere (domain) including it.* Furthermore, we have also shown that the asymptotic values of the cell phase functions, $\underline{\bar{C}}^{\Omega}$ and $\underline{\bar{S}}^{\Omega}$, can be obtained from those of the sphere by means of the relations,

$$\underline{\bar{C}}^{\Omega} = \underline{\bar{C}}^{S} - k\int_{M} |N(\mathbf{r}')\rangle V^{M}(\mathbf{r}')\langle\psi^{S}(\mathbf{r}')|\,\mathrm{d}^3r', \tag{3.143}$$

and

$$\underline{\bar{S}}^{\Omega} = \underline{\bar{S}}^{S} + k\int_{M} |J(\mathbf{r}')\rangle V^{M}(\mathbf{r}')\langle\psi^{S}(\mathbf{r}')|\,\mathrm{d}^3r'. \tag{3.144}$$

As already mentioned, these formal results can have important computational ramifications. We note that occasionally it may be easier to solve for ψ^S than for ψ^Ω. For example, this is the case in the empty lattice, consisting of a constant shifted potential throughout space, where ψ^S is spherically symmetric. There, $\underline{\bar{C}}^{\Omega}$ and $\underline{\bar{S}}^{\Omega}$ can be found by means of Eqs.(3.143) and (3.144) alleviating the need to solve cumbersome coupled-channel equations. Referring to Eqs. (3.71) and (3.72), this procedure amounts to the use of the exact basis function under the integral sign, with all internal summations carried to convergence. This process would be expected to provide much faster rates of convergence than a direct solution of the coupled-channel equations. Alternatively, expressing the wave function as a sum over phase functions may lead to divergent behavior as one attempts to determine the phase functions for values of L that approach the maximum value used in the expansion of the sum. This is another manifestation of the importance of carrying to convergence all internal sums. How far the internal index should be taken for a given preassigned convergence criterion, of course, depends (among other things) on the value of the outer index. However, carrying the internal sum to about twice the value of the outer index, for sufficiently large values of the latter, may displace divergences far enough and lead to acceptable convergence rates, and savings of computational effort. Such savings in effort may indeed materialize in some realistic applications, or in test cases, in which the potential inside the sphere can be described by an analytic function. In general, optimal use of Eqs.(3.143) and (3.144) can be made in those cases in which one can define a potential in the moon region of a cell that leads to an easy determination of ψ^S.

References

[1] Leonard I. Schiff, *Quantum Mechanics* (McGraw-Hill Book Company, New York, 1955).
[2] Charles J. Joachain, *Quantum Collision Theory* (North Holland, New York, 1983).
[3] Albert Messiah, *Quantum Mechanics* (John Wiley and Sons, New York, 1966).
[4] Eugene Merzbacher, *Quantum Mechanics* (John Wiley and Sons, New York, 1970).
[5] E. U. Condon and G. H. Shortley, *The Theory of Atomic Spectra* (Cambridge University Press, Cambridge, 1976).
[6] F. Calogero, *Variable Phase Approach to Potential Scattering* (Academic, New York, 1967).
[7] A. Gonis, X. -G. Zhang, and D. M. Nicholson, Phys. Rev. B**40**, 947 (1989).
[8] A. Gonis, Erik C. Sowa, and P. A. Sterne, Phys. Rev. Lett. **66**, 2207 (1991).
[9] R. K. Nesbet, Phys. Rev. B**41**, 4984 (1990).
[10] R. G. Brown and M. Ciftan, in *Applications of Multiple Scattering Theory to Materials Science*, W. H. Butler, P. H. Dederichs, A. Gonis, and R. L. Weaver, eds. Materials Research Society Conference Proceedings, Vol. 253, 1992.

4

Formal Development of MST

In this chapter, we present a brief overview of some elements of formal multiple scattering theory that will be useful in our subsequent discussion. As is customary, we assume that the Hamiltonian, H_0, for a noninteracting system has only a continuous spectrum, while the Hamiltonian, $H = H_0 + V$, for the physical (interacting) system has the same continuous spectrum, but may possess a discrete spectrum of bound states below the continuum. Thus, it is assumed that V does not support bound states in the continuous part of the spectrum. These assumptions are unlikely to limit significantly the application of MST in the context of electronic structure calculations. More detailed discussions of these and other commonly made assumptions in scattering theory may be found in the literature. A summary of some formal concepts of scattering theory is given in Appendix F.

4.1 Scattering Theory for a Single Potential

Given a Hamiltonian, H_0, with eigenstates [1, 2] $|\chi_\alpha\rangle$,

$$H_0|\chi_\alpha\rangle = E_\alpha|\chi_\alpha\rangle, \tag{4.1}$$

for an unperturbed system, the scattered-wave solutions, $|\psi_\alpha^\pm\rangle$, of an interacting (perturbed) system with the Hamiltonian, $H = H_0 + V$,

$$H|\psi_\alpha^\pm(V)\rangle = E_\alpha|\psi_\alpha^\pm(V)\rangle \tag{4.2}$$

satisfy the Lippmann–Schwinger equations [3],

$$|\psi_\alpha^\pm(V)\rangle = |\chi_\alpha\rangle + (E - H_0 \pm i\epsilon)^{-1} V |\psi_\alpha^\pm(V)\rangle, \tag{4.3}$$

where ϵ is a positive infinitesimal. The states $|\psi_\alpha^+\rangle$ or $|\psi_\alpha^-\rangle$ are those that in the remote past or future, respectively, coincide with the free-particle state, $|\chi_\alpha\rangle$, and correspond to the same eigenvalue, E_α.

4.1.1 The S-Matrix and the t-Matrix

In a scattering experiment, free states are transformed into scattered states. In the language of quantum mechanics, this transformation is described by the *scattering operator*, or *S-operator*.[1] Such an operator can be postulated for any physical system characterized by a collision process. In a particular representation, such as that of angular momentum, the S-operator is expressed as the *S-matrix*.

We can construct a representation of the S-matrix in terms of the eigenstates of H_0 and H. First, we note that the Lippmann–Schwinger equation, Eq. (4.3), can also be written in the form,

$$|\psi_\alpha^\pm\rangle = |\chi_\alpha\rangle + (E_\alpha - H \pm \mathrm{i}\epsilon)^{-1} V |\chi_\alpha\rangle. \tag{4.4}$$

From the operator identity,

$$\frac{1}{E - H + \mathrm{i}\epsilon} - \frac{1}{E - H - \mathrm{i}\epsilon} = -2\pi\mathrm{i}\delta(E - H), \tag{4.5}$$

we obtain the following interesting relation between the eigenvectors $|\psi_\alpha^+\rangle$ and $|\psi_\alpha^-\rangle$,

$$|\psi_\alpha^+\rangle - |\psi_\alpha^-\rangle = -2\pi\mathrm{i}\delta(E_\alpha - H) V |\chi_\alpha\rangle. \tag{4.6}$$

We now define the elements $S_{\alpha\alpha'}$ of the S-matrix in the basis of the eigenvectors $|\chi_\alpha\rangle$ of H_0,

$$S_{\alpha\alpha'} = \langle\chi_\alpha|S|\chi_{\alpha'}\rangle \equiv \langle\psi_\alpha^-|\psi_\alpha^+\rangle. \tag{4.7}$$

Through the use of Eq.(4.5), we can also write,

$$\begin{aligned} S_{\alpha\alpha'} &= \left[\langle\psi_\alpha^+| - 2\pi\langle\chi_\alpha|V\delta(E_\alpha - H)\right] |\psi_{\alpha'}^+\rangle \\ &= \langle\psi_\alpha^+|\psi_{\alpha'}^+\rangle - 2\pi\mathrm{i}\delta(E_\alpha - E_{\alpha'})\langle\chi_\alpha|V|\psi_{\alpha'}^+\rangle, \end{aligned} \tag{4.8}$$

or

$$S_{\alpha\alpha'} = \langle\psi_\alpha^-|\psi_{\alpha'}^-\rangle - 2\pi\mathrm{i}\delta(E_\alpha - E_{\alpha'})\langle\psi_\alpha^-|V|\chi_{\alpha'}\rangle, \tag{4.9}$$

where the fact that the states $|\psi^\pm\rangle$ are eigenstates of H has been used in replacing H with its eigenvalue in the argument of the δ-function. It is

---

[1]The symbol S has been used in Chapter 3 to denote a phase function. Hopefully, the content of the discussion will make clear which quantity is being represented by this symbol in a particular case.

shown in Appendix F that the eigenvectors of H have the same normalization as those of H_0 so that the first two terms in the last two expressions are equal. Thus Eq.(4.9) implies that the S-matrix can be written as a sum of a potential-independent and a potential-dependent term. The potential dependent term is called the t-matrix. Thus, when $E_\alpha = E_{\alpha'} = E$, we have

$$\langle \chi_\alpha | V | \psi^+_{\alpha'} \rangle = \langle \psi^-_\alpha | V | \chi_{\alpha'} \rangle = T_{\alpha\alpha'}. \tag{4.10}$$

These expressions define the on-the-energy-shell ($E_\alpha = E_{\alpha'}$) t-matrix element, $T_{\alpha\alpha'}$. It follows that the matrix elements $T_{\alpha\alpha'}$ are symmetric with respect to the indices α and α'. Because they are restricted to lie on the energy shell, these matrix elements do not define an operator. Their off-shell extensions, however, do define proper operators through the relations

$$T^+_{\alpha\alpha'} = \langle \chi_\alpha(E_\alpha) | T^+ | \chi_{\alpha'}(E_{\alpha'}) \rangle = \langle \chi_\alpha(E_\alpha) | V | \psi^+_{\alpha'}(E_{\alpha'}) \rangle \tag{4.11}$$

and

$$T^-_{\alpha\alpha'} = \langle \chi_\alpha(E_\alpha) | T^- | \chi_{\alpha'}(E_{\alpha'}) \rangle = \langle \psi^-_\alpha(E_\alpha) | V | \chi_{\alpha'}(E_{\alpha'}) \rangle. \tag{4.12}$$

On the energy shell, the matrices $T^+_{\alpha\alpha'}$ and $T^-_{\alpha\alpha'}$ coincide with the t-matrix $T_{\alpha\alpha'}$,

$$T^+_{\alpha\alpha'} = T^-_{\alpha\alpha'} = T_{\alpha\alpha'} \quad \text{when} \quad E_\alpha = E_{\alpha'} = E. \tag{4.13}$$

Now, the S-matrix can be written explicitly in terms of the t-matrix,

$$S_{\alpha\alpha'} = \rho(E_\alpha)^{-1} \delta_{\hat{\alpha}\hat{\alpha}'} \delta(E_\alpha - E_{\alpha'}) - 2\pi \mathrm{i} T_{\alpha\alpha'}(E_\alpha), \tag{4.14}$$

where the normalization indicated in Eq.(F.21) has been used. The t-matrix often provides a more convenient quantity than the S-matrix for the discussion of a scattering experiment. Note that it is the t-matrix part of the S-matrix, arising from the potential-dependent part of the latter, that causes change in the system. If $V = 0$, then the t-matrix will vanish and the S-matrix will describe the development of the system under H_0. In fact, it is the on-the-energy-shell t-matrix, (also called the transition matrix), which is usually made the basis of developments of multiple scattering theory.

Explicit expressions for the t-matrix in connection with spherically symmetric as well as non spherical cell potentials were derived in the last chapter. Because of its formal and computational importance, in the following subsection we summarize some useful properties of the t-matrix. (It is common practice in the literature of multiple scattering theory to refer to the t-matrix as the scattering matrix. For the sake of consistency we shall follow this practice, taking care to make the appropriate distinctions when the possibility of confusion may arise.)

4.1.2 t-Matrices and Green Functions

The free-particle Green function operator, G_0, and the full Green function operator, G, are given by the expressions,

$$G_0^{\pm}(E) = (E - H_0 \pm i\epsilon)^{-1}, \tag{4.15}$$

and

$$G^{\pm}(E) = (E - H \pm i\epsilon)^{-1}, \tag{4.16}$$

which are associated with the Hamiltonians H_0 and H, respectively. Using these expressions, we can cast Eq.(4.3) into a variety of equivalent forms,

$$\begin{aligned} |\psi_\alpha^{\pm}(V)\rangle &= |\chi_\alpha\rangle + G_0^{\pm}(E)V|\psi_\alpha^{\pm}(V)\rangle \\ &= |\chi_\alpha\rangle + G^{\pm}(E)V|\chi_\alpha\rangle \\ &= |\chi_\alpha\rangle + G_0^{\pm}(E)T^{\pm}(V)|\chi_\alpha\rangle, \end{aligned} \tag{4.17}$$

where the t-matrix operators, $T^{\pm}$,are defined by (Eqs.(4.11) and (4.12))

$$T_{\alpha\alpha'}^{+} = \langle\chi_\alpha|T^{+}|\chi_{\alpha'}\rangle = \langle\chi_\alpha|V|\psi_{\alpha'}^{+}(V)\rangle \tag{4.18}$$

and

$$T_{\alpha\alpha'}^{-} = \langle\chi_\alpha|T^{-}|\chi_{\alpha'}\rangle = \langle\chi_\alpha|V|\psi_{\alpha'}^{-}(V)\rangle. \tag{4.19}$$

Scattering theory can be described fully in terms of on-the- energy-shell elements of T^{+} or T^{-}. On the energy shell, $E_\alpha = E_{\alpha'} = E$, and we have,

$$T_{\alpha\alpha'}(E) = T_{\alpha\alpha'}^{+} = T_{\alpha\alpha'}^{-}. \tag{4.20}$$

Now, from a simple iteration of the first line of Eq.(4.17) and comparison with the third line, we find that T is given formally by the expression,

$$T(V) = V + VG_0V + VG_0VG_0V + \cdots, \tag{4.21}$$

where energy arguments have been suppressed. If this series converges, $T(V)$ satisfies the *Dyson equation*,

$$T(V) = V + VG_0T(V) \tag{4.22}$$

which has the formal solution,

$$\begin{aligned} T(V) &= (1 - VG_0)^{-1}V \\ &= (V^{-1} - G_0)^{-1}. \end{aligned} \tag{4.23}$$

Even in those instances in which the series in Eq.(4.21) does not converge, the relation, Eq.(4.22), between T and V is usually still valid, especially for the types of potentials of primary interest here, namely, those that are spatially bounded and vanish identically outside a sphere of finite radius. It follows from Eq.(4.17) that V and $T(V)$ satisfy the relation,

$$V|\psi_\alpha(V)\rangle = T(V)|\chi_\alpha\rangle. \tag{4.24}$$

We note that the Lippmann–Schwinger equations can also be expressed in terms of the Green functions,

$$\begin{aligned} G &= G_0 + G_0 V G \\ &= \left[G_0^{-1} - V\right]^{-1} \\ &= G_0 + G_0 T(V) G_0, \end{aligned} \tag{4.25}$$

which leads to the Green-function analogue of Eq.(4.24),

$$VG = T(V)G_0. \tag{4.26}$$

4.2 Two-Potential Scattering

In this section we generalize the Lippmann–Schwinger equations to the case in which two potentials act simultaneously. The following formalism is useful in deriving the equations of multiple scattering theory in their most general form, namely, when scattering events take place in the presence of a potential rather than in free space. In addition, the operator algebra that underlies the following development projects out the fundamental notions of multiple scattering theory and is helpful in enhancing an intuitive understanding of the formalism of MST.

The Lippmann–Schwinger equations, Eq.(4.3),

$$|\psi_\alpha^\pm(V)\rangle = |\chi_\alpha\rangle + (E - H_0 \pm i\epsilon)^{-1} V |\psi_\alpha^\pm(V)\rangle, \tag{4.27}$$

yield the solutions $|\psi_\alpha^+\rangle$ and $|\psi_\alpha^-\rangle$ of the Schrödinger equation corresponding to a potential, V, which evolve out of the unperturbed state $|\chi_\alpha\rangle$ at the same energy, E. In this equation, H_0 is the Hamiltonian of the unperturbed system. We seek a generalization of the Lippmann- Schwinger equations in which the unperturbed Hamiltonian is that of a reference system, $H_r = H_0 + U$. Then V would describe a perturbation superimposed on the reference system. On purely intuitive grounds, we might expect that the Lippmann–Schwinger equations can be used to describe the scattering caused by V in the field of U. This turns out to be the case provided that the role of the free-space solutions, $|\chi_\alpha\rangle$, is played by the solutions $|\phi_\alpha^\pm\rangle$ of the Lippmann–Schwinger equations corresponding to the potential U. We will now show that in the presence of two potentials we can write the Lippmann–Schwinger equations in the form

$$\begin{aligned} |\psi_\alpha^\pm\rangle &= |\phi_\alpha^\pm\rangle + (E_\alpha - H_0 - U \pm \mathrm{i}\epsilon)^{-1} V |\psi_\alpha^\pm\rangle \\ &= |\phi_\alpha^\pm\rangle + G_r^\pm V |\psi_\alpha^\pm\rangle, \end{aligned} \tag{4.28}$$

where

$$G_r^\pm = (E_\alpha - H_0 - U \pm \mathrm{i}\epsilon)^{-1} \tag{4.29}$$

is the Green function for the reference medium, and the states $|\phi_\alpha^\pm\rangle$ are the solutions of the Lippmann–Schwinger equations corresponding to the potential U acting alone. In order to derive Eq.(4.28), we recall that the transition matrix, $T_{\alpha\beta}$, corresponding to the potential $U+V$ is given by the expression (see Eq.(4.10)),

$$T_{\alpha\beta} = \langle\chi_\alpha|U+V|\psi_\beta^{\pm}\rangle, \tag{4.30}$$

where the states $|\psi_\alpha^{\pm}\rangle$ evolve out of the free states $|\chi_\alpha\rangle$ and satisfy the Lippmann–Schwinger equations,

$$|\psi_\alpha^{\pm}(U+V)\rangle = |\chi_\alpha\rangle + (E-H_0 \pm i\epsilon)^{-1}(U+V)|\psi_\alpha^{\pm}(U+V)\rangle. \tag{4.31}$$

We now introduce the states $|\phi^{\pm}\rangle$ which are the solutions of the Lippmann–Schwinger equations corresponding to the potential U,

$$|\phi_\alpha^{\pm}(U)\rangle = |\chi_\alpha\rangle + (E-H_0 \pm i\epsilon)^{-1}U|\phi_\alpha^{\pm}(U)\rangle. \tag{4.32}$$

When this equation is solved for the state $\langle\chi|$ in terms of the states $\langle\phi^-|$ and the result is substituted into Eq.(4.30) one obtains [2],

$$\begin{aligned} T_{\alpha\beta} &= \langle\phi_\alpha^-|U|\psi_\beta^+\rangle + \langle\phi_\alpha^-|V|\psi_\beta^+\rangle \\ &\quad - \langle\phi_\alpha^-|U(E_\beta - H_0 + \mathrm{i}\epsilon)^{-1}(U+V)|\psi_\beta^+\rangle \\ &= \langle\phi_\alpha^-|V|\psi_\beta^+\rangle + \langle\phi_\alpha^-|U|\chi_\beta\rangle, \end{aligned} \tag{4.33}$$

where the last line follows from Eq.(4.31). Using the defining Eqs.(4.10) for the on-the-energy-shell t-matrix, we can write the last term in Eq.(4.33) in the form,

$$\langle\phi_\alpha^-|U|\chi_\beta\rangle = \langle\chi_\alpha|U|\phi_\beta^+\rangle \equiv T_{\alpha\beta}(U). \tag{4.34}$$

Now, the expression for the total scattering matrix takes the form,

$$\begin{aligned} T_{\alpha\beta} &= \langle\chi_\alpha|U|\phi_\beta^+\rangle + \langle\phi_\alpha^-|U|\psi_\beta^+\rangle \\ &= T_{\alpha\beta}(U) + \tilde{T}_{\alpha\beta}(V), \end{aligned} \tag{4.35}$$

where the matrix element $T_{\alpha\beta}(U)$ determines the probability of transitions from state $|\chi_\alpha\rangle$ to state $|\chi_\beta\rangle$ under the action of the potential U alone, while the matrix element $\tilde{T}_{\alpha\beta}(V)$ describes the effects of any further scattering induced by V.

In order to clarify the meaning of this last term, we use the operator identity

$$A^{-1} - B^{-1} = A^{-1}(A-B)B^{-1}, \tag{4.36}$$

with $A = (E_\alpha - H_0 + \mathrm{i}\epsilon)$ and $B = (E_\alpha - H + \mathrm{i}\epsilon)$, and the relation $-H_0 = -H + V$ to rewrite the Lippmann–Schwinger equation, Eq.(4.31), in the closed form,

$$|\psi_\alpha^+\rangle = |\chi_\alpha\rangle + (E_\alpha - H_0 - U - V + \mathrm{i}\epsilon)^{-1}(U+V)|\chi_\alpha\rangle. \tag{4.37}$$

Similarly, we have

$$|\phi_\alpha^+\rangle = |\chi_\alpha\rangle + (E_\alpha - H_0 - U + \mathrm{i}\epsilon)^{-1}U|\chi_\alpha\rangle. \tag{4.38}$$

It is important to emphasize that in spite of their appearance these equations *do not* provide a closed form solution of the Lippmann–Schwinger equations in terms of the free states $|\chi_\alpha\rangle$. This is so because they involve the Green function, G, for the entire interacting system, a quantity that is still unknown. However, Eqs.(4.37) and (4.38) can be used to advantage in a formal sense. Upon subtracting Eq.(4.38) from Eq.(4.37) and using again the operator identity, Eq.(4.36), we obtain the expression,

$$\begin{aligned}|\psi_\alpha^+\rangle &= |\phi_\alpha^+\rangle + (E_\alpha - H_0 - U - V + \mathrm{i}\epsilon)^{-1} \\ &\quad \times V\left[|\chi_\alpha\rangle + (E_\alpha - H_0 - V + \mathrm{i}\epsilon)^{-1} V|\chi_\alpha\rangle\right].\end{aligned} \tag{4.39}$$

Now, from Eq.(4.38) the terms inside the square brackets are equal to $|\phi_\alpha^+\rangle$ so that we can write,

$$|\psi_\alpha^+\rangle = |\phi_\alpha^+\rangle + (E_\alpha - H_0 - U - V + \mathrm{i}\epsilon)^{-1} V|\phi_\alpha^+\rangle, \tag{4.40}$$

or equivalently (refer to the derivation of Eqs.(4.37) and (4.38)),

$$|\psi_\alpha^+\rangle = |\phi_\alpha^+\rangle + (E_\alpha - H_0 - U + \mathrm{i}\epsilon)^{-1} V|\psi_\alpha^+\rangle. \tag{4.41}$$

This is Eq.(4.28). An analogous expression can be derived in terms of the states $|\phi_\alpha^-\rangle$ and $|\psi_\alpha^-\rangle$ yielding the desired expressions for the generalized Lippmann–Schwinger equations, Eq.(4.28). Thus, the term

$$\begin{aligned}\tilde{T}_{\alpha\beta}(V/U) &\equiv \langle\phi_\alpha^-|V|\psi_\beta^+\rangle \\ &= \langle\phi_\alpha^-|\tilde{T}(V/U)|\phi_\beta^+\rangle\end{aligned} \tag{4.42}$$

can be interpreted as the probability amplitude for the transition caused by a perturbation, V, between the states $|\phi_\alpha^\pm\rangle$ and $|\phi_\beta^\pm\rangle$ already scattered (distorted) by the reference potential U. We see easily that in the case of $U = 0$, Eq.(4.42) reduces to the ordinary form of the t-matrix for scattering produced by a single potential, Eq.(4.12).

Let us now summarize the most important results of two-potential scattering theory. In the presence of two potentials, $U + V$, acting simultaneously the Lippmann–Schwinger equations for the state vectors take the form, (suppressing the superscripts $\pm$),

$$\begin{aligned}|\psi_\alpha(U+V)\rangle &= |\chi_\alpha\rangle + G_0[U+V]|\psi_\alpha(U+V)\rangle \\ &= |\chi_\alpha\rangle + G_0 T(U+V)|\chi_\alpha\rangle.\end{aligned} \tag{4.43}$$

It can be readily shown that corresponding expressions can be written for the Green function,

$$\begin{aligned}G &= G_0 + G_0[U+V]G \\ &= G_0 + G_0 T(U+V) G_0.\end{aligned} \tag{4.44}$$

Equivalently, we can also write,

$$\begin{aligned}|\psi_\alpha(U+V)\rangle &= |\phi_\alpha(U)\rangle + G(U)V|\psi_\alpha(U+V)\rangle \\ &= |\phi_\alpha(U)\rangle + G(U)\tilde{T}(V/U)|\phi_\alpha(U)\rangle\end{aligned} \tag{4.45}$$

and

$$
\begin{aligned}
G &= G(U) + G(U)VG \\
&= G(U) + G(U)\tilde{T}(V/U)G(U),
\end{aligned}
\tag{4.46}
$$

where $|\phi_\alpha(U)\rangle$ and $G(U)$ are the wave function and Green function, respectively, associated with the potential U alone, and $T(V/U)$ denotes the additional scattering produced by V in a wave already distorted by the presence of U. This last quantity provides the proper description of the scattering by V in the field of U. The use of Eq.(4.22) followed by a combination of terms yields the relation,

$$
T(U+V) = T(U) + [1 + T(U)G_0]V[1 + G_0T(U+V)]. \tag{4.47}
$$

This expression will be used below to derive the equations of multiple scattering theory. It should also be noted that Eq.(4.47) can be derived directly from Eq.(4.21) by replacing V by $V+U$, expanding, and collecting terms.

4.2.1 *An Integral Equation for the t-Matrix*

Before embarking on the derivation of the MST equations, we note in passing that the analysis of two-potential scattering presented above can be used to derive an alternative and computationally convenient integral equation for the t-matrix. Writing $U + V = U + \delta U$ and using Eq.(4.33), we have in the limit $\delta U \to 0$,

$$
\begin{aligned}
t(U+\delta U) - t(U) &= \delta t(U) \\
&= \langle \psi(U)|\delta U|\psi(U)\rangle
\end{aligned}
\tag{4.48}
$$

where $|\psi\rangle$ satisfies the Lippmann–Schwinger equation with respect to the potential U. This functional equation holds for all variations of U with respect to any chosen parameter, such as the length of the radius vector, r. In the angular-momentum representation, and for the choice of normalization made there, it leads directly to Eq.(3.105).

4.3 The Equations of Multiple Scattering Theory

Straightforward iteration of Eq. (4.47) yields the series,

$$
\begin{aligned}
T(U+V) = T(U) + T(V) &+ T(U)G_0T(V) + T(V)G_0T(U) \\
&+ T(U)G_0T(V)G_0T(U) + \ldots,
\end{aligned}
\tag{4.49}
$$

where $T(V)$ is the t-matrix corresponding to the potential V acting alone. Note one characteristic property of Eq.(4.49), namely, that no two successive scattering events correspond to the same potential, that is, no term like

$T(U)G_0T(U)$ appears in this expression. This feature, which was encountered in our intuitive discussion of multiple scattering theory in Chapter 2, reflects the fact that $T(U)$ and $T(V)$ describe the *complete* scattering from the potentials U and V, respectively. Equation (4.49) can be readily generalized to the case in which a number of potentials are acting simultaneously. We may consider one of the potentials, say U, as being the sum of two other potentials, and one of those new potentials as the sum of two other ones, and so on, to arrive at the expression corresponding to Eq.(4.49) when the potential is given as a sum of terms, $V = \sum_{i=1} V^i$. In this case Eq.(4.21) assumes the form,

$$T \equiv T\left[\sum_i V^i\right] = V + VG_0V + VG_0VG_0V + \cdots$$
$$= \sum_i V^i + \left[\sum_i V^i\right] G_0 \left[\sum_j V^j\right] + \cdots. \tag{4.50}$$

Now, the sum of all repeated, consecutive products with the same potential (cell) index, i, such as $V^i + V^iG_0V^i + V^iG_0V^iG_0V^i + ...$, can be grouped together and replaced by the cell t-matrix, $t(V^i) \equiv t^i$.This allows one to cast Eq.(4.50) in the form,

$$T\left[\sum_i V^i\right] = \sum_i t^i + \sum_i \sum_{j\neq i} t^i G_o t^j + \cdots, \tag{4.51}$$

which is clearly a generalization of Eq.(4.49) to the case in which a number of potentials are acting together. As was the case in Eq.(4.49), no two consecutive scattering events in Eq.(4.51) refer to the same potential. Applied to the case of nonoverlapping cell potentials, this expression contains no consecutive scattering terms corresponding to the same cell. Equation (4.51) is the basic equation of multiple scattering theory that was derived in Chapter 2, Eq.(2.6), along intuitive grounds based on Huygens' principle. It shows that *the t-matrix of a scattering assembly is made up of scattering sequences, each sequence involving scattering at individual potential cells with free-particle propagation between scattering events.*

One may proceed from Eq.(4.51) in the same manner as from Eq.(2.27). Denoting by T^{ij} the sum of all terms (scattering sequences) in Eq.(4.51) that start with t^i and end with t^j leads to the expression,

$$T = \sum_{i,j} T^{ij} \equiv \left[\left(\sum_i V^i\right)^{-1} - G_0\right]^{-1}. \tag{4.52}$$

It can be shown through direct iteration that the quantities T^{ij} satisfy the *equation of motion*,

$$T^{ij} = t^i \delta_{ij} + t^i G_0 \sum_{k \neq i} T^{kj}. \tag{4.53}$$

Equations (4.52) and (4.53) constitute a set of fundamental equations of multiple scattering theory and are equivalent to those arrived at through alternative derivations [4]. Furthermore, Eq.(4.52) exhibits a fundamental property of the total t-matrix, T; the t-matrix (scattering matrix) corresponding to an assembly of scattering potential cells depends on the *total* potential, $\sum_i V^i$, being entirely independent of the shape, extent, or overlap of individual cells. In the case of nonoverlapping potentials, certain simplifications become possible. In this case, we can introduce in a unique manner the site off-diagonal elements, G_0^{ij}, of the free-particle propagator between cells (potentials) i and j, (that part of G_0 which connects t^i to t^j in Eq.(4.51)) and write Eq.(4.53) in the form,

$$T^{ij} = t^i \delta_{ij} + t^i \sum_{k \neq i} G_0^{ik} T^{kj}. \tag{4.54}$$

Thus, the T^{ij} can be obtained as the inverse of a matrix M with matrix elements,

$$M^{ij} = m^i \delta_{ij} - G_0^{ij}(1 - \delta_{ij}), \tag{4.55}$$

where $m^i \equiv (t^i)^{-1}$ is the (formal) inverse of the t-matrix associated with cell Ω_i.

It is clear that the matrix elements, T^{ij}, and the associated Green-function matrix elements, G_0^{ij}, depend on the particular partition of a potential into cells and on the choice of cell centers. In other words, these matrix elements are functions of the representation chosen to describe the potential. On the other hand, as already mentioned, the full scattering matrix, T, and the associated Green function are invariant with respect to representation. This invariance can be easily demonstrated in an alternative way by means of the expression

$$V = \sum_i V^i = \sum_i [T(V^i)^{-1} + G_0]^{-1}, \tag{4.56}$$

which follows from Eq.(4.52), and exhibits the invariance of the total potential, V, with respect to partitioning into cells. This invariance is preserved in the angular momentum representation of multiple scattering theory.

4.3.1 The Wave Functions of Multiple Scattering Theory

It is particularly instructive to interpret the equations of multiple scattering theory in terms of waves being incident upon and scattered by the presence of a potential. To fix ideas, let us recall Eq.(4.24), which for a single potential, V^i, in vacuum takes the form,

$$V^i|\psi\rangle = t^i|\chi\rangle, \tag{4.57}$$

with state indices, α, suppressed. We can think of the state $|\chi\rangle$ in this equation as describing an incident wave which, upon interaction with the potential V, is transformed to an outgoing wave, $|\psi\rangle$. This state is the solution of the Lippmann–Schwinger equations for the entire system including the scatterer. We wish to carry this interpretation over to the case of an assembly of scattering cells. To this end, we introduce [5] the state $|\psi^{\mathrm{in},i}\rangle$ which describes the wave incident on cell i, (analogous to $|\,\chi\rangle$ in Eq.(4.57)), in the presence of all other scatterers. In formal analogy with Eq.(4.57) we can write

$$V^i|\psi\rangle = t^i|\psi^{\mathrm{in},i}\rangle, \tag{4.58}$$

where once again $|\psi\rangle$ denotes the solution of the Lippmann–Schwinger equations for the entire system, including the scatterer at i. It remains only for us to derive a recognizable form for $|\psi^{\mathrm{in},i}\rangle$.

From Eq.(4.24) we have the relation,

$$\begin{aligned}\sum_i V^i|\psi\rangle &= \sum_{i,j} T^{ij}|\chi\rangle \\ &= \sum_i T^i|\chi\rangle,\end{aligned} \tag{4.59}$$

where we have defined the quantity,

$$T^i = \sum_{j\neq i} T^{ij}, \tag{4.60}$$

describing all multiple scattering events emanating from cell i. It follows from Eq.(4.54) that

$$T^i = t^i\left[1 + G_0\sum_{k\neq i} T^k\right]. \tag{4.61}$$

Now, from Eq.(4.59) we obtain the effect of V^i on the state $|\psi\rangle$,

$$\begin{aligned}V^i|\psi\rangle &= T^i|\chi\rangle \\ &= t^i\left[1 + G_0\sum_{k\neq i} T^k\right]|\chi\rangle \\ &= t^i|\psi^{\mathrm{in},i}\rangle,\end{aligned} \tag{4.62}$$

which leads to the identification,

$$|\psi^{\text{in},i}\rangle = \left[1 + G_0 \sum_{k \neq i} T^k\right] |\chi\rangle. \tag{4.63}$$

Through the use of Eq.(4.62), the last equation can also be written in the form

$$|\psi^{\text{in},i}\rangle = |\chi\rangle + G_0 \sum_{k \neq i} t^k |\psi^{\text{in},k}\rangle. \tag{4.64}$$

This equation describes the incoming wave at cell (site) i as the superposition of waves that are incident on all other sites, j, are scattered there and propagate via G_0 toward site i. To this set of waves, we must often add an independent overall incident wave, $|\chi\rangle$, if one is present. Note, however, that such an incident wave is *not* necessary in order for $|\psi^{\text{in},i}\rangle$ to exist (as is the case in determining the bound states of a potential).

Now, the wave function for the entire system can be written in the form,

$$\begin{aligned} |\psi\rangle &= |\chi\rangle + G_0 T |\chi\rangle \\ &= |\chi\rangle + G_0 \sum_i t^i |\psi^{\text{in},i}\rangle, \end{aligned} \tag{4.65}$$

which expresses the total wave function as a *multicenter expansion* in terms of waves incident on the sites (cells) of the system. This interpretation is valid even in the absence of an overall incident wave, $|\chi\rangle$.

It is also useful to define an outgoing wave, $|\psi^{\text{out},i}\rangle$, for site i as

$$|\psi^{\text{out},i}\rangle = G_0 t^i |\psi^{\text{in},i}\rangle, \tag{4.66}$$

so that we can write Eq.(4.65) in the form,

$$\begin{aligned} |\psi\rangle &= |\chi\rangle + \sum_{j \neq i} G_0 t^j |\psi^{\text{in},j}\rangle + G_0 t^i |\psi^{\text{in},i}\rangle \\ &= |\chi\rangle + \sum_{j \neq i} G_0 t^j |\psi^{\text{in},j}\rangle + |\psi^{\text{out},i}\rangle \\ &= |\psi^{\text{in},i}\rangle + |\psi^{\text{out},i}\rangle. \end{aligned} \tag{4.67}$$

This equation gives a *one-center* representation of the state $|\psi\rangle$. It states that the wave function for the *entire system* can be thought of as the superposition of the incident and outgoing waves associated with site i. This one-center expansion of the total wave function is equivalent to the multicenter expansion, Eq.(4.65), with the connecting bridge between them being given by Eq.(4.64).

Up to this point we have been able to express the total wave function for a scattering assembly in both one-center, Eq.(4.67), and multicenter, Eq.(4.65), expansions which are connected by Eq.(4.64). This last equation can also be used to obtain a secular equation for the bound states of the system, that is, the states that have a nonvanishing amplitude even in the absence of an overall incident wave, $|\chi\rangle$. To see this, we write Eq.(4.64) in the form

$$\sum_j [\delta_{ij} - G_0 t^j (1 - \delta_{ij})] |\psi^{\text{in},j}\rangle = |\chi\rangle, \tag{4.68}$$

which can be viewed as a decomposition, or linear combination, of the free wave, $|\chi\rangle$, in terms of the waves incident at all the sites i of the system. In the absence of a free wave, Eq.(4.68) represents a system of linear, homogeneous equations for the amplitudes, $|\psi^{\text{in},j}\rangle$. As is well known, nontrivial solutions for these amplitudes can exist only if the following condition is satisfied:

$$|[\delta_{ij} - G_0 t^j (1 - \delta_{ij})]| = 0. \tag{4.69}$$

In a particular representation, the expression on the left-hand side of the last equation becomes a determinant, and we regain the well-known result that nontrivial solutions of a linear homogeneous equation exist only if the determinant of the coefficients vanishes.

This is one form of the secular equation often encountered in applications of multiple scattering theory. Its solutions determine the bound states of a system of scatterers, for example, the Bloch states of a periodic solid. Provided that $\det[t^i] \neq 0$, we can write[2] instead of Eq.(4.69),

$$\det[M] = \det[m^i - G_0(1 - \delta_{i,j})] = 0. \tag{4.70}$$

We recognize M as the matrix exhibited in Eq.(4.55). Either of the forms of the secular equation, Eq.(4.69) or Eq.(4.70), can and have been used in realistic applications of multiple scattering theory. We note in passing that the formal removal of $\det[t^i]$ implies that in practice one deals with square matrices. The extent to which this is justified in the case of generally shaped potentials will be commented upon in a following chapter.

It is worth pointing out that the discussion of this section gives formal expression to the intuitive interpretation of multiple scattering theory which allows one to view all multiply scattered waves as emanating to infinity between successive scattering events. Let us compare Eqs. (4.57) and (4.58). The wave function $|\psi^{\text{in},i}\rangle$ can be thought of as proceeding from infinity, in exactly analogous fashion as the free wave, $|\chi\rangle$. At the same time, this formal analogy must be used with care. Note that the t-matrix occurring in Eq.(4.58) does not bear the same relation to $|\psi^{\text{in},i}\rangle$ as it does to $|\chi\rangle$. This is because the t-matrix is defined strictly with respect to free-particle states, whereas the states $|\psi^{\text{in},i}\rangle$ contain the influence of a potential. However, in spite of such formal differences the intuitive description of multiple scattering theory given in Chapter 2 remains a useful mental construct. We

---

[2]Assuming the existence of $\underline{t}^{-1} = \underline{m}$.

will have occasion to use this construct when we examine the significance of the interstitial region (regions of free space between cells), and the form of the scattered wave function near cell boundaries.

4.4 Representations

4.4.1 The Coordinate Representation

The equations of multiple scattering theory derived in abstract operator space in the previous sections find immediate realizations in the coordinate representation. With the usual notation, $\psi(\mathbf{r}) \equiv \langle \mathbf{r}|\psi\rangle$ and $G(\mathbf{r},\mathbf{r}') \equiv \langle \mathbf{r}|G|\mathbf{r}'\rangle$, we obtain the expressions,

$$\psi(\mathbf{r}) = \chi(\mathbf{r}) + \int G_0(\mathbf{r},\mathbf{r}')V(\mathbf{r}')\psi(\mathbf{r}')\mathrm{d}^3r', \tag{4.71}$$

$$\psi(\mathbf{r}) = \chi(\mathbf{r}) + \int G_0(\mathbf{r},\mathbf{r}')T(\mathbf{r}',\mathbf{r}'')\chi(\mathbf{r}'')\mathrm{d}^3r'\,\mathrm{d}^3r'', \tag{4.72}$$

and

$$T(\mathbf{r},\mathbf{r}') = V(\mathbf{r})\left[\delta(\mathbf{r}-\mathbf{r}') + \int G_0(\mathbf{r},\mathbf{r}'')T(\mathbf{r}'',\mathbf{r}')\mathrm{d}^3r''\right], \tag{4.73}$$

corresponding to Eqs.(4.3), (4.17), and (4.22), respectively. Similar expressions can be obtained for the Green function, for example,

$$\begin{aligned} G(\mathbf{r},\mathbf{r}') &= G_0(\mathbf{r},\mathbf{r}') + \int G_0(\mathbf{r},\mathbf{r}_1)V(\mathbf{r}_1)G(\mathbf{r}_1,\mathbf{r}')\mathrm{d}^3r_1 \\ &= G_0(\mathbf{r},\mathbf{r}') + \iint G_0(\mathbf{r},\mathbf{r}_1)T(\mathbf{r}_1,\mathbf{r}_2)G_0(\mathbf{r}_2,\mathbf{r}')\mathrm{d}^3r_1\mathrm{d}^3r_2. \end{aligned} \tag{4.74}$$

It follows from Eq.(4.73) that $T(\mathbf{r},\mathbf{r}')$ vanishes when either one of its arguments lies outside the region of the potential, $V(\mathbf{r})$. In our applications of multiple scattering theory, we will use the specific expression for the free-particle propagator that corresponds to outgoing waves, (Eq.(B.3)) (suppressing the superscript (+) here and subsequently),

$$G_0(\mathbf{r},\mathbf{r}') \equiv G_0(\mathbf{r}-\mathbf{r}') = -\frac{1}{4\pi}\frac{e^{ik|\mathbf{r}-\mathbf{r}'|}}{|\mathbf{r}-\mathbf{r}'|}. \tag{4.75}$$

As was the case in the abstract formulation of multiple scattering theory, it is convenient to identify cell-diagonal and cell-off-diagonal elements of $G_0(\mathbf{r},\mathbf{r}')$ according to whether the vectors $\mathbf{r}$ and $\mathbf{r}'$ lie in the same cell or in different cells. It is also convenient to consider any two adjacent potential cells as being separated by a thin strip of zero potential whose width can be allowed to vanish at the conclusion of formal arguments. We can now obtain the coordinate representation of the equation of motion of multiple scattering theory, Eq.(4.54),

$$T^{ij}(\mathbf{r},\mathbf{r}') = t^i(\mathbf{r},\mathbf{r}') + \sum_{k\neq i} \int t^i(\mathbf{r},\mathbf{r}_1) G_0^{ik}(\mathbf{r}_1 - \mathbf{r}_2) \times T^{kj}(\mathbf{r}_2,\mathbf{r}')\mathrm{d}^3r_1\,\mathrm{d}^3r_2. \tag{4.76}$$

Here, $G_0^{ik}(\mathbf{r}-\mathbf{r}')$ is that cell-off-diagonal element of the free-particle propagator that corresponds to the vectors $\mathbf{r}$ and $\mathbf{r}'$ being confined to cells i and k, respectively. A similar interpretation holds for the elements $T^{ij}(\mathbf{r},\mathbf{r}')$.

The essence of multiple scattering theory is readily discernible upon iteration of Eq.(4.76). This reveals the makeup of the total scattering matrix, $T = \sum_{ij} T^{ij}$, as the sum of all multiple scattering events associated with individual cells, described by the cell scattering matrices, $t^i(\mathbf{r},\mathbf{r}')$, with free propagation from cell to cell via $G^{ik}(\mathbf{r}-\mathbf{r}')$ between individual scattering events.

This separation of structure and potential is possibly the most outstanding and useful property of MST. This property is retained when MST is expressed within the angular-momentum representation, discussed immediately below.

4.4.2 The Angular-Momentum Representation

Informative as they may be, the integral equations of multiple scattering theory in the coordinate representation are very difficult to handle computationally. However, in most cases these integral equations can be transformed into equations between matrices whose elements are labeled by the index $L = (\ell, m)$ of angular momentum eigenstates. The resulting *angular momentum representation* has been found to be of great value in numerical applications of multiple scattering theory. In this representation, the total scattering matrix takes the form

$$T_{LL'} = \int\int J_L(\mathbf{r})T(\mathbf{r},\mathbf{r}')J_{L'}(\mathbf{r}')\mathrm{d}^3r\,\mathrm{d}^3r', \tag{4.77}$$

where $J_L(\mathbf{r})$ denotes the product of the spherical Bessel function, $j_\ell(kr)$, with a spherical harmonic, as defined in Appendix C. This expression shows clearly that the t-matrix is symmetric with respect to the indices L and L'. Using the vector and matrix notation introduced in Appendix D, we can write Eq. (4.77) in the form

$$\underline{T} = \int\int |J(\mathbf{r})\rangle T(\mathbf{r},\mathbf{r}')\langle J(\mathbf{r}')|\,\mathrm{d}^3r\mathrm{d}^3r', \tag{4.78}$$

with underlines denoting matrices in angular-momentum space. Upon breaking the regions of integration in Eq.(4.78) into integrations over the interiors of individual cells, and using the expansion properties of the spherical functions, $J_L(\mathbf{r})$, given in Appendix D, we can cast Eq. (4.78) into the form,

$$\begin{aligned} \underline{\mathbf{T}} &= \sum_{i,j} \underline{g}(-\mathbf{R}_i)\underline{T}^{ij}\underline{g}(\mathbf{R}_j) \\ &= \langle \underline{\mathbf{g}} \,|\underline{\mathbf{T}}|\, \underline{\mathbf{g}}^{\dagger} \rangle, \end{aligned} \tag{4.79}$$

where underlined boldfaced symbols denote matrices in both site (cell) and angular-momentum spaces. In Eq.(4.79), $\underline{T}^{ij}$ is given by the expression,

$$\underline{T}^{ij} = \int_{\Omega_i} \int_{\Omega_j} |J(\mathbf{r})\rangle T^{ij}(\mathbf{r}, \mathbf{r}')\langle J(\mathbf{r}')|\, \mathrm{d}^3 r\, \mathrm{d}^3 r', \tag{4.80}$$

in which the integrals and the cell vectors $\mathbf{r}$ and $\mathbf{r}'$ are measured from the centers and are confined to the interiors of the cells i and j. The quantities $[\underline{g}(\mathbf{R})]_{LL'}$ are the matrix elements of the translation operator in the angular momentum representation and are given explicitly in Eq.(D.5). We point out that Eq.(4.79) is the angular-momentum representation of the abstract Eq.(4.52), and that it is completely general, being valid for any set of nonoverlapping scattering potentials, regardless of their shape or the choice of cell centers.

It follows from Eq.(4.77) that the two expressions of a matrix quantity indexed by L obtained with two different choices for the center of a cell, that is, a change in the point about which the L-expansions are being constructed, are connected by a matrix transformation that depends only on the vector distance between the two centers of expansion. If we let $\underline{\mathbf{A}}$ denote a general intercell quantity expressed in the angular-momentum representation, and $\underline{\mathbf{A}}(\mathbf{a}, \mathbf{b})$ denote the same quantity evaluated when the centers of the cells displaced by the vectors $\mathbf{a}$ and $\mathbf{b}$, respectively, we have

$$\underline{\mathbf{A}}(\mathbf{a}, \mathbf{b}) = \underline{\mathbf{g}}(\mathbf{a})\underline{\mathbf{A}}\underline{\mathbf{g}}^{\dagger}(\mathbf{b}). \tag{4.81}$$

4.4.3 Representability of the Green Function and the Wave Function

We are now in a position to justify our use of Eq.(3.66) in which the solution, $\psi(\mathbf{r})$, of the Schrödinger equation,

$$[\nabla^2 + E - V(\mathbf{r})]\psi(\mathbf{r}) = 0, \tag{4.82}$$

is represented as a linear combination,

$$\psi(\mathbf{r}) = \sum_L A_L \psi_L(\mathbf{r}), \tag{4.83}$$

of solutions, or basis functions, corresponding to specific values of angular momentum. The validity of Eq.(4.83) for the case of the single, spatially bounded scatterer will be important in the development of multiple scattering theory in subsequent chapters.

First, we show that the Green function, $G(\mathbf{r},\mathbf{r}')$, defined as the solution of the equation (for a proof by direct construction see Appendix G),

$$[\nabla^2 + E - V(\mathbf{r})]G(\mathbf{r},\mathbf{r}') = \delta(\mathbf{r}-\mathbf{r}') \tag{4.84}$$

has a convergent expansion of the form,

$$G(\mathbf{r},\mathbf{r}') = \sum_L \psi_L(\mathbf{r}_<)F_L(\mathbf{r}_>), \tag{4.85}$$

where $\psi_L(\mathbf{r})$ is equal to a regular solid harmonic, $J_L(\mathbf{r})$, near the origin, and $F_L(\mathbf{r})$ is proportional to an irregular solid harmonic $H_L(\mathbf{r}_>)$ outside the range of the potential (outside the sphere bounding the potential). Elsewhere, these two functions are, respectively, regular and irregular solutions of the Schrödinger equation. In particular, $\psi_L(\mathbf{r})$ is given by Eq.(3.70).

To get a sense of the validity of Eq.(4.85) consider the Green function, $G(\mathbf{r},\mathbf{r}')$, associated with an arbitrarily shaped but spatially bounded potential, $V(\mathbf{r})$, with both arguments outside a sphere circumscribing the potential. From Eqs.(4.25) and (3.110) we can write

$$\begin{aligned} G(\mathbf{r},\mathbf{r}') &= -ik\sum_L J_L(\mathbf{r}_<)H_L(\mathbf{r}_>) + (-ik)^2\sum_L\sum_{L'} H_{L'}(\mathbf{r})t_{L'L}H_L(\mathbf{r}') \\ &= -ik\sum_L \left[J_L(\mathbf{r}) - ik\sum_{L'} H_{L'}(\mathbf{r})t_{L'L}\right] H_L(\mathbf{r}') \\ &= -ik\left[\langle J(\mathbf{r})| - ik\langle H(\mathbf{r})|\, \underline{t}\,\right]|H(\mathbf{r}')\rangle. \end{aligned} \tag{4.86}$$

Here, we have set $r' > r$, and have made use of the vector/matrix notation of Appendix D. Note that the expression inside the brackets is that given by Eq.(3.111) and represents the regular solution of the Schrödinger equation outside the bounding sphere. From the definition of the t-matrix, Eq.(3.91), we obtain

$$G(\mathbf{r},\mathbf{r}') = \langle\psi(\mathbf{r}_<)|F(\mathbf{r}_>)\rangle, \tag{4.87}$$

where

$$\langle\psi(\mathbf{r})| = \langle J(\mathbf{r})|(\underline{C} - \mathrm{i}\underline{S}) + i\langle H(\mathbf{r})|\underline{S} \tag{4.88}$$

and

$$\begin{aligned} |F(\mathbf{r})\rangle &= -ik(\underline{C} - \mathrm{i}\underline{S})^{-1}|H(\mathbf{r})\rangle \\ &= -ik\underline{\tilde{C}}^{-1}|H(\mathbf{r})\rangle. \end{aligned} \tag{4.89}$$

This demonstrates that the Green function for the Schrödinger equation with an arbitrarily shaped potential can, when both arguments lie outside the range of the potential, be written in the "factored" form of Eq.(4.87). Because the Green function must be continuous everywhere as a function of either of its arguments, as these arguments cross the surface of the bounding sphere the same expansion holds inside the sphere, but with the functions

$|\psi(\mathbf{r})\rangle$ and $|F(\mathbf{r})\rangle$ now being the regular and irregular solutions, respectively, of the equation in the potential region. In particular, the function that matches smoothly onto the asymptotic form given in Eq.(4.88) on the surface of the bounding sphere is given by an equation analogous to Eq.(3.70), but with $|N(\mathbf{r})\rangle$ in that equation replaced with $|H(\mathbf{r})\rangle$,

$$|\psi(\mathbf{r})\rangle = \tilde{\underline{c}}(r)|J(\mathbf{r})\rangle + i\tilde{\underline{s}}(r)|H(\mathbf{r})\rangle. \tag{4.90}$$

The phase functions $\tilde{\underline{c}}(r)$ and $\tilde{\underline{s}}(r)$ are such that on the surface of the bounding sphere they become equal to $\tilde{\underline{C}}$ and $\tilde{\underline{S}}$, respectively.

We note that Eq.(4.87) is completely analogous to Eq.(3.110) for the free-particle propagator. Both of these equations express the Green function in a factored form involving the regular and irregular solutions of the corresponding Schrödinger equation. As shown directly in Appendix G, the Green function in the form of Eq.(4.85) satisfies the defining Eq.(4.84).

Now, because $G(\mathbf{r}, \mathbf{r}')$ satisfies Eq.(4.84) and $\psi(\mathbf{r})$ is a solution of the Schrödinger equation corresponding to $V(\mathbf{r})$, we have the identity

$$\begin{aligned}\psi(\mathbf{r}) = & \int d\mathbf{r}'[\nabla'^2 + E - V(\mathbf{r}')]G(\mathbf{r}, \mathbf{r}')\psi(\mathbf{r}') \\ & - \int d\mathbf{r}' G(\mathbf{r}, \mathbf{r}')[\nabla'^2 + E - V(\mathbf{r}')]\psi(\mathbf{r}').\end{aligned} \tag{4.91}$$

Through the use of Green's theorem, this expression can be converted into the integral

$$\psi(\mathbf{r}) = \int_{S_n} dS'\, \hat{n} \cdot [\nabla' G(\mathbf{r}, \mathbf{r}') - G(\mathbf{r}, \mathbf{r}')\nabla']\psi(\mathbf{r}'), \tag{4.92}$$

over the surface of any sphere, S_n, bounding the potential. Using the expansion (4.85) for points $\mathbf{r}$ *inside* the sphere, we obtain immediately Eq.(4.83), where the coefficients A_L are given by the expression

$$A_L = \int_{S_n} dS'\, \hat{n} \cdot [\nabla' F_L(\mathbf{r}') - F_L(\mathbf{r}')\nabla']\psi(\mathbf{r}'). \tag{4.93}$$

In the limit as $l \to \infty$, we have $\psi_L \to J_L(\mathbf{r})$ and $F_L(\mathbf{r}) \to H_L(\mathbf{r})$ (since the centrifugal term in the Schrödinger equation overwhelms the potential in that limit). Because $r' > r$, the argument of the irregular solution, $H_L(\mathbf{r}')$, is larger than that of the regular solution, $J_L(\mathbf{r})$, and the sum converges absolutely. This justifies the use of the expansion in Eq.(4.83), which will be used often in our subsequent discussion.

At this point, it should be pointed out that the expansion in (4.83) is *not* a completeness relation in the usual sense. Such a relation would manifest itself as a sum (or integral) over the *energy* variable, and would allow one to expand *any* function in terms of the eigensolutions of the Schrödinger equation. By contrast, the expansion in (4.83) *represents a solution of the Schrödinger equation at a given energy as a linear combination of basis functions (partial solutions) at the same energy.* Hence the necessity of demonstrating explicitly the validity of Eq.(4.83).

It is useful to compare certain basic features of the expansion for the wave function and the Green function derived above. Let us note that the Schrödinger equation with the boundary conditions placed on it is a homogeneous equation for the wave function, thus determining that quantity only within a normalization constant (see Section 3.3.4.) On the other hand, the Green function satisfies an inhomogeneous equation and is determined uniquely by a differential equation and its boundary conditions. Thus, the Green function can be used to normalize the wave function. Alternatively, one cannot choose independently the normalizations of the regular and irregular solutions of the Schrödinger equation. These must be chosen so that they are consistent with the normalization of the corresponding Green function.

As an example, consider the expansion in Eq.(4.87), with $\psi(\mathbf{r})$ given by Eq.(4.88). This expansion corresponds to a regular solution which outside the bounding sphere is given by Eq.(4.88), and which behaves as $\langle J(\mathbf{r})|$ near the origin. As near the origin the Green function must have the form $\langle J(\mathbf{r}_<)|H(\mathbf{r}_>)\rangle$, it follows that the irregular solution that outside the bounding sphere is given by Eq.(4.89) must approach $|H(\mathbf{r})\rangle$ as $\mathbf{r} \to 0$. As a specific demonstration of this argument, take the center of the potential (the center of expansion) to lie in free space. Then, outside the bounding sphere the Green function is given by Eq.(4.87), while near the origin we have $G(\mathbf{r},\mathbf{r}') = G_0(\mathbf{r},\mathbf{r}') = \langle J(\mathbf{r}_<)|H(\mathbf{r}_>)$. Similarly, if outside the bounding sphere the Green function is given by Eq.(4.86), then the regular solution is given by the term inside the brackets in that equation and approaches $\langle J(\mathbf{r})|\underline{\tilde{C}}^{-1}$ at the origin. (See the discussion in Section 3.4.1.) It follows that the irregular solution that in the asymptotic expansion in Eq.(4.86) behaves as $|H(\mathbf{r})\rangle$, must approach $\underline{\tilde{C}}|H(\mathbf{r})\rangle$ at the origin, thus guaranteeing that the Green function there is given by the expansion $\langle J(\mathbf{r}_<)|H(\mathbf{r}_>)$.

4.4.4 Example of Representability

By way of illustration of Eq.(4.83), let us consider in some detail the role played by and the form of the coefficients, A_L, in that equation for a specific case. These coefficients are to be chosen so that the wave function, $\psi(\mathbf{r})$, satisfies the proper set of boundary conditions imposed on the Schrödinger equation, even though the basis functions, $\psi_L^n(\mathbf{r})$, are determined without regard to the boundary conditions of the problem.

Consider the elementary problem of a cubic box of side L with a potential of the form

$$V(\mathbf{r}) = \begin{cases} -V_0, & \mathbf{r} \text{ inside the box } (0 < x,\ y,\ z < L) \\ \infty, & \text{otherwise} \end{cases} \tag{4.94}$$

The wavefunction inside the box that corresponds to a positive energy, E, and that vanishes at the sides of the box, has the form

$$\psi(\mathbf{r}) = C \sin k_x x' \sin k_y y' \sin k_z z', \tag{4.95}$$

where the origin of coordinates is taken at the point (0,0,0), in one of the corners of the box. The components of the $\mathbf{k}$ vector satisfy the relation

$$k_x^2 + k_y^2 + k_z^2 = E + V_0, \tag{4.96}$$

and each component is given by the expression

$$k_\alpha = \frac{n_\alpha \pi}{L}, \quad \alpha = x, y, z, \quad n_\alpha = 1, 2, 3 \ldots. \tag{4.97}$$

These values for k_α guarantee that the wave function vanishes identically at the sides of the box. They also yield immediately the discrete eigenenergies of the system in the form

$$E = \frac{\pi^2}{L^2}[n_x^2 + n_y^2 + n_z^2] - V_0. \tag{4.98}$$

It is easy to show that the box wave function, $\psi(\mathbf{r})$, can be written in the form of Eq.(4.83) and derive explicit expressions for the coefficients A_L. In Eq.(4.95), we write each sine function as the sum of two exponentials and obtain the expression

$$\psi(\mathbf{r}) = \frac{\mathrm{i}C}{8} \sum_{n=1}^{8} (-1)^{n-1} \mathrm{e}^{\mathrm{i}\mathbf{k}_n \cdot \mathbf{r}'}, \tag{4.99}$$

where we have defined the vectors

$$\begin{aligned}
\mathbf{k}_1 &= k_x \mathbf{e}_1 + k_y \mathbf{e}_2 + k_z \mathbf{e}_3, & \mathbf{k}_5 &= -k_x \mathbf{e}_1 + k_y \mathbf{e}_2 - k_z \mathbf{e}_3, \\
\mathbf{k}_2 &= k_x \mathbf{e}_1 + k_y \mathbf{e}_2 - k_z \mathbf{e}_3, & \mathbf{k}_6 &= -k_x \mathbf{e}_1 + k_y \mathbf{e}_2 + k_z \mathbf{e}_3, \\
\mathbf{k}_3 &= k_x \mathbf{e}_1 - k_y \mathbf{e}_2 - k_z \mathbf{e}_3, & \mathbf{k}_7 &= -k_x \mathbf{e}_1 - k_y \mathbf{e}_2 + k_z \mathbf{e}_3, \\
\mathbf{k}_4 &= k_x \mathbf{e}_1 - k_y \mathbf{e}_2 + k_z \mathbf{e}_3, & \mathbf{k}_8 &= -k_x \mathbf{e}_1 - k_y \mathbf{e}_2 - k_z \mathbf{e}_3.
\end{aligned} \tag{4.100}$$

Here, the symbols $\mathbf{e}_\alpha$, $\alpha = 1, 2, 3$ denote unit vectors along the three Cartesian axes, x, y, and z.

Because the angular momentum expansion is usually performed about an origin at the center of the cell, we shift the origin of the wave function in Eq.(4.99) by use of the translation $\mathbf{r} = \mathbf{r}' - \mathbf{r}_0$ where $\mathbf{r}_0$ is the center of the cell, $(L/2)(\mathbf{e}_1 + \mathbf{e}_2 + \mathbf{e}_3)$. Thus, the wave function can be written as

$$\psi(\mathbf{r}) = \frac{\mathrm{i}C}{8} \sum_{n=1}^{8} (-1)^{n-1} \mathrm{e}^{\mathrm{i}\mathbf{k}_n \cdot (\mathbf{r} + \mathbf{r}_0)}, \tag{4.101}$$

We now expand the $\mathrm{e}^{\mathrm{i}\mathbf{k}_n \cdot \mathbf{r}}$ in Eq.(4.101) in terms of spherical functions using Bauer's identity, Eq.(D.1), and write

$$\psi(\mathbf{r}) = \sum_L A_L J_L(\mathbf{r}), \tag{4.102}$$

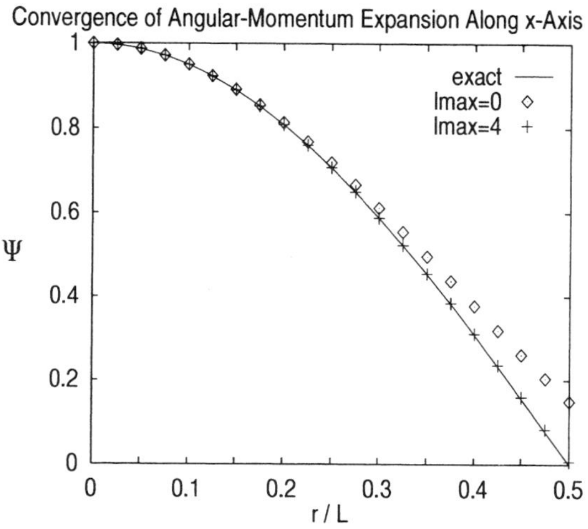

FIGURE 4.1. Wave function plotted along a line from the center of the box to the center of a side. The exact wave function is represented well using only $l = 0$ and $l = 4$ angular-momentum states.

where $J_L(\mathbf{r}) = j_l(kr)Y_L(\hat{r})$ with $k^2 = k_x^2 + k_y^2 + k_z^2$ and the coefficients in the expansion are given by

$$A_L = \frac{\mathrm{i}\pi C}{2} \sum_{n=1}^{8} (-1)^{n-1} \mathrm{e}^{\mathrm{i}\mathbf{k}_n \cdot \mathbf{r}_0} i^l Y_L(\mathbf{k}_n). \tag{4.103}$$

The coefficients of Y_L in the above expression have unit modulus. Thus the solution to the Schrödinger equation can be expanded in a convergent series of local solutions of definite angular momentum.

Figures 4.1 and 4.2 display the convergence of the angular momentum expansions for the lowest energy state ($n_x = n_y = n_z = 1$). The convergence is rapid with only two terms ($l = 0, l = 4$) needed in the direction toward the center of the side of the box. Three terms are needed in the direction of the corner of the box ($l = 0, l = 4, l = 6$). The $l = 2$ term does not appear because of the cubic symmetry of this state.

4.4.5 The Representability Theorem

Fortified by the formalism and the above illustrations, we are now ready to state the results of our discussion in the form of a theorem:

The Representability Theorem: The global solution of the Schrödinger equation in a domain, D, can be expanded in a convergent series of local solutions at the same energy defined over subdomains, d, and which have the limiting form $J_L(\mathbf{r})$ as $r \to 0$ provided the potential has no singularities in d other than a possible singularity at the origin weaker than r^{-2}.

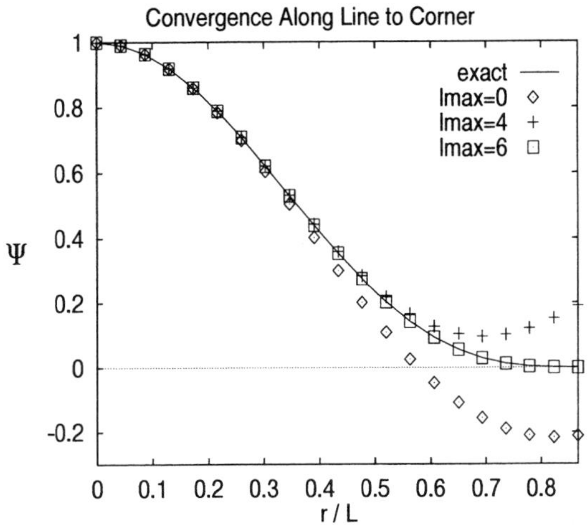

FIGURE 4.2. Wave function plotted along a line from the center of the box to one of the corners. The exact wave function is represented well using $l = 0$, $l = 4$, and $l = 6$ angular-momentum states.

It is to be noted that the theorem may fail in cases in which the potential contains nonintegrable singularities, in which case neither $\psi(\mathbf{r})$ nor the basis functions $\psi_L(\mathbf{r})$ may exist. The theorem, however, holds in cases in which the potential contains integrable singularities such as a step function.

The representability theorem makes a very strong statement regarding the solutions of the Schrödinger equation. As we will see, it gives a prescription for constructing the solution of the Schrödinger equation for a collection of cells from the solutions associated with individual cells. When the number of cells becomes very large, in principle infinite, then the direct determination of the coefficients A_L becomes impossible. In this case, and also in general for any number of scatterers, multiple scattering theory provides a powerful procedure for the determination of these coefficients. *The coefficients of the linear superpositions of the local cell functions are found as the eigenvectors of the secular equation of MST, as is discussed in the following and subsequent chapters.* It should, however, be kept in mind that the determination of these coefficients may require the use of double or multiple, conditional sums in order to assure convergence.

Finally, we mention that the representability theorem can be extended to apply to the irregular solution of the Schrödinger equation, as well as the Green function itself.

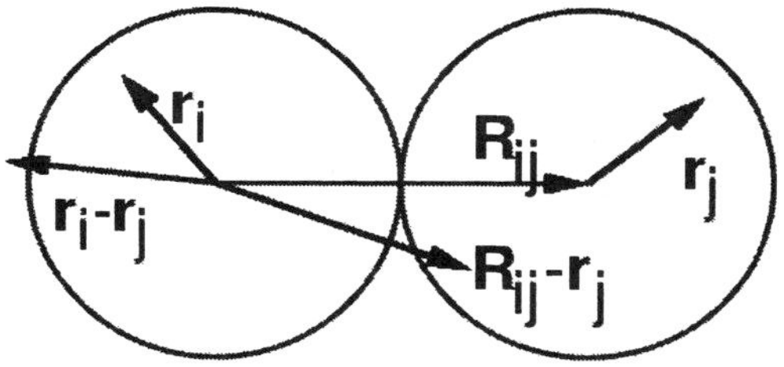

FIGURE 4.3. Several important inequalities are satisfied for muffin-tin potentials.

4.5 Muffin-Tin Potentials

In the muffin tin (MT) approximation to the potential, one assumes that the cells constituting a scattering assembly are positioned far enough apart so that each can be circumscribed by a sphere that does not intersect any sphere circumscribing another cell. In numerical applications, it has been customary to assume also that the potentials are spherically symmetric, but it is not necessary for us to make this assumption in our discussion. As a simple application of the formalism presented above, in this section we derive explicit expressions in the L-representation for the t-matrix and the secular equation for an assembly of MT scatterers. An extended discussion of MST applied to MT potentials is given in the next chapter.

In the MT case, the cell vectors $\mathbf{r}_i = \mathbf{r} - \mathbf{R}_i$ are confined inside nonoverlapping spheres and together with the intercell vectors $\mathbf{R}_{ij} = \mathbf{R}_j - \mathbf{R}_i$ satisfy the conditions,

$$|\mathbf{R}_{ij} - \mathbf{r}_j| > |\mathbf{r}_i|, \quad |\mathbf{R}_{ij} + \mathbf{r}_i| > |\mathbf{r}_j|, \tag{4.104}$$

and

$$|\mathbf{R}_{ij}| > |\mathbf{r}_j|, \quad |\mathbf{R}_{ij}| > |\mathbf{r}_j|, \quad |\mathbf{R}_{ij}| > |\mathbf{r}_j - \mathbf{r}_i|. \tag{4.105}$$

In this case, the cell-off-diagonal elements of the free-particle propagator, $G_0(\mathbf{r} - \mathbf{r}') \equiv G_0(-\mathbf{r}_i + \mathbf{R}_{ij} + \mathbf{r}_j)$ can be expanded in terms of spherical functions as is indicated in Eqs.(D.18) and (D.19),

$$\begin{aligned} G_0(\mathbf{r}_i - \mathbf{R}_{ij} - \mathbf{r}_j) &= -ik \sum_L J_L(\mathbf{r}_i) H_L(\mathbf{r}_j + \mathbf{R}_{ij}) \\ &= -ik \sum_{LL'} J_L(\mathbf{r}_i) G_{LL'}(\mathbf{R}_{ij}) J_{L'}(\mathbf{r}_j) \\ &= -ik \langle J(\mathbf{r}_i) \left| \underline{G}(\mathbf{R}_{ij}) \right| J(\mathbf{r}_j) \rangle, \end{aligned} \tag{4.106}$$

where $\underline{G}(\mathbf{R}_{ij})$ is a real-space structure constant as defined in Eq.(D.10). Because of the MT conditions between the various vectors involved in the last expressions, the sums over L and L' can be carried out irrespective of order, and always lead to converged results [6]. Upon multiplying Eq.(4.76) from the left and from the right by $J_L(\mathbf{r})$ and $J_{L'}(\mathbf{r}')$, respectively, replacing G_0 by its "factored" form, Eq.(4.106), and carrying out the indicated integrals using the definition given in Eq. (4.80), we obtain the equation of motion for the t-matrix in the form,

$$\underline{T}^{ij} = \underline{t}^i \delta_{ij} + \underline{t}^i \sum_{k \neq i} \tilde{\underline{G}}^{ik} \underline{T}^{kj}, \tag{4.107}$$

where $\tilde{\underline{G}}^{ik} = -ik\underline{G}(\mathbf{R}_{ik})$. It follows immediately from Eq. (4.107) that $\underline{T}^{ij}$ is the (i,j)th matrix element of the inverse of a matrix $\underline{\mathbf{M}}$ whose elements are given by the expression,

$$\underline{M}^{ij} = \underline{m}^i \delta_{ij} - \tilde{\underline{G}}^{ij}(1 - \delta_{ij}). \tag{4.108}$$

For reference purposes, we shall refer to Eqs. (4.107) and (4.108) as the *MT-form of multiple scattering theory.*

We note that in the MT form both Eqs.(4.107) and (4.108), written in terms of matrices in the angular-momentum representation, have the same form as the abstract operator Eqs.(4.54) and (4.55). Furthermore, a straightforward iteration of Eq.(4.107) yields an expansion of $\underline{T}^{ij}$, or of $\underline{\mathbf{M}}^{-1}$, in terms of products of the individual cell t-matrices which converges unconditionally [6]. We will see later that this last property is usually not preserved in generalizations of multiple scattering theory to space-filling cell potentials. In order to conform with established notation, we shall denote the inverse of the matrix $\underline{\mathbf{M}}$ by $\underline{\tau}$, so that $[\underline{\mathbf{M}}^{-1}]^{ij} = \underline{\tau}^{ij}$.

In the MT case, the quantity $\underline{\tau}^{ij}$ is the on-the-energy-shell scattering-path operator (SPO) introduced by Györffy and Stott [7], and in the form of Eq. (4.108) it is particularly convenient for computational purposes[3]. This form clearly indicates the separation of the potential, embodied in the cell-scattering matrices, from the structural aspects, represented by the structure constants, of a scattering assembly. The structure constants, $\underline{G}^{ij}$, are the real-space analogues of the reciprocal-space quantities introduced by Korringa [8], and by Kohn and Rostoker [9] (KKR) for the study of periodic solids.

The MT form of multiple scattering theory has been used extensively in calculations of the electronic structure of periodic solid materials. In this case, the electronic structure (band structure), $E(\mathbf{k})$, is determined by the poles of $\underline{\tau}(\mathbf{k})$, which is the Fourier transform of the $\underline{\tau}^{ij}$, with $\mathbf{k}$ being a vector in the first Brillouin zone of the reciprocal lattice. The poles of

[3]Because τ is not an operator in the proper sense of the word, a possibly more appropriate name would be the scattering-path matrix. In our discussion, we will use the terms scattering path operator (SPO) and scattering path matrix interchangeably.

$\underline{M}(\mathbf{k})$ are determined through the solutions of the secular equation,

$$\det\left[\underline{m} - \underline{\tilde{G}}(\mathbf{k})\right] = 0, \tag{4.109}$$

where the $\underline{G}(\mathbf{k})$ are the well-known KKR structure constants. This secular equation determines the bound states of the system, which in this case are the familiar Bloch states. As Eq.(4.68) indicates, the vanishing of the determinant in Eq. (4.109) is also the condition for the existence of solutions of the equations of multiple scattering theory in the absence of an externally incident wave. Much of our discussion having to do with the generalization of multiple scattering theory to space-filling cells will revolve around the validity of Eq. (4.109), and other formally analogous expressions, in the case of nonMT potentials.

References

[1] Roger G. Newton, *Scattering Theory of Waves and Particles* (Springer-Verlag, New York, 1966, 2nd Edition.)

[2] Marvin L. Goldberger and Kenneth M. Watson, *Collision Theory* (Robert E. Krieger Publishing Co., Huntington, NY, 1975.)

[3] A thorough discussion of the Lippmann–Schwinger equation for spherically symmetric potentials at a basic quantum-mechanical level is given by A. Böhm, *Quantum Mechanics* (Springer-Verlag, New York, 1979).

[4] J. S. Faulkner, J. Phys. C., Solid State Phys. **10**, 4661 (1977).

[5] P. J. Braspenning, *A Multiple Scattering Treatment of Dilute Metal Alloys*, Thesis, University of Amsterdam (1982), unpublished.

[6] R. Zeller, J. Phys. C **20**, 2347 (1987).

[7] B. L. Györffy and M. J. Stott, in *Band Structure Spectroscopy of Metals and Alloys*, edited by D. J. Fabian and L. M. Watson (Academic Press, London, 1972).

[8] J. Korringa, Physica, **13**, 392 (1947).

[9] W. Kohn and N. Rostoker, Phys. Rev. **94**, 1111 (1954).

5

MST for Muffin-Tin Potentials

In this chapter, we begin our study of multiple scattering theory (MST) within the angular-momentum representation. This representation allows one to reduce the integral equations of MST to much more convenient matrix equations. For historical reasons, as well as for the sake of clarity, we consider first the case of muffin-tin potentials, that is, potentials that can be separated from one another by nonoverlapping spheres and that, in most applications, are also taken to be spherically symmetric. The last condition, that of spherical symmetry (although helpful in computations) provides only minimal simplification in the development of MST, but the *geometric* condition that the potentials be separated by nonoverlapping spheres greatly simplifies the expansion of the free-particle propagator in angular-momentum states and will be used explicitly throughout this chapter. In order to prepare the ground for subsequent developments, we will use notation which is sufficiently general to encompass nonspherical potentials. The contact with the spherically symmetric case can be made by considering all appropriate matrix quantities, such as the cell phase functions and t-matrices, to be diagonal in the angular-momentum indices. The completely general case of nonoverlapping potentials of arbitrary, space-filling shape will be considered in following chapters.

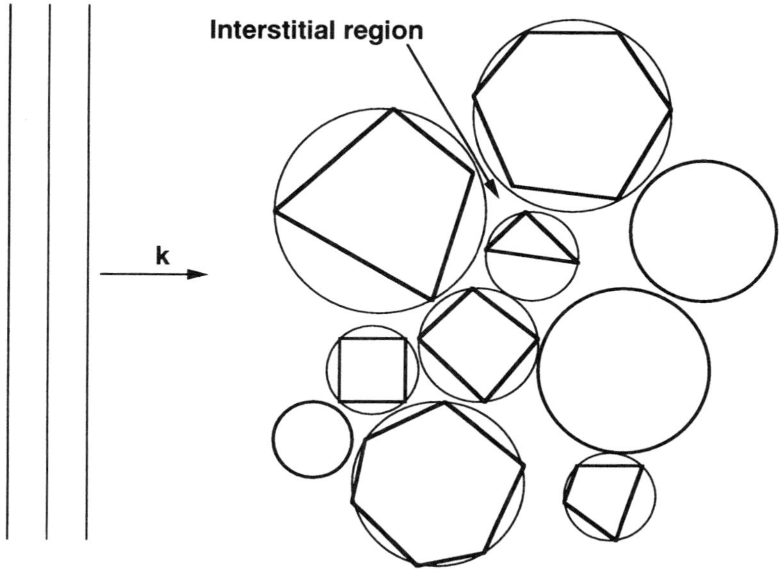

FIGURE 5.1. A plane wave incident on a collection of MT scatterers.

5.1 Multiple Scattering Series

Consider a plane wave incident upon a collection of spatially bounded scattering centers as is indicated schematically in Fig. 5.1. The region of free space (or of zero potential) outside all bounding spheres is commonly referred to as the *interstitial region.* As we saw in Section 4.4, the Lippmann–Schwinger equation which describes the propagation of the incident wave through the scattering assembly can be written in the form,

$$\begin{aligned}\psi(\mathbf{r}) &= \chi(\mathbf{r}) + \int G_0(\mathbf{r},\mathbf{r}')V(\mathbf{r}')\psi(\mathbf{r}')\mathrm{d}^3r' \\ &= \chi(\mathbf{r}) + \int G_0(\mathbf{r},\mathbf{r}')T(\mathbf{r}',\mathbf{r}'')\chi(\mathbf{r}'')\mathrm{d}^3\,r'\mathrm{d}^3r'', \end{aligned} \tag{5.1}$$

where $T(\mathbf{r},\mathbf{r}')$ is the scattering matrix for the assembly of scatterers considered as a single unit. Similarly, the system Green function can be written in the form of Eq.(4.74). The scattering matrix, T, describes the complete effect of scattering from the assembly and in principle is determined through the use of Eq.(4.23) with the potential $V(\mathbf{r})$ in that equation being the potential of the entire system.

This direct method for evaluating T is generally quite cumbersome so we would like to have a method for evaluating T that makes explicit use of the t-matrices or wave functions associated with individual cells. Such a method can be obtained following the intuitive discussion of Chapter 2 and the formal development of Chapter 4 which show that the wave function,

$\psi(\mathbf{r})$, can be thought of as being gradually built up by a process of *multiple scattering* from individual scattering centers, starting from an incident wave, $|\chi_\alpha\rangle$, that satisfies the Helmholtz equation, that is, the Schrödinger equation with $V(\mathbf{r}) = 0$. Thus, as discussed in Chapter 4, the total wave amplitude incident on site α is the sum of the original incident wave plus waves scattered from all of the other sites,

$$|\psi_\alpha^{in}\rangle = |\chi_\alpha\rangle + G_0^+ \sum_{\beta\neq\alpha} t^\beta |\psi_\beta^{in}\rangle. \tag{5.2}$$

Because the wave scattered from site α is $t^\alpha|\psi_\alpha^{in}>$, the total wave function in the vicinity of site α, that is, the sum of the incident wave and the scattered waves coming from all of the sites, will be

$$|\psi\rangle = |\chi_\alpha\rangle + G_0^+ \sum_{\beta} t^\beta |\psi_\beta^{in}\rangle. \tag{5.3}$$

These two equations were derived in the preceding chapter, Eqs.(4.64) and (4.65). Now, a straightforward iteration of Eqs. (5.2) and (5.3) shows that the total wave function can be written in the form,

$$|\psi\rangle = |\chi_\alpha\rangle + G_0^+ T|\chi\rangle, \tag{5.4}$$

where the total t-matrix is given by the *multiple scattering* series,

$$T = \sum_\alpha t^\alpha + \sum_\alpha \sum_{\beta\neq\alpha} t^\alpha G_0 t^\beta + \cdots, \tag{5.5}$$

and where the superscripts $(+)$ again are suppressed in order to simplify our notation. This result can be compared with Eq.(4.53). The corresponding expression for the Green function is readily obtained from Eq.(4.25)

$$G = G_0 + G_0 \sum_\alpha t^\alpha G_0 + G_0 \sum_\alpha t^\alpha G_0 \sum_{\beta\neq\alpha} t^\beta G_0 + \cdots. \tag{5.6}$$

There are two important points to be noted in Eqs.(5.5) and (5.6). First, as noted previously, the exclusions in the summations prevent two successive scattering events from occurring at the same site. Second, Eq.(5.4) is valid regardless of the spatial arrangement or form of the individual scatterers provided that the individual scattering matrices can be defined, and that individual scattering events can be decoupled and treated as independent from one another. It is not difficult to see that these assumptions are valid for MT potentials. We note that under the usual definitions, $\psi(\mathbf{r}) = \langle\mathbf{r}|\psi\rangle$ and $\chi(\mathbf{r}) = \langle\mathbf{r}|\chi\rangle$, Eq.(5.1) is the coordinate representation of the abstract Eq.(5.4).

As we saw in the previous chapter, Eq.(4.76), the coordinate representation of the site matrix elements of the t-matrix takes the form,

$$T(\mathbf{r},\mathbf{r}') = \sum_{i,j} T^{ij}(\mathbf{r},\mathbf{r}') \tag{5.7}$$

$$T^{ij}(\mathbf{r},\mathbf{r}') = t^i(\mathbf{r},\mathbf{r}')\delta_{ij} + \sum_{k\neq i}\int_{\Omega_i} \mathrm{d}^3r_1 \int_{\Omega_k} \mathrm{d}^3r_2\, t^i(\mathbf{r},\mathbf{r}_1) \times G_0^{ik}(\mathbf{r}_1,\mathbf{r}_2)T^{kj}(\mathbf{r}_2,\mathbf{r}'), \tag{5.8}$$

or

$$\begin{aligned} T^{ij}(\mathbf{r},\mathbf{r}') = \;& t^i(\mathbf{r},\mathbf{r}')\delta_{ij} + (1-\delta_{ij})\int_{\Omega_i} \mathrm{d}^3r_1 \int_{\Omega_j} \mathrm{d}^3r_2\, t^i(\mathbf{r},\mathbf{r}_1) \\ & \times G_0^{ij}(\mathbf{r}_1,\mathbf{r}_2)t^j(\mathbf{r}_2,\mathbf{r}') \\ & + \sum_{k\neq i}\int_{\Omega_i} \mathrm{d}^3r_1 \int_{\Omega_k} \mathrm{d}^3r_2 \int_{\Omega_k} \mathrm{d}^3r_3 \int_{\Omega_j} \mathrm{d}^3r_4 \\ & \times t^i(\mathbf{r},\mathbf{r}_1)G_0^{ik}(\mathbf{r}_1,\mathbf{r}_2)t^k(\mathbf{r}_2,\mathbf{r}_3)G_0^{kj}(\mathbf{r}_3,\mathbf{r}_4)t^j(\mathbf{r}_4,\mathbf{r}') + \cdots, \end{aligned} \tag{5.9}$$

which clearly corresponds to Eq.(4.51). Here, it is to be understood that the variables $\mathbf{r}$ and $\mathbf{r}'$ in any expression of the form $A^{ij}(\mathbf{r},\mathbf{r}')$ are confined to the cells Ω_i and Ω_j, respectively, and that integrations extend over the domains of individual cells (more precisely over the range of the potentials associated with the cells).

5.1.1 The Angular-Momentum Representation

Although Eqs.(5.8) and (5.9) give an intuitively appealing picture of how the total scattering described by $T^{ij}(\mathbf{r},\mathbf{r}')$ is related to the elementary scattering events described by $t^i(\mathbf{r},\mathbf{r}')$, they are not very useful for computations. In this section, we derive equations better suited for carrying out calculations by specializing to the angular-momentum representation. For MT potentials the individual cell potentials can be bounded by spheres that do not overlap one another. Thus, the cell vectors $\mathbf{r}_1$ and $\mathbf{r}_2$ are measured from the centers and confined to the interiors of two cells, Ω_1 and Ω_2, respectively, and the intercell vector $\mathbf{R}_{12}$ connecting the centers of the cells (the centers of the bounding spheres) satisfy the inequalities,

$$|\mathbf{R}_{12}| > |\mathbf{r}_1|, \quad |\mathbf{R}_{12}| > |\mathbf{r}_2|, \quad \text{and} \quad |\mathbf{R}_{12}| > |\mathbf{r}_1 - \mathbf{r}_2|, \tag{5.10}$$

along with

$$|\mathbf{R}_{12} - \mathbf{r}_2| > |\mathbf{r}_1| \quad \text{and} \quad |\mathbf{R}_{12} + \mathbf{r}_1| > |\mathbf{r}_2|. \tag{5.11}$$

In order to obtain the desired expansions of T in angular-momentum eigenstates, we use the well-known expansion of the free-particle propagator, Eq.(D.6), (the bra and ket notation used here and often through the book is discussed and explained in Appendix D),

$$G_0(\mathbf{r}-\mathbf{r}') = -\mathrm{i}k\langle H(\mathbf{r})|J(\mathbf{r}')\rangle \quad \text{for} \quad r > r'. \tag{5.12}$$

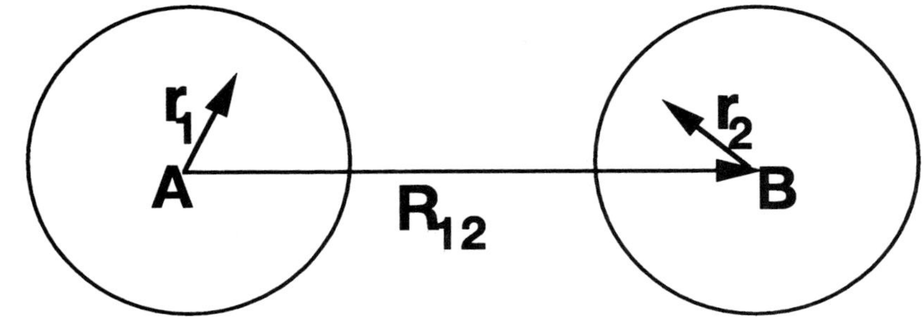

FIGURE 5.2. Two potential cells that satisfy the MT conditions, intra-cell vectors $\mathbf{r}_1$ and $\mathbf{r}_2$, and the intercell vector $\mathbf{R}_{12}$

In the MT case we have $\mathbf{r} - \mathbf{r}' = \mathbf{r}_1 - \mathbf{R}_{12} - \mathbf{r}_2$ and we can write Eq.(5.12) in the form

$$\begin{aligned} G_0(\mathbf{r}_1 - \mathbf{R}_{12} - \mathbf{r}_2) &= -ik\langle J(\mathbf{r}_1)|G(\mathbf{R}_{12})|J(\mathbf{r}_2)\rangle \\ &= -ik \sum_L \sum_{L'} J_L(\mathbf{r}_1) G_{LL'}(\mathbf{R}_{12}) J_{L'}(\mathbf{r}_2). \end{aligned} \tag{5.13}$$

Figure 5.2 shows a schematic diagram of the MT geometry. For $\mathbf{R}_{12}$ a direct lattice vector, the expansion coefficients $G_{LL'}(\mathbf{R}_{12})$ are the real-space representatives of the well-known KKR *structure constants* introduced by Korringa [1] and by Kohn and Rostoker [2] in their pioneering work on the application of the Green function method to the electronic structure of solids.

Now, for $\mathbf{r}_1$ and $\mathbf{r}_2$ confined inside different nonoverlapping spheres, the sums over L and L' can be performed irrespective of order.[1] Although angular-momentum expansions can be used in connection with more general geometries, the corresponding sums in general must be treated with greater care. *This is perhaps the most important point to keep in mind in our subsequent discussion in which MST is extended to space-filling potentials.*

With these considerations in mind, let us examine the single-scattering term in Eq.(5.6), which we write in the form [3],

$$\begin{aligned} \langle \mathbf{r}|G_0 t^i G_0|\mathbf{r}'\rangle &= \int \mathrm{d}^3 y \int \mathrm{d}^3 z G_0(\mathbf{r} - \mathbf{y}) \\ &\quad \times t^i(\mathbf{y} - \mathbf{R}_i, \mathbf{z} - \mathbf{R}_i) G_0(\mathbf{z} - \mathbf{r}') \\ &= \int \mathrm{d}^3 y' \int \mathrm{d}^3 z' G_0(\mathbf{r} - \mathbf{R}_i - \mathbf{y}') \end{aligned}$$

[1]Note that this independence is preserved even if the third inequality in (5.10) is violated. Thus, for the case of two adjacent cells, Eq.(5.13) remains valid for all cell vectors that do not lie in the region of overlap of the spheres bounding the cells.

$$\times\, t^i(\mathbf{y}' - \mathbf{z}')G_0(\mathbf{z}' + \mathbf{R}_i - \mathbf{r}'), \tag{5.14}$$

where $\mathbf{R}_i$ labels the center of scattering cell (MT) at site i, and $\mathbf{y}'$ and $\mathbf{z}'$ are the relative coordinates of the vectors $\mathbf{y}$ and $\mathbf{z}$ with respect to the $\mathbf{R}_i$. Let us consider specifically the case in which both arguments of $G(\mathbf{r}, \mathbf{r}')$ lie outside all bounding spheres, that is, in the interstitial region and still satisfy the MT condition. Then, because of the inequalities, $|\mathbf{r} - \mathbf{R}_i| > y'$ and $|\mathbf{r}' - \mathbf{R}_i| > z'$, we have

$$\begin{aligned}\langle \mathbf{r}|G_0 t^i G_0|\mathbf{r}'\rangle &= \int \mathrm{d}^3y' \int \mathrm{d}^3z' \sum_L (-\mathrm{i}k) H_L(\mathbf{r} - \mathbf{R}_i) J_L(\mathbf{y}') t^i(\mathbf{y}', \mathbf{z}') \\ &\quad \times \sum_{L'} (-\mathrm{i}k) J_{L'}(\mathbf{z}') H_{L'}(\mathbf{r}' - \mathbf{R}_i) \\ &= \sum_{LL'} (-\mathrm{i}k)^2 H_L(\mathbf{r} - \mathbf{R}_i) t^i_{LL'} H_{L'}(\mathbf{r}' - \mathbf{R}_i),\end{aligned} \tag{5.15}$$

where the definition of $t^i_{LL'}$, Eq.(4.77), has been used. Similarly, the two-scatterer term in the expansion of the Green function can be written in the form,

$$\begin{aligned}\langle \mathbf{r}|G_0 t^i G_0 t^j G_0|\mathbf{r}'\rangle &= \sum_{LL_1} \sum_{L_2 L'} (-\mathrm{i}k)^3 H_L(\mathbf{r} - \mathbf{R}_i) t^i_{LL_1} \\ &\quad \times G_{L_1 L_2}(\mathbf{R}_{ij}) t^j_{L_2 L'} H_{L'}(\mathbf{r}' - \mathbf{R}_j).\end{aligned} \tag{5.16}$$

Proceeding in this manner, we obtain for $\mathbf{r}$ in the interstitial region near cell m and $\mathbf{r}'$ in the interstitial region near cell n, the expression

$$\begin{aligned}G(\mathbf{r}, \mathbf{r}') = G_0(\mathbf{r} - \mathbf{r}') &+ \sum_{m,n} \sum_{L_1 L_2} (-\mathrm{i}k)^2 H_{L_1}(\mathbf{r} - \mathbf{R}_m) \tau^{mn}_{L_1 L_2}(E) \\ &\times H_{L_2}(\mathbf{r}' - \mathbf{R}_n).\end{aligned} \tag{5.17}$$

Here we have used the *scattering-path matrix* [4][2](SPM), $\tau^{mn}_{LL'}(E)$, introduced following Eq.(4.108) to denote the matrix elements of the on-the-energy shell part of the scattering matrix, T. The SPM is a function of the energy and as follows from Eq.(5.9) satisfies the expansion,

$$\tau^{mn}_{LL'} = t^m_{LL'} \delta_{mn} + (1 - \delta_{mn}) \sum_{L_1 L_2} t^m_{LL_1} \tilde{G}_{L_1 L_2}(\mathbf{R}_{mn}) t^n_{L_2 L'} + \cdots. \tag{5.18}$$

Here we have used the symbol $\tilde{G}_{L_1 L_2}$ to denote $-ikG_{L_1 L_2}$. This will simplify some of the MST expressions.

[2]This quantity is more often called the scattering-path *operator* in the literature, but it is not technically an operator because it is only defined on the energy shell. We shall use both terms interchangeably.

The SPM, $\underline{\tau}$ satisfies the *equation of motion*,

$$\tau_{LL'}^{mn} = t_{LL'}^{m}\delta_{mn} + \sum_{k\neq m}\sum_{L_1L_2} t_{LL_1}^{m}\tilde{G}_{L_1L_2}(\mathbf{R}_{mk})\tau_{L_2L'}^{kn}, \tag{5.19}$$

which can also be written in terms of matrices in L space,

$$\underline{\tau}^{mn} = \underline{t}^{m}\left[\delta_{mn} + \sum_{k\neq m}\underline{\tilde{G}}(\mathbf{R}_{mk})\underline{\tau}^{kn}\right]. \tag{5.20}$$

We now see that the last equation can be solved formally in the form

$$\underline{\tau}^{mn} = \left[\underline{M}^{-1}\right]^{mn}, \tag{5.21}$$

where the matrix $\underline{M}$ is defined by the elements

$$\begin{aligned}\underline{M}^{ij} &= \left[\underline{t}^{i}\right]^{-1}\delta_{ij} - \underline{\tilde{G}}(\mathbf{R}_{ij})(1-\delta_{ij}) \\ &= \underline{m}^{i}\delta_{ij} - \underline{\tilde{G}}(\mathbf{R}_{ij})(1-\delta_{ij}),\end{aligned} \tag{5.22}$$

and is, as expected, of the MT form.

The formal similarity of the last matrix expression with Eq.(4.108) in operator space is to be noted. Provided that the expansion of $\underline{\tau}$ given in Eq.(5.20) converges, Eqs.(5.21) and (5.22) constitute the formal solution in real space of the SPM for an arbitrary assembly of MT scatterers in terms of the t-matrices of the individual cells. Thus, we have succeeded in our attempt to express the on-the-energy shell scattering matrix of a collection of nonoverlapping MT potentials in terms of the t-matrices of the constituent parts and the structure constants of the system. It is important to emphasize that the derivation of Eq.(5.22) depends on the MT geometry which allows the expansion of the free-particle propagator in Eq.(5.12).

5.1.2 *Electronic Structure of a Periodic Solid*

Before proceeding with further consideration of Eq.(5.20), let us illustrate how this equation can be used to obtain the E vs. $\mathbf{k}$ dispersion relation, (band structure) of a translationally invariant material. We consider the case in which identical scatterers are arranged on the sites of a simple Bravais lattice, so that $\underline{t}^{i} = \underline{t}$ for all i. Then, the SPM, $\underline{\tau}$, is diagonal in reciprocal, $\mathbf{k}$, space and Eq.(5.20) can be solved by means of lattice Fourier transforms. Introducing the Fourier transforms,

$$\underline{\tau}(\mathbf{k}) = \frac{1}{N}\sum_{n,m} e^{i\mathbf{k}\cdot(\mathbf{R}_m-\mathbf{R}_n)}\underline{\tau}^{mn}, \tag{5.23}$$

$$\begin{aligned}\underline{\tau}^{mn} &= \frac{1}{N}\sum_{\mathbf{k}} e^{-i\mathbf{k}\cdot(\mathbf{R}_m-\mathbf{R}_n)}\underline{\tau}(\mathbf{k}) \\ &= \frac{1}{\Omega_{BZ}}\int_{BZ} d^3k\, e^{-i\mathbf{k}\cdot(\mathbf{R}_m-\mathbf{R}_n)}\underline{\tau}(\mathbf{k}),\end{aligned} \tag{5.24}$$

and

$$\underline{\tilde{G}}(\mathbf{k}) = \frac{1}{N} \sum_{n,m} e^{i\mathbf{k}\cdot(\mathbf{R}_m - \mathbf{R}_n)} \underline{G}(\mathbf{R}_{mn}), \tag{5.25}$$

we obtain from Eq.(5.20) the solution

$$\underline{\tau}(\mathbf{k}) = \left[\underline{m} - \underline{\tilde{G}}(\mathbf{k})\right]^{-1}. \tag{5.26}$$

Both $\underline{t}$ and $\underline{\tilde{G}}(\mathbf{k})$ in the last expression are functions of the energy, E, but $\underline{t}$ depends *only* on the potential of a cell, whereas $\underline{\tilde{G}}(\mathbf{k})$ depends *only* on the structure of the lattice. Thus, Eq.(5.26) reflects the complete separation of the potential and the structural aspects of the multiple scattering assembly within the computationally convenient angular-momentum representation. The quantities $\underline{\tilde{G}}(\mathbf{k})$ were introduced by Korringa [1] and by Kohn and Rostoker [2], and are commonly referred to as the KKR structure constants. These structure constants need to be calculated only once for a given lattice, stored and used as it becomes necessary in connection with different MT potentials on the lattice. Methods for the evaluation of the KKR structure constants, often also called the *structural Green function*,[3] can be found in the literature [5].

But, let us now return to the calculation of the electronic structure. Because the band structure is given by the poles of the scattering matrix, we see that at a given energy, E, the allowed eigenvalues, $E(\mathbf{k})$, can be obtained from the solutions of the so-called *secular equation*,[4]

$$\det\left|\underline{\tau}(\mathbf{k})\right| = \det\left|\underline{m} - \underline{\tilde{G}}(\mathbf{k})\right| = 0. \tag{5.27}$$

This equation is evidently the angular-momentum representation of the abstract operator Eq.(4.109). Alternatively, Eq.(5.27) can be solved as a function of the energy for a fixed value of $\mathbf{k}$ in the first Brillouin zone of the reciprocal lattice. These two alternative modes of searching for the roots of the secular equation are referred to as the "k-search" and "E-search" modes, respectively. Either one leads to the determination of the band structure, $E_n(\mathbf{k})$, associated with the various bands, n, of the periodic solid.

It is to be noted that the secular equation Eq.(5.27) (as well as Eqs.(5.21) and (5.22)) involves matrices that are in principle infinite-dimensional, labeled by the angular-momentum index, L, which stands for both the orbital quantum number, $l = 0, 1, 2, \ldots$, and the azimuthal quantum number,

---

[3] The term "structural Green" function is also used in the literature to denote a quantity closely related to the scattering path matrix.

[4] The secular equation can be written in a number of different forms, which in the case of MT potentials lead to identical results. However, in nonMT cases, these forms may exhibit different numerical behavior, as is discussed in the following two chapters.

$-l \le m \le l$. In realistic calculations, however, it is necessary to truncate l at some relatively small number, say $l = 2$, $l = 3$, or $l = 4$. Such truncations have been found to be well justified in most cases of physical interest, such as that of metallic systems on close-packed lattices. In these cases the potentials are very nearly spherically symmetric, which justifies the MT approximation, and a truncation at low l values, (e.g. $l = 2$), has proved adequate. The MT approximation may not be very realistic in the case of systems with low rotational and translational symmetry, for example, surfaces and highly anisotropic charge distributions such as occur in covalently bonded materials. Thus, there is a need to extend multiple scattering theory to arbitrarily shaped, space-filling cell potentials, a task that will be undertaken beginning with the next chapter.

If the potentials are taken to be spherically symmetric, then it follows from Eq.(3.57) that the secular equation (5.27) takes the simpler form

$$\det\left|\underline{m} - \underline{\tilde{G}}(\mathbf{k})\right| = \det\left|-k\cot\delta_\ell - k\underline{B}(\mathbf{k}) + ik\underline{g}(\mathbf{k})\right| = 0, \tag{5.28}$$

where δ_ℓ is the phase shift for the ℓth partial wave introduced in Chapter 3, and $\underline{B}$ and $\underline{g}$ are the expansion coefficients for the Neumann and Hankel functions, respectively, which are given in Appendix D. We shall show later that an alternative formulation of MST in which G_0 is expanded in terms of Neumann rather than Hankel functions corresponding to the assumption of standing wave boundary conditions yields a secular matrix of the form $\det|\cot\delta_\ell + \underline{B}(\mathbf{k})| = 0$. We shall also show how the solutions of the secular equation lead to the determination of the eigenstates and wave functions.

Figure 5.3 shows the band structure, the $E(\mathbf{k})$ vs $\mathbf{k}$ relation, for elemental Cu along various directions in the Brillouin zone, obtained [6] in the MT approximation. The figure shows an electronic structure that is typical of noble and transition metals. There is a band reminiscent of nearly "free" electrons that forms a parabola in the vicinity of the Γ point ($\mathbf{k} = \mathbf{0}$). The extension of this parabola can be seen again in the vicinity of the Fermi energy (solid horizontal line near $E = 0.62$ Ry). Between these two pieces of the "free electron" parabola one can see relatively flat bands derived from the atomic d-states that hybridize with the free electron-like states.

5.2 The Green Function in MST

The one-electron Green function is a very useful quantity because most of the observables that one might want to calculate can be obtained from it. In this subsection we develop Eq.(5.6) for the Green function into a form suitable for computations using an angular-momentum basis. We continue to consider systems of MT potentials.

Again restricting the arguments of $G(\mathbf{r}, \mathbf{r}')$ to lie outside the bounding spheres but inside a given Wigner–Seitz cell, say the one at the origin, we

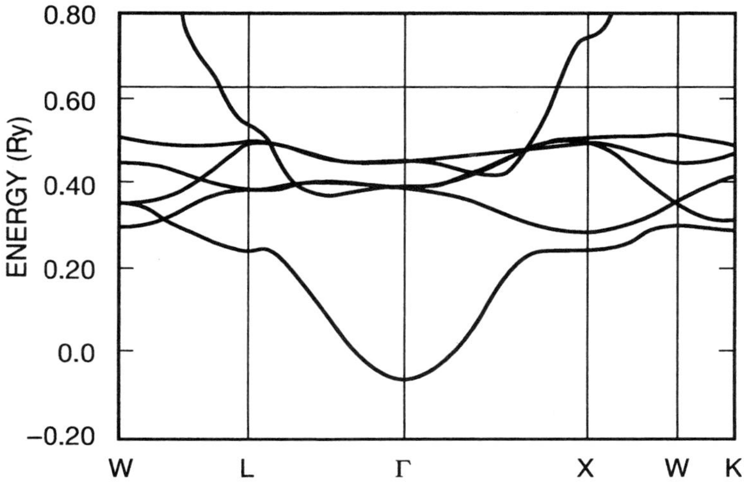

FIGURE 5.3. The band structure of elemental fcc Cu in the MT approximation. V. L. Moruzzi, J. F. Janak, and A. R. Williams, *Calculated Electronic Properties of Metals* (Pergamon, New York, 1978).

can write Eq.(5.17) in the form [3],

$$
\begin{aligned}
G(\mathbf{r},\mathbf{r}') = G_0(\mathbf{r},\mathbf{r}') &+ (-ik)^2 \langle H(\mathbf{r})|\underline{\tau}^{00}|H(\mathbf{r}')\rangle \\
&- ik \sum_{n\neq 0} \langle H(\mathbf{r})|\underline{\tau}^{0n}\underline{\tilde{G}}(\mathbf{R}_n)|J(\mathbf{r}')\rangle \\
&- ik \sum_{m\neq 0} \langle J(\mathbf{r})|\underline{\tilde{G}}(\mathbf{R}_m)\underline{\tau}^{m0}|H(\mathbf{r}')\rangle \\
&+ \sum_{n\neq 0}\sum_{m\neq 0} \langle J(\mathbf{r})|\underline{\tilde{G}}(\mathbf{R}_m)\underline{\tau}^{mn}\underline{\tilde{G}}(\mathbf{R}_n)|J(\mathbf{r}')\rangle,
\end{aligned} \tag{5.29}
$$

where the vector/matrix notation introduced in Appendix D has been used. We can now use the expressions

$$
\underline{\tau}^{00} = \underline{t}^0 + \underline{t}^0 \sum_{m\neq 0} \underline{\tilde{G}}^{0m}\underline{\tau}^{m0} \tag{5.30}
$$

and

$$
\underline{\tau}^{00} = \underline{t}^0 + \sum_{n\neq 0} \underline{\tau}^{0n}\underline{\tilde{G}}^{n0}\underline{t}^0, \tag{5.31}
$$

which imply

$$
\sum_{m\neq 0} \underline{\tilde{G}}(\mathbf{R}_m)\underline{\tau}^{m0} = \underline{m}^0\underline{\tau}^{00} - 1, \tag{5.32}
$$

$$\sum_{m\neq 0} \underline{\tau}^{0m}\underline{\tilde{G}}(\mathbf{R}_m) = \underline{\tau}^{00}\underline{m}^0 - 1, \tag{5.33}$$

and

$$\sum_{m\neq 0}\sum_{n\neq 0} \underline{\tilde{G}}(\mathbf{R}_m)\underline{\tau}^{mn}\underline{\tilde{G}}(\mathbf{R}_n) = \underline{m}^0\underline{\tau}^{00}\underline{m}^0 - \underline{m}^0, \tag{5.34}$$

to cast Eq.(5.29) in the form,

$$\begin{aligned} G(\mathbf{r},\mathbf{r}') &= \langle J(\mathbf{r})\underline{m}^0 - ikH(\mathbf{r})|\underline{\tau}^{00}|\underline{m}^0 J(\mathbf{r}') - ikH(\mathbf{r}')\rangle \\ &\quad - \langle J(\mathbf{r})\underline{m}^0 - ikH(\mathbf{r})|J(\mathbf{r}')\rangle. \end{aligned} \tag{5.35}$$

Here, we used the expansion of the free-particle propagator, Eq.(5.12), and chose $\mathbf{r}'$ greater than $\mathbf{r}$. From the fact that the Green function is continuous with respect to either of its arguments, we can write an expression for the Green function even when the vectors $\mathbf{r}$ and $\mathbf{r}'$ lie inside the MT spheres,

$$G(\mathbf{r},\mathbf{r}') = G^{(s)}(\mathbf{r},\mathbf{r}') + \langle Z(\mathbf{r})|\underline{\tau}^{00}|Z(\mathbf{r}')\rangle, \tag{5.36}$$

where we have defined the singular part of the Green function corresponding to the cell at the origin,

$$G^{(s)}(\mathbf{r},\mathbf{r}') = -\langle Z(\mathbf{r})|\tilde{J}(\mathbf{r}')\rangle. \tag{5.37}$$

For $r > r'$, the singular term is

$$G^{(s)}(\mathbf{r},\mathbf{r}') = -\langle Z(\mathbf{r}')|\tilde{J}(\mathbf{r})\rangle. \tag{5.38}$$

The functions $|Z(\mathbf{r})\rangle$ and $|\tilde{J}(\mathbf{r})\rangle$ are those regular and irregular solutions, respectively, of the Schrödinger equation that at the radius of the MT sphere join smoothly to the functions,

$$|Z(\mathbf{r})\rangle \to \underline{m}|J(\mathbf{r})\rangle - ik|H(\mathbf{r})\rangle \tag{5.39}$$

and

$$|\tilde{J}(\mathbf{r})\rangle \to |J(\mathbf{r})\rangle. \tag{5.40}$$

Note that $|Z(\mathbf{r})\rangle$ corresponds to the wave function defined in Eq.(3.88). For real potentials and real energies the functions Z and $\tilde{J}$ can be chosen to be real.[5]

Although Eq.(5.36) was derived for $\mathbf{r}$ and $\mathbf{r}'$ outside the bounding sphere (but inside the cell at the origin) it satisfies the defining equation for the Green function,

$$\left[-\nabla^2 + k^2 + V\right] G(\mathbf{r},\mathbf{r}') = \delta(\mathbf{r}-\mathbf{r}') \tag{5.41}$$

[5]Caution should be used in using this form of the wave function with nonmuffin-tin potentials because of possible difficulties in inverting the t-matrix (or sine matrix).

and contains the proper boundary conditions so that it can readily be continued inside the sphere and the potential region. Thus, for $\mathbf{r}$ ($\mathbf{r}'$) inside no other MT spheres than $n(m)$ and for $r' > r$, we can write

$$G(\mathbf{r}, \mathbf{r}') = \langle Z^n(\mathbf{r}) | \underline{\tau}^{mn} | Z^m(\mathbf{r}') \rangle + \langle Z^n(\mathbf{r}) | \tilde{J}^n(\mathbf{r}') \rangle \delta_{nm}. \tag{5.42}$$

After $G(\mathbf{r}, \mathbf{r}')$ has been determined, all single-particle properties of the system under consideration can be calculated. For example, the charge density at position $\mathbf{r}$ inside cell (MT) n and at a real energy E can be obtained from the imaginary part of the Green function,

$$\begin{aligned} \rho^n(\mathbf{r}; E) &= -\frac{1}{\pi} \Im G(\mathbf{r}, \mathbf{r}; E) \\ &= -\frac{1}{\pi} \langle Z^n(\mathbf{r}) | \Im \underline{\tau}^{nn}(E) | Z(\mathbf{r}) \rangle \\ &= -\frac{1}{\pi} \sum_L \sum_{L'} Z_L^n(\mathbf{r}) \Im \tau_{LL'}^{nn}(E) Z_{L'}(\mathbf{r}), \end{aligned} \tag{5.43}$$

an expression which will be found to be valid for nonspherical as well as spherical potentials. (For complex energies the single-scatterer term, $G^{(1)}$, in Eq.(5.36) acquires an imaginary part and must be taken into consideration.) Thus, the density of states at a real energy E,

$$n(E) = \int \mathrm{d}^3 r \rho(\mathbf{r}, E), \tag{5.44}$$

is determined directly in terms of the imaginary part of $\underline{\tau}^{00}$. The charge density, $\rho(\mathbf{r})$, is given by an integral over the energy of the quantity $\rho(\mathbf{r}; E)$,

$$\rho(\mathbf{r}) = \int^{E_F} \mathrm{d}E \rho(\mathbf{r}, E) \tag{5.45}$$

where E_F is the Fermi level, that is, the energy of the highest occupied single-particle level in the material. Finally, the integrated density of states is given as an integral over $n(E)$,

$$N(E) = \int^{E} \mathrm{d}E' n(E'), \tag{5.46}$$

and denotes the total number of states lying below E.

Figure 5.4 depicts the charge density and DOS as well as integrated DOS of elemental fcc Cu obtained [6] in the MT approximation. The charge density is multiplied by the square of the radius in order to emphasize the structure at small values of the radius. The peaks in the charge density arise from the shell structure of the atom. The sharp structure with many peaks in the density of states arises from the d-bands.

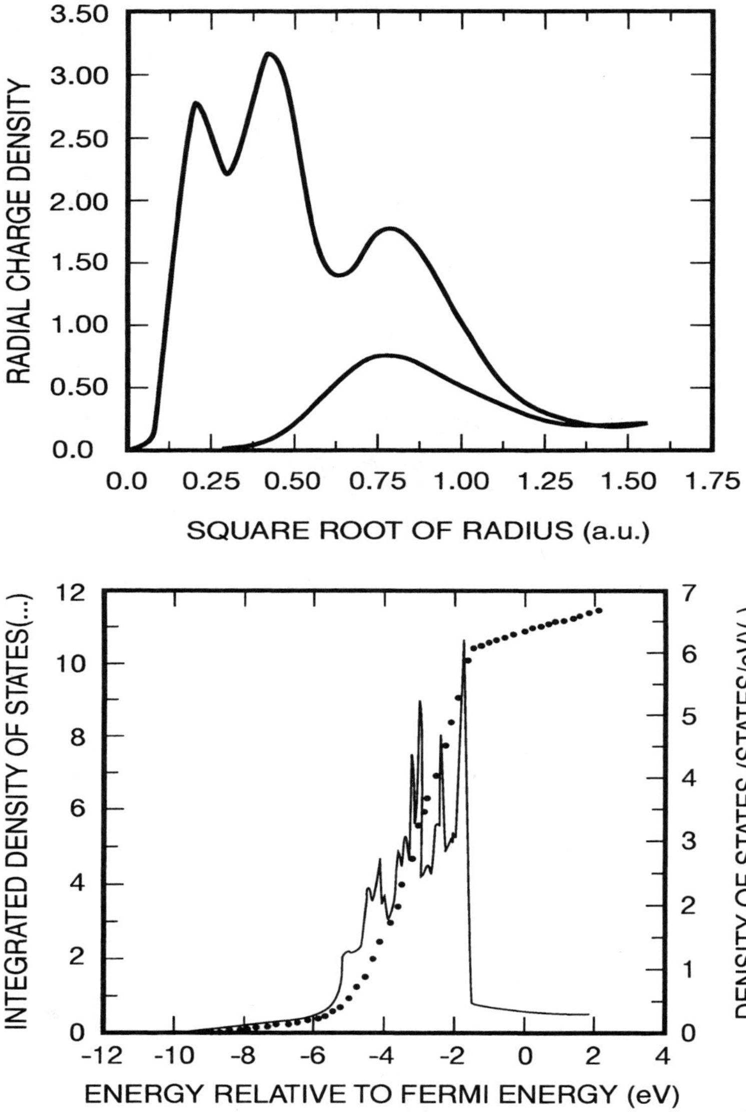

FIGURE 5.4. The charge density and valence charge density, upper panel, and density of states and integrated DOS, lower panel, for elemental fcc Cu[6].

5.3 Impurities in MST

After the Green function or the scattering path operator of any system is known it is relatively simple within MST to obtain the Green function or scattering path operator for a modified system in which one of the atomic potentials or a cluster of atomic potentials has been changed. This might be useful, for example, in calculating the electronic structure associated with a substitutional impurity in a periodic lattice.

As we saw in Section 5.1, Eqs. (5.21) and (5.22) for the scattering path operator can be written as the inverse of a matrix in angular-momentum and site indices,

$$\tau^{ij}_{LL'} = [M^{-1}]^{ij}_{LL'}, \tag{5.47}$$

where

$$M^{ij}_{LL'} = m^{i}_{LL'}\delta_{ij} - \tilde{G}_{LL'}(R_{ij})(1-\delta_{ij}). \tag{5.48}$$

If one of these potentials, say the one at the origin, is changed, $m^0 \to m^I$, the new scattering path operator will be obtained from the inverse of the matrix

$$M^{I;ij}_{LL'} = M^{ij}_{LL'} - \Delta m^0_{LL'}\delta_{i0}\delta_{j0}, \tag{5.49}$$

where

$$\Delta m^0_{LL'} = m^0_{LL'} - m^I_{LL'}, \tag{5.50}$$

so that the scattering path operator for the system with the impurity is given by the expression

$$\tau^{I;ij}_{LL'} = \sum_{L_1,k} \tau^{ik}_{LL_1} D^{kj}_{L_1L'} \tag{5.51}$$

where the matrix, $D^{ij}_{LL'}$ is given by $D^{ij}_{LL'} = [Y^{-1}]^{ij}_{LL'}$ with

$$Y^{ij}_{LL'} = \delta_{ij}\delta_{LL'} - \delta_{i0}\sum_{L_1}\Delta m^0_{LL_1}\tau^{0j}_{L_1L'}. \tag{5.52}$$

Note that the scattering path operator,$\tau^{I;00}_{LL'}$, for the site with the impurity and hence the Green function for that site is given entirely in terms of τ^{00} and Δm. This can be made more obvious if we rearrange Eq. (5.51) slightly to obtain the result,

$$\tau^{I;ij}_{LL'} = \tau^{ij}_{LL'} + \sum_{L_1L_2} \tau^{i0}_{LL_1} X^{00}_{L_1L_2}\tau^{0j}_{L_2L'}, \tag{5.53}$$

where

$$\begin{aligned} X^{00}_{LL'} &= \sum_{L_1} D^{00}_{LL_1}\Delta m^0_{L_1L'} \\ &= \Delta m^0_{LL'} + \sum_{L_1L_2}\Delta m^0_{LL_1}\tau^{00}_{L_1L_2}\Delta m^0_{L_2L'} \\ &\quad + \sum_{L_1L_2L_3L_4}\Delta m^0_{LL_1}\tau^{00}_{L_1L_2}\Delta m^0_{L_2L_L3}\tau^{00}_{L_3L_4}\Delta m^0_{L_4L'} + \cdots \end{aligned} \tag{5.54}$$

Thus there is an interesting analogy between the Eq. (5.53) and Eq. (5.6). Assuming that there is only one scatterer in free space, Eq. (5.6) becomes $G = G_0 + G_0 t G_0$ whereas Eq. (5.53) can be written as $\tau^I = \tau + \tau X \tau$. Here, τ, plays a role analogous to a free-space Green function, and X plays a role analogous to the t-matrix in free space. Thus, just as t can be used to make a Green function, G, which includes the effects of a scattering potential in free space from the elementary free-space Green function, (G_0); X can be used to make a scattering path operator, τ^I, which includes the effects of the impurity potential from the scattering path operator for a periodic system, (τ).

It is quite straightforward to generalize these results to a localized cluster of impurities. The Green function connecting sites within the impurity cluster can be written in terms of the parts of the scattering path operator of the original system which connect sites within the cluster. This feature of multiple scattering theory allows the calculation of the electronic structure of an infinite periodic system with impurities by first calculating the scattering path operator of the infinite periodic system, for example, in k-space using Eq. (5.26) and then using this scattering path operator to calculate the scattering path operator in the presence of the impurities. The size of the system of equations that has to be solved is given by the number of *impurity* sites times the number of angular-momentum states.

5.4 Coherent Potential Approximation

In Section 5.3 we observed that the quantity X^{00} defined in Eq. (5.54) gives the additional scattering caused by an impurity with inverse t-matrix, m^I in an otherwise perfect system with inverse t-matrix m^0. This result can be used to obtain an approximate mean field theory for the electronic structure of a substitutional alloy that is known as the coherent potential approximation [7] (CPA). The CPA is a rather general idea for treating disordered alloys that has been implemented within several formalisms. It is most often used within the context of a model tight-binding Hamiltonian. The implementation within multiple scattering theory, however, probably allows a more realistic treatment of the electronic structure of a physical alloy.

The problem of calculating the electronic structure of a random substitutional alloy is a classic problem in solid state physics. The CPA is often described as the best "single-site" solution to this problem. In the CPA one attempts to calculate the electronic structure, for example, the density of states on an atom in the alloy, by imagining the atom to be immersed in a periodic "medium" that behaves "on the average" like the remainder of the alloy. The "medium" is to be chosen self-consistently. If the electronic structure is calculated using multiple scattering theory, the only part of the theory that "knows" about the type of atom on a particular site is the t-matrix. Therefore we imagine that our "medium" consists of a system with a "coherent" t-matrix on each site, t_c. Corresponding to t_c there will be an inverse t-matrix, $m_c = t_c^{-1}$. Equation (5.54) can be used to calculate the additional scattering that occurs when an impurity atom of type "A" with t-matrix t^A is introduced into the lattice of "coherent" potentials with t-matrices, t_c,

$$\begin{aligned} X^A &= D^A \Delta m^A \\ D^A &= [1 - \tau_c \Delta m^A]^{-1} \\ \tau_c &= \frac{1}{\Omega_z} \int d\mathbf{k}[m_c - G(\mathbf{k})]^{-1}. \end{aligned} \tag{5.55}$$

Here the angular-momentum indices are suppressed and it is understood that all cite indices are 0 indicating the cite at the origin and $\Delta m^A = m^c - m^A$.

The CPA condition for obtaining the coherent medium or t-matrix is that, on the average, the additional scattering due to replacing a t_c at the origin by impurity t-matrices, t_A, t_B, etc. should vanish. Thus, for an alloy with n components with concentrations, c_i, $i = 1, \ldots, n$ we would determine t_c (or m_c) from the condition,

$$\sum_i c_i X^i = 0. \tag{5.56}$$

From Eq. (5.55) it is clear that Eq. (5.56) can also be written as

$$\sum_i c_i D^i = 1 \tag{5.57}$$

or as

$$\sum_i c_i \tau_c^i = \tau_c. \tag{5.58}$$

After the "coherent" t-matrix, t_c has been determined by solving either of the three equations, Eq. (5.56), (5.57), or (5.58), the electronic structure on a site occupied by a particular type of atom can be calculated from the Green function generated from the scattering path operator calculated for that type of atom treated as an impurity within the CPA — t_c medium,

$$\tau^A = \tau_c + \tau_c X^A \tau_c. \tag{5.59}$$

The CPA expressed in the context of multiple scattering theory [8, 9, 10, 11] has contributed significantly to our understanding of the properties of alloys [12, 13].

5.5 Screened MST

In Section 5.3 we took advantage of the Green function to calculate the changes in electronic structure caused by changes to one or a few of the potentials. One can also calculate the change in the Green function caused by making the same change to all of the potentials. One use for this might be to generate a reference medium with useful properties as was done in Section 5.4 where we derived a self-consistent medium to describe a disordered alloy. In this section we obtain a reference medium that gives better convergence for a periodic system.

We usually think of MST in terms of scattering by potentials with free electron propagation between the potentials. It may sometimes be useful, however, to imagine the electrons to be propagating in a medium filled with "reference" scatterers rather than uniform zero potential and scattering off modified scatterers which are equal to the difference between the t-matrix of the original scatterer and that of the reference scatterer. In this case the structure constants are replaced by modified structure constants which we can arrange to have desirable properties such as being more localized spatially than the standard structure constants for free electrons.

From Eq.(5.26) we can write the inverse of the Fourier transform of the scattering path operator of a periodic system with an inverse t-matrix, $\underline{m}$ on each site as

$$\underline{\tau}^{-1}(\mathbf{k}) = [\underline{m} - \underline{\tilde{G}}(\mathbf{k})]. \tag{5.60}$$

If the inverse t-matrix on each site were $\underline{m}_r$, from a "reference" potential, we would have

$$\underline{\tau}_r^{-1}(\mathbf{k}) = [\underline{m}_r - \underline{\tilde{G}}(\mathbf{k})]. \tag{5.61}$$

Thus we can write

$$\underline{\tau}^{-1}(\mathbf{k}) = [\underline{m} - \underline{m}_r + \underline{\tau}_r^{-1}(\mathbf{k})] = [\underline{1} + \Delta\underline{m}\ \underline{\tau}_r(\mathbf{k})]\underline{\tau}_r^{-1}(\mathbf{k}). \tag{5.62}$$

Thus the scattering path operator can be written as

$$\underline{\tau}^{ij} = \frac{1}{\Omega_z}\int d\mathbf{k}\ \underline{\tau}_r(\mathbf{k})[\underline{1} + \Delta\underline{m}\ \underline{\tau}_r(\mathbf{k})]^{-1} e^{i\mathbf{k}\cdot\mathbf{R}_{ij}} \tag{5.63}$$

One possible advantage of using a reference medium in a calculation is that the scattering path operator for the reference medium, which plays a role similar to the structure constants in the usual calculation, can be made to have a much smaller spatial range than the ordinary structure constants. One way to achieve a faster decay in space of these reference scattering path operators is to choose a repulsive potential for the reference medium that is strong enough to raise the bottom of its conduction band above the energy region of interest. Several groups [14, 15, 16, 17, 18] have used this technique to generate effective structure constants with much shorter spatial range than free electron structure constants. These effective structure constants have the added advantage that they are not singular at values of $\mathbf{k}$ and E that satisfy the free electron dispersion relation.

Figure 5.5 shows how the structure constants and the effective or "screened" structure constants decay with distance. The figure shows the "partial norms" which are proportional to $[\sum_{mm'} |\tau^{ij}_{r;lm,l'm'}(E)|^2]^{1/2}$. These partial norms are the same for values of R_{ij} corresponding to a given shell of neighbors. The calculations displayed here were performed by Moghadam and Stocks [18].

The topic of screened structure constants within MST is currently an active area of research. One possible application may be to very large systems consisting of tens or hundreds of different atoms. For such large systems it may be very advantageous to use the "screened" structure constants so that the matrices that must be inverted are relatively sparse. Numerical techniques are available that can invert sparse matrices in a time that scales as the first power of the number of atoms or number of equations rather than the third power as is required for general matrices.

5.6 Alternative Derivation of MST

In section 5.1 we derived the secular equation for MST by summing a multiple scattering series. In this section we shall attempt to gain further insight into the formalism of multiple scattering theory by deriving the secular equation of MST directly from the properties of the wave function as expressed in the Lippmann–Schwinger equation, Eq.(5.1),

$$\psi(\mathbf{r}) = \chi(\mathbf{r}) + \int G_0(\mathbf{r},\mathbf{r}')V(\mathbf{r}')\psi(\mathbf{r}')\mathrm{d}^3r'. \tag{5.64}$$

We are interested in obtaining the stationary states of the potential V and thus set $\chi(\mathbf{r}) = 0$. We, therefore, have to solve the homogeneous integral equation,

$$\psi(\mathbf{r}) = \int G_0(\mathbf{r},\mathbf{r}')V(\mathbf{r}')\psi(\mathbf{r}')\mathrm{d}^3r'. \tag{5.65}$$

Because the potential is the sum of atomic potentials confined within separate nonoverlapping muffin-tin spheres, $V(\mathbf{r}) = \sum_m v_m(\mathbf{r})$, we can write the integral over $\mathbf{r}'$ as the sum of integrals over the individual muffin-tins,

$$\psi(\mathbf{r}) = \int_n \mathrm{d}\mathbf{r}'G_0(\mathbf{r}-\mathbf{r}')V_n(\mathbf{r}')\psi(\mathbf{r}') + \sum_{m\neq n}\int_m \mathrm{d}\mathbf{r}'G_0(\mathbf{r}-\mathbf{r}')V_m(\mathbf{r}')\psi(\mathbf{r}'). \tag{5.66}$$

Let us begin by considering the case in which the point $\mathbf{r}$ is very near but just outside muffin-tin n (so that the MT conditions on the lengths of the intercell and intracell vectors hold as discussed in Section 4.5). We shall expand the wave function in the vicinity of this muffin-tin using the

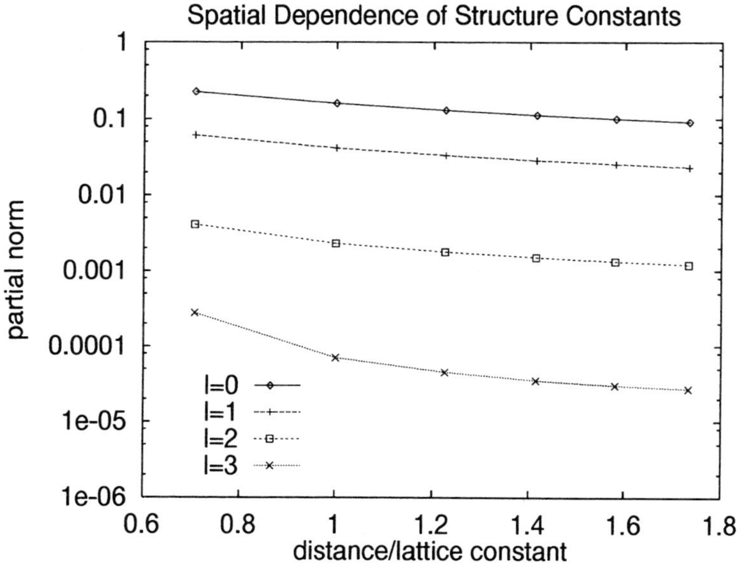

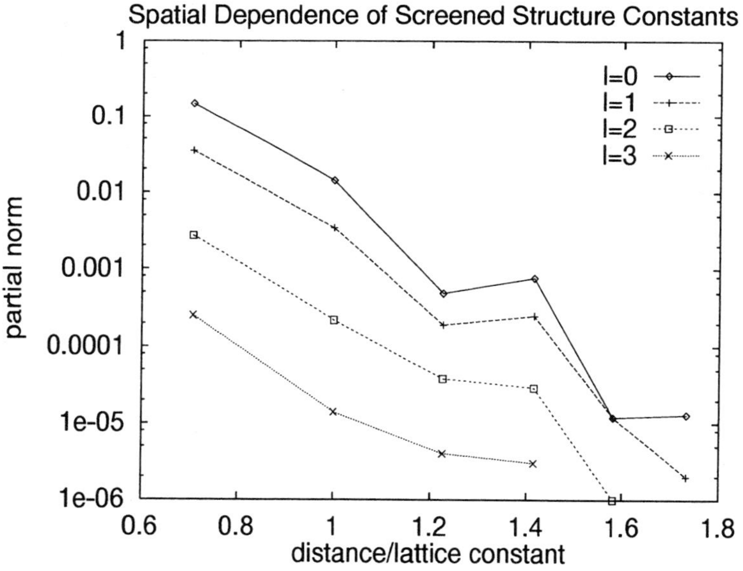

FIGURE 5.5. A comparison of free electron structure constants and effective "screened" structure constants.

basis functions defined in Section 3.4,

$$\psi(\mathbf{r}) = \sum_L \psi_L^n(\mathbf{r}_n) a_L^n. \tag{5.67}$$

Any of the forms for the basis functions that were obtained in Section 3.4 could be used. However, the subsequent derivation in which we use standing wave boundary conditions is slightly simpler if we use basis functions of the form,

$$\psi_L^n(\mathbf{r}_n) = \sum_{L'} [J_{L'}(\mathbf{r}_n) C_{L'L}^n - N_{L'}(\mathbf{r}_n) S_{L'L}^n]. \tag{5.68}$$

Standing wave boundary conditions are invoked by using the form $-\frac{\cos\sqrt{E}|\mathbf{r}-\mathbf{r}'|}{4\pi|\mathbf{r}-\mathbf{r}'|}$ for the free-particle Green function. This gives a different secular equation from that obtained from the multiple scattering expansion of Eq.(4.74) because that expansion used outgoing wave boundary conditions that entered through the free-particle Green function, $-\frac{\exp i\sqrt{E}|\mathbf{r}-\mathbf{r}'|}{4\pi|\mathbf{r}-\mathbf{r}'|}$. Substituting the expansion for the basis function into the Lippmann–Schwinger equation (5.66) and expanding the free-particle Green functions we obtain

$$\begin{aligned}
&\sum_{LL'} [J_{L'}(\mathbf{r}_n) C_{L'L}^n - N_{L'}(\mathbf{r}_n) S_{L'L}^n] a_L^n = \\
&\quad k \sum_{L'} N_{L'}(\mathbf{r}_n) \int_n d\mathbf{r}'_n J_{L'}(\mathbf{r}'_n) V(\mathbf{r}'_n) \sum_L \psi_L^n(\mathbf{r}'_n) a_L^n \\
&\quad + \sum_{m\neq n} k \sum_{L'} N_{L'}(\mathbf{r}_m) \int_m d\mathbf{r}'_m J_{L'}(\mathbf{r}'_m) V(\mathbf{r}'_m) \sum_L \psi_L^m(\mathbf{r}'_m) a_L^m
\end{aligned} \tag{5.69}$$

From the definition $S_{L'L}^n = -k \int_n d\mathbf{r} J_{L'}(\mathbf{r}) V_n(\mathbf{r}) \psi_L(\mathbf{r})$, we obtain

$$\sum_{LL'} J_{L'}(\mathbf{r}_n) C_{L'L}^n a_L^n = -\sum_{m\neq n} \sum_{L'} N_{L'}(\mathbf{r}_m) \sum_L S_{L'L}^m a_L^m. \tag{5.70}$$

The Neumann function, $N_{L'}(\mathbf{r}_m) = N_{L'}(\mathbf{r}_n - R_{nm})$ can be expanded using

$$N_{L'}(\mathbf{r}_n - \mathbf{R}_{nm}) = \sum_{L''} J_{L''}(\mathbf{r}_n) B_{L''L'}(\mathbf{R}_{nm}), \tag{5.71}$$

so that we obtain

$$\sum_L C_{L'L}^n a_L^n + \sum_{m\neq n} \sum_{L''L} B_{L'L''}(\mathbf{R}_{nm}) S_{L''L}^m a_L^m = 0, \tag{5.72}$$

which may be compared with Eq.(5.28). If we had used the form of the free-space Green function corresponding to outgoing waves in Eq. (5.65), we could have obtained the secular equation of Eq.(5.28). For an infinite periodic system, it can be shown that the two secular equations have the same eigenvalues.

5.7 Korringa's Derivation

We present here a final derivation of MST due to Korringa [1] which is based on achieving continuity of wave function value and derivative at each of the muffin-tin spheres. This derivation also leads naturally to the Muffin-Tin Orbital picture which is discussed in the next session. For simplicity we shall restrict ourselves to spherical scatterers in this section.

Inside the nth muffin-tin we can expand the wave function in the form

$$\psi^n(\mathbf{r}_n) = \sum_L \phi_L^n(\mathbf{r}_n) a_L^n, \tag{5.73}$$

where subscript L stands for (ℓ, m), $\mathbf{r}_n$ is a vector from the center, $\mathbf{R}_n$, of atom n (i.e., $\mathbf{r}_n = \mathbf{r} - \mathbf{R}_n$), and where the the local basis function $\phi_L^n(\mathbf{r}_n)$ is the product of a radial wave function and a spherical harmonic

$$\phi_L^n(\mathbf{r}_n) = R_\ell^n(r_n) Y_L(\hat{r}_n). \tag{5.74}$$

Each of the basis functions $\phi_L^n(\mathbf{r}_n)$ is an exact local solution to the Schrödinger equation, and, because the potential vanishes outside the muffin-tin, it joins smoothly to a combination of regular and irregular solid harmonics, $(r_n > r_{\mathrm{MT}})$

$$\psi^n(\mathbf{r}_n) = \sum_L \left\{ J_L(\mathbf{r}_n) a_L^n + H_L(\mathbf{r}_n) b_L^n \right\} = \sum_L \left\{ J_L(\mathbf{r}_n) + H_L(\mathbf{r}_n) t_\ell^n \right\} a_L^n, \tag{5.75}$$

where $J_L(\mathbf{r}_n) = j_\ell(\sqrt{E} r_n) Y_L(\hat{r}_n)$ and $H_L(\mathbf{r}_n) = -i\sqrt{E} h_\ell(\sqrt{E} r_n) Y_L(\hat{r}_n)$. The t-matrix, $t_\ell^n = b_L^n / a_L^n$, is determined by the condition that the logarithmic derivative of $\phi_L^n(\mathbf{r})$ be continuous across the muffin-tin boundary. Note that we can choose the (arbitrary) normalization of $\phi_L^n(\mathbf{r})$ so that

$$\phi_L^n(\mathbf{r}) = J_L(\mathbf{r}_n) + H_L(\mathbf{r}_n) t_\ell^n \quad \text{for } r = r_{MT}. \tag{5.76}$$

This ensures that the coefficients a_L^n are the same in Eqs.(5.73) and (5.75).

One of the fundamental postulates of multiple scattering theory [1] is the assumption that (for the bound states of the system) the wave function in the region outside all of the muffin-tins can be represented as a sum of outgoing waves, coming from each scatterer, each regular at infinity,

$$\psi^o(\mathbf{r}) = \sum_{n'L'} H_{L'}(\mathbf{r}_{n'}) b_{L'}^{n'}. \tag{5.77}$$

The MST secular equation can be obtained by requiring that the two expansions be consistent in the vicinity of the muffin-tin radius of potential n,

$$\sum_L \left\{ J_L(\mathbf{r}_n) a_L^n + H_L(\mathbf{r}_n) b_L^n \right\} = \sum_{n'L'} H_{L'}(\mathbf{r}_{n'}) b_{L'}^{n'}. \tag{5.78}$$

This leads to the condition,

$$\sum_L J_L(\mathbf{r}_n)a_L^n = \sum_{n'\neq n,L'} H_{L'}(\mathbf{r}_{n'})b_{L'}^{n'}, \tag{5.79}$$

which is simply the statement that the incoming wave on site n is the sum of the outgoing waves from all of the other sites. Replacing $b_{L'}^{n'}$ by $a_{L'}^{n'}t_{\ell'}^{n'}$, and expanding the irregular solid harmonic centered on site n' by a sum of regular solid harmonics centered on site n using the MST structure constants,

$$H_{L'}(\mathbf{r}_{n'}) = \sum_L J_L(\mathbf{r}_n)G_{LL'}^{nn'} \tag{5.80}$$

$$G_{LL'}^{nn'} = 4\pi \sum_{L''} \mathrm{i}^{\ell-\ell'-\ell''} C(LL'L'')H_{L'}(\mathbf{R}_{nn'}), \tag{5.81}$$

with $C(LL'L'')$ being a Gaunt number (integral of three spherical harmonics), we obtain

$$\sum_L J_L(\mathbf{r}_n)a_L^n = \sum_{n'\neq n,L,L'} J_L(\mathbf{r}_n)G_{LL'}^{nn'}t_{\ell'}^{n'}a_{L'}^{n'} \tag{5.82}$$

which gives us the muffin-tin MST equation

$$a_L^n = \sum_{n'\neq n,L,} G_{LL'}^{nn'}t_{\ell'}^{n'}a_{L'}^{n'}. \tag{5.83}$$

From this expression, the MT secular equation of MST, Eq.(5.28), follows by the replacement, $b_L^n = t_L^n a_L^n$.

5.8 Relation to Muffin-Tin Orbital Theory

In this section we show the relation between the wave function defined in previous sections and the muffin-tin orbital idea of Andersen [19]. In muffin-tin orbital (MTO) theory the system wave function is written as a linear superposition of basis functions $\chi_L^n(E, \mathbf{r})$,

$$\psi^{\mathrm{mto}}(\mathbf{r}) = \sum_{nL} \chi_L^n(\mathbf{r}_n)a_L^n. \tag{5.84}$$

These basis functions, which Andersen calls muffin-tin orbitals or MTOs may be defined (using a notation and choice of normalization that emphasizes the relationship with MST-KKR theory) by

$$\chi_L^n(\mathbf{r}_n) = \begin{cases} H_L(\mathbf{r}_n)t_\ell^n & \text{everywhere outside muffin-tin } n; \\ \phi_L^n(\mathbf{r}_n) - J_L(\mathbf{r}_n) & \text{inside muffin-tin } n. \end{cases} \tag{5.85}$$

Note that the MTOs are smooth and continuous at the muffin-tin boundary because of Eq.(5.76). Each of the MTOs satisfies the Schrödinger equation outside its muffin-tin, but does not satisfy it inside because of the extra term, $-J_L(\mathbf{r}_n)$. The condition for ψ^{mto} (Eq.5.84) to be a valid solution to the Schrödinger equation is that the "tails" of the MTO's from all of the sites other than n cancel the term $-J_L(\mathbf{r}_n)$. Thus the additional terms added to the "true" solution inside muffin-tin n are,

$$-\sum_L J_L(\mathbf{r}_n)a_L^n + \sum_{n'\neq n}\sum_{L'} H_{L'}(\mathbf{r}_{n'})t_{\ell'}^{n'}a_{L'}^{n'}. \tag{5.86}$$

If we require that these terms sum to zero we obtain the same consistency relation, Eq. (5.79), that we derived for MST and thus the same secular equation. Now Eq.(5.86) cannot, in general, be exactly satisfied for an arbitrary ℓ truncation of the secular equation. Use of the secular equation Eq. (7.3) in the expression for the additional terms inside muffin-tin n, Eq. (5.86), yields $\sum_{L>\ell_{\text{max}}} J_L(\mathbf{r}_n)a_L^n$. We point out that $\psi^{\text{mto}}(\mathbf{r})$ is the same function as the so-called *augmented wave function* which can be defined within MST and is discussed in Chapter 7, except that it is represented as a multicenter expansion,

$$\psi_a^n(\mathbf{r}_n) = \sum_{n'L}^{\ell_{\text{max}}} \chi_L^{n'}(\mathbf{r}_{n'})a_L^{n'}. \tag{5.87}$$

The detailed discussion of augmented MST wave functions in Chapter 7 hopefully will further clarify the analogy to MT-orbital theory.

5.9 MST for $E < 0$

In this section we present a detailed study of MST for bound states [20]. We shall also present numerical results for the convergence of the energy and the wave function as a function of ℓ_{max}, the maximum value of orbital angular momentum used in the expansion of the wave function. In order to clarify some of the concepts introduced thus far, we present a detailed study of the multiple scattering treatment of a system of two spherically symmetric potentials in three-dimensional space [20]. In this presentation, we amplify much of the formalism presented in previous sections of this chapter, attempting to clarify such notions as one-center and multicenter expansions and the construction of the wave function within MST. We also show in this section how the MST formalism is modified when the energies of interest are smaller than the value of the potential in the interstitial region and illustrates the use of "scattering" theory even when there are no propagating states. Similar studies of two MT scatterers in one and two dimensions can be found in the literature [21, 22].

The study of the two-MT problem presented here serves an additional purpose as well: It provides a useful foundation and a point of comparison against which the convergence and accuracy of MST when applied to non-MT, space-filling cell potentials can be judged. Such a generalized formalism is presented in the following chapters. Specifically, we study numerically, and derive analytic expressions for, the rate of convergence of the angular-momentum expansions of both the wave function and the secular equation. These studies indicate that MST provides an exact solution to the single particle Schrödinger equation for an assembly of spherically symmetric MT potentials, and that the errors made due to the truncation of the angular-momentum expansions can be made arbitrarily small by use of a sufficiently large number of angular-momentum states.

5.9.1 The Two-Scatterer Problem in Three Dimensions

Our objective is to solve the time-independent Schrödinger equation which (in atomic units) is given by

$$[-\nabla^2 + V(\mathbf{r}) - E]\Psi(\mathbf{r}, E) = 0 \tag{5.88}$$

for the special case where the potential $V(\mathbf{r})$ has the "muffin-tin" form; that is, it consists of a sum of nonoverlapping, spherically symmetric, potentials,

$$V(\mathbf{r}) = \sum_n v_n(\mathbf{r} - \mathbf{R}_n). \tag{5.89}$$

We shall solve this problem by extending Korringa's approach to the case of $E < 0$.

We will be interested in the bound states of a potential consisting of two MT, spherically symmetric potential cells, which we denote by A and B. We now think of space as being divided into two regions in such a way that region I consists of those points "inside" one of the potentials, that is, those points where $V(\mathbf{r})$ is nonzero, and region II consists of those points "outside" all of the potentials, that is, those points where $V(\mathbf{r})$ vanishes. Region I, for the particular case of two scatterers is composed of two pieces, IA inside potential A and IB inside potential B, as is illustrated in Fig. 5.6. The basic idea of MST is to solve the wave equation separately in the region inside each scatterer (regions IA and IB) and in the region outside all scatterers (region II) and then to fit these partial solutions together into a solution that is acceptable over all of space by matching the pieces in magnitude and derivative at the boundaries of each region. If this can be accomplished, then it follows from the theory of second-order partial differential equations that the resulting wave function represents the exact solution of the Schrödinger equation for the system at hand. Thus, in analogy with Eq.(3.66) we write the solutions for regions IA and IB as,

$$\Psi_{In} = \sum_L c_L^n R_\ell^n(r_n) Y_L(\hat{r}_n), \qquad n = (A, B), \tag{5.90}$$

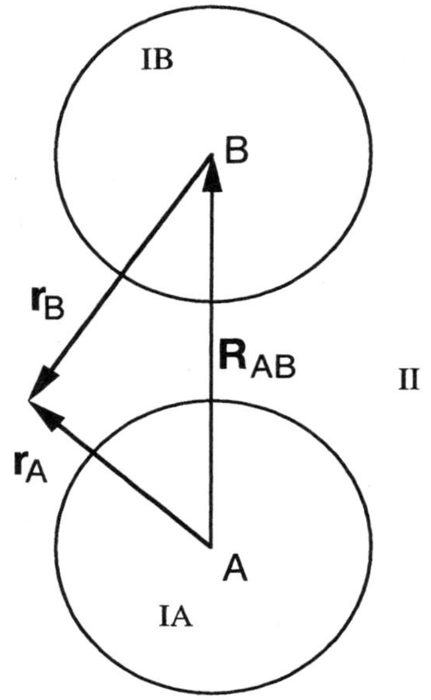

FIGURE 5.6. Two spherical scatterers in three dimensions.

and for region II as (see Eq.(5.77)),

$$\Psi_{II} = \sum_{n=A,B} \sum_{L} b_L^n \tilde{h}_\ell(z_n) Y_L(\hat{r}_n). \tag{5.91}$$

This last expression is a special way of writing the wave function outside the MT sphere by setting $\sum_L c_L t_{LL'} = b_{L'}$,where $\underline{t}$ is the t-matrix in Eq.(3.112). The notations r_n and $\hat{r}_n$ indicate, respectively, distances and directions measured from the center of site n, while the notation z_n is used as a shorthand for κr_n where $\kappa = \sqrt{-E}$. Because we are looking for bound states of the two scatterer problem we must look for states at negative energy relative to the energy zero of region II.

As discussed in Chapter 3, the "inside" functions R_ℓ^n used in region I are solutions of the "radial" part of the wave equation for a single scatterer

$$[-r_n^{-1}(\mathrm{d}^2/\mathrm{d}r_n^2)r_n + \ell(\ell+1)/r_n^2 + v_n(r_n) - E]R_\ell^n(E, r_n) = 0. \tag{5.92}$$

and behave near the origin as r_n^ℓ. The "outside" functions, $\tilde{h}_\ell$, used in region II are solutions of the free-space radial equation Eq.(5.92) with $v_n = 0$ or Eq.(3.17). At negative energies, the functions $\tilde{h}_\ell(z)$ are obtained by analytic continuation of the spherical Hankel functions given in Eqs.(3.31) and (3.32) to imaginary argument. We choose $\tilde{h}_\ell(z) = -i^\ell h_\ell(iz)$. These functions behave as $\tilde{h}_\ell(z_n) \to e^{-z_n}/z_n$ in the limit $r_n \to \infty$, and as $(2\ell - 1)!!/z_n^{\ell+1}$ when r approaches the origin.

The MST equations result from the requirement that the wave functions Ψ_{IA} and Ψ_{IB} match smoothly and continuously onto Ψ_{II} at the boundary between regions I and II. In order to satisfy this matching condition it is convenient to define auxiliary functions in region II, Ψ_{IIA} and Ψ_{IIB}, that are single-center expansions about the centers of potentials A and B respectively:

$$\Psi_{IIn} = \sum_L a_\ell^n \tilde{j}_\ell(z_n) Y_L(\hat{r}_n) + \sum_L b_L^n \tilde{h}_\ell(z_n) Y_L(\hat{r}_n), \qquad n = (A, B). \quad (5.93)$$

Here, $\tilde{j}_\ell$ is obtained from the analytic continuation of the spherical Bessel function to imaginary argument, $\tilde{j}_\ell(z) = i^{-\ell} j_\ell(iz)$. The function $\tilde{j}_\ell(z_n)$ behaves near the origin as $z_n^\ell/(2\ell+1)!!$ but is irregular as $r_n \to \infty$, behaving there as $e^{z_n}/2z_n$.

It is a relatively simple matter to match Ψ_{IA} to Ψ_{IIA} and to match Ψ_{IB} to Ψ_{IIB}. The coefficient c_L^n can be used to match the magnitude of each partial wave at the muffin-tin radius, r_n^0, and the ratio of a_L^n to b_L^n can be chosen to ensure that the logarithmic derivatives match. This latter condition determines the t-matrix, $t_\ell(E)$, although for our purposes it is more convenient to work with the inverse of the t-matrix which we denote by $m_\ell(E)$[6]. Thus m_L^n, defined by the relation

$$a_L^n = m_L^n b_L^n, \quad (5.94)$$

is given by the expression,

$$m_L^n(E) = m_\ell^n(E) = -\frac{z_n^0 \tilde{h}'_\ell(z_n^0) - r_n^0 \gamma_\ell^n \tilde{h}_\ell(z_n^0)}{z_n^0 \tilde{j}'_\ell(z_n^0) - r_n^0 \gamma_\ell^n \tilde{j}_\ell(z_n^0)}, \quad (5.95)$$

where γ_ℓ^n is the logarithmic derivative of the radial wave function evaluated at the muffin-tin radius, r_n^0,

$$\gamma_\ell^n(E) = \left[\frac{dR_\ell^n(E, r_n)/dr_n}{R_\ell^n(E, r_n)}\right]_{r_n = r_n^0}, \quad (5.96)$$

and where $z_n^0 = \kappa r_n^0$.

It is clear that neither Ψ_{IIA} nor Ψ_{IIB} is an acceptable solution for all of region II because they both contain terms that grow exponentially as $r_n \to \infty$. For the purposes of MST, however, it is only necessary that they be able to accurately represent Ψ_{II} in the vicinity of their respective scatterers. This condition will be satisfied if the following relation holds in the vicinity of scatterer A:

$$\sum_L m_\ell^A b_L^A \tilde{j}_\ell(z_A) Y_L(\hat{r}_A) = \sum_L b_L^B \tilde{h}_\ell(z_B) Y_L(\hat{r}_B) \quad (5.97)$$

[6]The function m_ℓ defined by Eqs.(5.94) and (5.95) is related to the t-matrix of scattering theory by $m_\ell = (-)^\ell t_\ell^{-1}/\sqrt{-E}$.

and if an analogous relation holds in the vicinity of scatterer B:

$$\sum_L m_\ell^B b_L^B \tilde{j}_\ell(z_B) Y_L(\hat{r}_B) = \sum_L b_L^A \tilde{h}_\ell(z_A) Y_L(\hat{r}_A). \tag{5.98}$$

The final step in deriving the MST equations requires the existence of the convergent expansions

$$\tilde{h}_{\ell'}(z_B) Y_{L'}(\hat{r}_B) = \sum_L \tilde{j}_\ell(z_A) Y_L(\hat{r}_A) G_{L,L'}(E, \mathbf{R}_{AB}), \tag{5.99}$$

and

$$\tilde{h}_{\ell'}(z_A) Y_{L'}(\hat{r}_A) = \sum_L \tilde{j}_\ell(z_B) Y_L(\hat{r}_B) G_{L,L'}(E, \mathbf{R}_{BA}), \tag{5.100}$$

where $\mathbf{R}_{AB} = \mathbf{R}_B - \mathbf{R}_A$. It is shown in Appendix D that such expansions exist if $r_A < R_{AB}$ in Eq.(5.99) and $r_B < R_{BA}$ in Eq.(5.100). The expansion coefficients, $G_{L,L'}(E, \mathbf{R})$ are the real-space structure constants defined in Eq.(D.11). If $\mathbf{R}_{AB}$ is chosen along the z-axis, the z-component of the angular momentum is a good quantum number allowing the expansion coefficients to be simplified somewhat;

$$\begin{aligned} G_{L,L'}(E, R\hat{z}) &= G^m_{\ell,\ell'}(E, R)\delta_{m,m'} \\ &= \sum_{\ell''=|\ell-\ell'|}^{\ell+\ell'} (2\ell''+1) d^m(\ell,\ell',\ell'') \tilde{h}_{\ell''}(\kappa R)(\pm 1)^{\ell''} \\ &\quad \times (-1)^\ell \sqrt{(2\ell+1)(2\ell'+1)}, \end{aligned} \tag{5.101}$$

where the $+$ $(-)$ sign is taken if $\mathbf{R}$ lies along the $+(-)z$-axis and where the "reduced" Gaunt numbers, $d^m(\ell,\ell',\ell'')$ vanish unless $\ell+\ell'+\ell''$ is an even integer. In that case they are given, respectively, for the $m=0$ (σ) and the $m=1$ (π) states by the expressions

$$d^\sigma(\ell,\ell',\ell'') = \int_0^1 dx\, P_\ell(x), P_{\ell'}(x) P_{\ell''}(x) \tag{5.102}$$

and

$$d^\pi(l,l',l'') = \frac{\int_0^1 dx\,(1-x^2)\, P'_\ell(x)\, P'_{\ell'}(x)\, P_{\ell''}(x)}{\sqrt{\ell\ell'(\ell+1)(\ell'+1)}}. \tag{5.103}$$

Substitution of these equations into Eqs.(5.97) and (5.98) yields a set of homogeneous linear equations for the coefficients, b_ℓ,

$$m_\ell^A(E) b_\ell^A = \sum_{\ell'} G^m_{\ell,\ell'}(E, \mathbf{R}_{AB}) b_{\ell'}^B \tag{5.104}$$

$$m_\ell^B(E) b_\ell^B = \sum_{\ell'} G^m_{\ell,\ell'}(E, \mathbf{R}_{BA}) b_{\ell'}^A. \tag{5.105}$$

These are the equations of multiple scattering theory determining the coefficients b_l.

The energies for which these equations have a solution are the eigenvalues of Eq.(5.88) and the solution vector b_L^n determines the overall solution through Eq.(5.91) in region II and through Eqs.(5.90) and (5.93) in region I. If the scatterers are identical, $m^A = m^B$, further simplification is possible because, in that case, the solutions for the A and B scatterers are related by symmetry, being either symmetric or antisymmetric with respect to inversion through the midpoint of the line joining their centers. States that are symmetric with respect to inversion are denoted as "gerade" by the spectroscopists, and states that are anti-symmetric are labeled "ungerade" [23]. Thus the coefficients of the expansions about the A and B sites are related by

$$b_\ell^B = (\pm 1)\,(-1)^\ell b_\ell^A, \tag{5.106}$$

where the $(+)$ sign is appropriate for the symmetric (g) states and the $(-)$ sign is appropriate for the antisymmetric (u) states. This simplification allows the MST equations for a homonuclear diatomic system to be written as

$$\sum_{\ell'}[m_\ell^A(E)\delta_{\ell,\ell'} - G_{\ell,\ell'}^m(E,S)(\pm 1)(-1)^{\ell'}]b_{\ell'}^A = 0. \tag{5.107}$$

Here we have used S to represent $|\mathbf{R}_{AB}|$, the separation between the scatterers. In practice, when these equations are solved numerically, it is necessary to truncate the infinite system of linear equations. When this is done the energy and the coefficients will depend on $\ell_{\max}$ the maximum value of ℓ used in the solution of the symmetric, homogeneous, linear system,

$$\sum_{\ell'=0}^{\ell_{\max}}[m_\ell^A(E_{\ell_{\max}})\delta_{\ell,\ell'} - G_{\ell,\ell'}^m(E_{\ell_{\max}},S)(\pm 1)(-1)^{\ell'}]b_{\ell'}^A(\ell_{\max}) = 0,$$

$$(\ell = 0, \ldots, \ell_{\max}) \tag{5.108}$$

This is a specific case of the secular Eq.(5.28) for the two-potential scattering system.

5.9.2 *Arbitrary Number of MT Potentials*

The wave-matching argument of the previous section can be readily generalized to an arbitrary number of MT potentials. Inside the nth muffin-tin we can expand the wave function in the form

$$\psi^n(\mathbf{r}_n) = \sum_L a_L^n \psi_L^n(\mathbf{r}_n), \tag{5.109}$$

where the subscript L stands for (ℓ, m), $\mathbf{r}_n$ is a vector from the center, $\mathbf{R}_n$, of cell n (i.e., $\mathbf{r}_n = \mathbf{r} - \mathbf{R}_n$), and where the the local basis function $\psi_L^n(\mathbf{r}_n)$

is the product of a radial wave function and a spherical harmonic

$$\psi_L^n(\mathbf{r}_n) = R_\ell^n(r_n) Y_L(\hat{r}_n). \tag{5.110}$$

Each of the basis functions $\psi_L^n(\mathbf{r}_n)$ is an exact local solution to the Schrödinger equation, which, because the potential vanishes outside the muffin-tin, joins smoothly to a combination of regular and irregular solid harmonics,

$$\psi^n(\mathbf{r}_n) = \sum_L \left\{ a_L^n J_L(\mathbf{r}_n) + b_L^n H_L(\mathbf{r}_n) \right\} = \sum_L a_L^n \left\{ J_L(\mathbf{r}_n) + t_\ell^n H_L(\mathbf{r}_n) \right\}, \tag{5.111}$$

for $(r_n > r_{MT})$, where $J_L(\mathbf{r}_n) = j_\ell(\sqrt{E} r_n) Y_L(\hat{r}_n)$ and $H_L(\mathbf{r}_n) = -i\sqrt{E} h_\ell(\sqrt{E} r_n) Y_L(\hat{r}_n)$. The coefficients a_L^n are to be determined as the eigenstates of the secular matrix of MST. The t-matrix, $t_\ell^n = b_L^n / a_L^n$, is determined by the condition that the logarithmic derivative of $\psi_L^n(\mathbf{r})$ be continuous across the muffin-tin boundary.

Outside the muffin-tins we can use a multicenter expansion for the wave functions,

$$\psi^o(\mathbf{r}) = \sum_{n'L'} b_{L'}^{n'} H_{L'}(\mathbf{r}_{n'}). \tag{5.112}$$

The MST secular equation can be obtained by requiring that the two expansions be consistent in the vicinity of the muffin-tin radius of potential n,

$$\sum_L \left\{ a_L^n J_L(\mathbf{r}_n) + b_L^n H_L(\mathbf{r}_n) \right\} = \sum_{n'L'} b_{L'}^{n'} H_{L'}(\mathbf{r}_{n'}), \tag{5.113}$$

which leads to the relation

$$\sum_L a_L^n J_L(\mathbf{r}_n) = \sum_{n' \neq n, L'} b_{L'}^{n'} H_{L'}(\mathbf{r}_{n'}). \tag{5.114}$$

Replacing $b_{L'}^{n'}$ by $a_{L'}^{n'} t_{\ell'}^{n'}$, and expanding the irregular solid harmonic centered on site n' by a sum of regular solid harmonics centered on site n through the use of the structure constants, Eq.(D.10), we obtain

$$\sum_L J_L(\mathbf{r}_n) a_L^n = \sum_{n' \neq n, L, L'} J_L(\mathbf{r}_n) G_{LL'}(\mathbf{R}_{nn'}) t_{\ell'}^{n'} a_{L'}^{n'} \tag{5.115}$$

which gives us the muffin-tin MST secular equation

$$a_L^n = \sum_{n' \neq n, L'} G_{LL'}(\mathbf{R}_{nn'}) t_{\ell'}^{n'} a_{L'}^{n'}. \tag{5.116}$$

With $\underline{m}^i = (\underline{t}^i)^{-1}$, Eq.(5.116) is the generalization of Eq.(5.106) to the case of an arbitrary number of MT potentials. It is easy to see that for periodic systems Eq.(5.116) leads to the secular Eq.(5.27), which determines the band structure of the system. Note that the a_L^n are expressed in terms of the t-matrices (and hence the basis functions) of cells other than cell n, and thus provide an expansion of the wave function in cell n in terms of basis functions associated with other cells. This is consistent with the representability theorem discussed in Section 4.4. In the case of MT potentials, the sum over L' in the last equation converges absolutely. As we will see in further discussion, this sum can also be performed in the case of nonMT, space-filling cells, and can even be applied to the irregular solutions of the Schrödinger equation. This last statement is proved in Appendix H.

It is interesting to point out a particular feature of Eq.(5.115), namely, for $\mathbf{r}_n$ outside a sphere that touches a neighboring MT, the single-site expansion, that is, the sum on the left-hand side of the equation, may diverge. This can be shown from the estimates of the $\ell_{\max}$ dependence of of a_ℓ and b_ℓ obtained in Section 5.10. On the other hand, the multicenter expansion, Eq.(5.112), will converge at least to values of r_n that are smaller than the intercell distance.

5.9.3 Convergence and Accuracy of MST (*)

In this section, we discuss in some detail the solution of the MST equations for the case introduced in Section 5.9.1 of two spherical-well potentials. Our objective will be to obtain a better understanding of the convergence of muffin-tin MST with respect to the truncation of angular-momentum sums. Results similar to those described below have also been obtained for the truncated Coulomb potential [20].

For a spherical-well potential of depth V, Eq.(5.96) for the logarithmic derivative at the muffin-tin radius becomes

$$\gamma_\ell(E) = \sqrt{E+V} j_\ell'(\sqrt{E+V} r^0)/j_\ell(\sqrt{E+V} r^0), \tag{5.117}$$

where r^0 is the radius of the spherical well. For the problem of two identical spherical wells we can take r^0 to be unity without loss of generality if the separation, S, is measured in units of r^0 and if E and V are measured in units of $1/{r^0}^2$. Table 5.1 gives some of the calculated bound-state energies obtained by solving the MST equations.

The $1\sigma_g$ and $1\sigma_u$ states of Table 5.1 are those corresponding to a potential of depth $V = 5$. An isolated potential of this depth has only one bound state.

Figure 5.7 shows how the energy converges as a function of the number of angular-momentum states for various values of the separation for the $1\sigma_g$ and $1\sigma_u$ states. Figure 5.8 shows how the wave function and its normal derivative converge for these same states. In the case of the energy, the numbers plotted are the differences between $E(\ell_{\max})$ and $E(\infty)$, which in practice is obtained by increasing $\ell_{\max}$ until $E(\ell_{\max})$ is constant within the precision of the computer. For the wave function and its derivative, the plotted numbers are the relative root mean square differences between Ψ_{IA}, Eq.(5.90) and Ψ_{II}, Eq.(5.91), averaged over the surface of scatterer

S	$1\sigma_g$	$1\sigma_u$
2.0	-1.326675638612688	$-.5079864354759665$
2.4	-1.144684789098379	$-.6835567377508935$
3.0	-1.026932973585779	$-.8219491981993462$
4.5	$-.9469747391316194$	$-.9151273063088851$
6.0	$-.9324137204726934$	$-.9286027142803350$
∞	$-.9314261194176701$	$-.9314261194176701$

TABLE 5.1. Energies of the $1\sigma_g$ and $1\sigma_u$ states calculated for two spherical-well potentials. The depth of the potential, V, is $5/r^{0\,2}$. The energy is measured in units of $1/r^{0\,2}$ and S is measured in units of r^0.

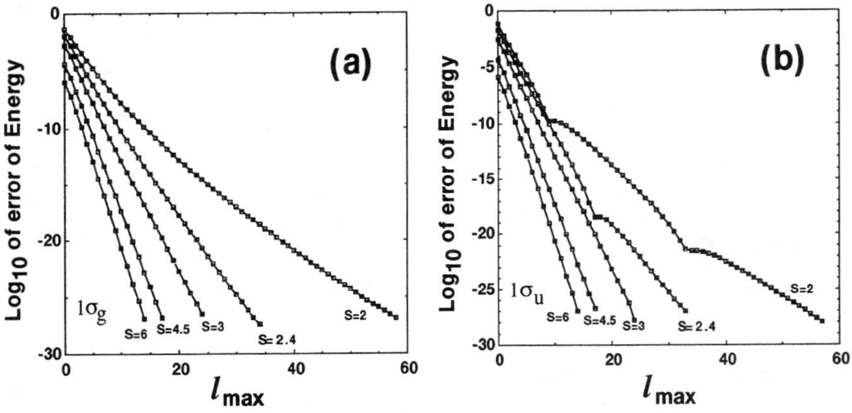

FIGURE 5.7. Convergence of the bound-state energy for a molecule consisting of two spherical-well potentials. S labels the separation between the centers of the potentials in units of the muffin-tin radius, r^0, so that $S = 2$ indicates touching spheres. The ordinate is the logarithm (base 10) of the difference between $E_{\ell_{\max}}$ and the converged bound-state energy. (a) $1\sigma_g$ state for $V = 5$. (b) $1\sigma_u$ state for $V = 5$.

A. The relative root mean square error of the wave function referred to in Fig. 5.8, for example, is

$$\left[\frac{\int dS_A(\Psi_{IA} - \Psi_{II})^2}{\int \Psi_{II}^2 dS_A}\right]^{1/2}. \tag{5.118}$$

The corresponding quantity with the wave functions replaced by their normal derivatives with respect to r is also plotted in Fig. 5.8.

The vanishing of the differences in the wave function and its normal derivative at the boundary between regions I and II suffices to determine the correct solution to the Schrödinger equation. We recall that Ψ_{In} is a superposition of terms, each of which satisfies the Schrödinger equation

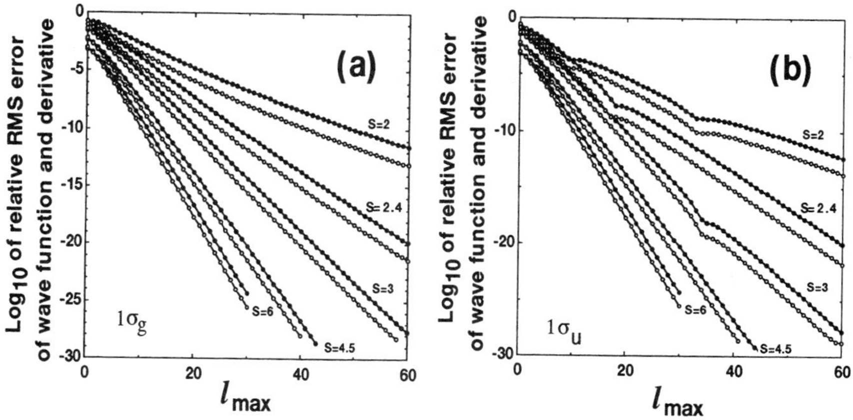

FIGURE 5.8. Logarithm (base 10) of the relative root mean square error of the wave function (open circles) and of the normal derivative of the wave function (filled circles) for a molecule consisting of two spherical-well potentials. The square of the relative error in the wave function or normal derivative at the muffin-tin radius has been averaged over the surface of one of the scatterers. It is the logarithm of the square root of this quantity that is plotted. The parameters are the same as for the previous figure.

by construction in region $In(n =$ A or B) and is regular in that region. Similarly, each term of Ψ_{II} satisfies the Schrödinger equation in all of region II, is regular in that region and satisfies the boundary condition at $r \to \infty$. Because the Schrödinger equation is a linear partial differential equation, a sum of solutions is also a solution. Therefore, if the wave function and its derivative are continuous everywhere then the wave function must be an acceptable solution to the Schrödinger equation [24]. The only points at which the continuity and smoothness of the wave function could be questioned are those at the surfaces that separate regions I and II. As is seen in Figs. 5.7 and 5.8, the energy, the wave function, and the normal derivative of the wave function all converge in an approximately exponential manner with increasing $\ell_{\max}$. The variation in slope of some of the curves at low values of ℓ, as well as the somewhat strange behavior of the $1\sigma_u$ state exhibited in these plots, can be understood, respectively, in terms of the dependence on the potential and the antisymmetric nature of the state $1\sigma_u$ [21].

5.10 The Convergence Properties of MST (*)

In this section we discuss the analytic behavior of the errors in the energies and the wave functions obtained within MST as functions of the angular-momentum cut-off, $\ell_{\max}$. Although the analysis is limited to bound states and to the case of two scatterers, the estimate of the errors in the two-scatterer problem should provide a good indication of the accuracy of MST. (The following discussion is somewhat involved and could be skipped by a reader not interested in such mathematical details.)

5.10.1 Energy Convergence

We first consider the accuracy of the energy of a bound state calculated with an angular-momentum truncation, $\ell_{\max}$. We denote the truncation error of this bound state energy by $E_{\ell_{\max}} - E_\infty$, where E_∞ is the bound-state energy given by the MST equations in the limit $\ell_{\max} \to \infty$. It is convenient to rewrite Eq.(5.108) as

$$M^{\ell_{\max}}(E,S)\; B^A(\ell_{\max}) = 0, \tag{5.119}$$

where $B^A(\ell_{\max})$ is a vector consisting of $b_\ell^A(\ell_{\max})$, and $M^{\ell_{\max}}(E,S)$ is a matrix with elements,

$$M^{\ell_{\max}}_{\ell,\ell'}(E,S) = m_\ell^A(E)\delta_{\ell,\ell'} - G^m_{\ell,\ell'}(E,S)(\pm 1)(-1)^{\ell'}, \qquad (\ell,\ell' = 0,\ \ldots\ \ell_{\max}). \tag{5.120}$$

Obviously, $B^A(\ell_{\max})$ is the eigenvector of $M^{\ell_{\max}}(E,S)$ corresponding to the eigenvalue 0. We now define an $\ell_{\max}\times\ell_{\max}$ matrix, P, which contains all eigenvectors of $M^{\ell_{\max}}(E,S)$. Because $M^{\ell_{\max}}(E,S)$ is symmetric, we have

$$P^{-1} = P^T, \tag{5.121}$$

and

$$\sum_{\ell,\ell'} P_{\ell,i}\; M^{\ell_{\max}}_{\ell,\ell'}(E,S)\; P_{\ell',j} = \alpha_i^{\ell_{\max}}(E)\delta_{i,j}, \tag{5.122}$$

where $\alpha_i^{\ell_{\max}}(E)$ are the eigenvalues of $M^{\ell_{\max}}(E,S)$, and $\alpha_0^{\ell_{\max}}(E_{\ell_{\max}}) = 0$. The error in the energy can be estimated through the expression,

$$E_{\ell_{\max}} - E_\infty = \frac{\alpha_0^\infty(E_{\ell_{\max}})}{\left(\frac{d\alpha_0^\infty(E)}{dE}\right)_{E_{\ell_{\max}}}}. \tag{5.123}$$

It now remains to find $\alpha_0^\infty(E_{\ell_{\max}})$. To this end, we partition the matrix, $M^\infty(E,S)$, into block matrices

$$M^\infty(E,S) = \begin{pmatrix} M^{\ell_{\max}}(E,S) & M_a^{\ell_{\max}} \\ M_b^{\ell_{\max}} & M_c^{\ell_{\max}} \end{pmatrix} \tag{5.124}$$

and multiply it by the matrices P and P^T,

$$\begin{pmatrix} P^T & 0 \\ 0 & I \end{pmatrix} \begin{pmatrix} M^{\ell_{\max}}(E,S) & M_a^{\ell_{\max}} \\ M_b^{\ell_{\max}} & M_c^{\ell_{\max}} \end{pmatrix} \begin{pmatrix} P & 0 \\ 0 & I \end{pmatrix} = \begin{pmatrix} \alpha^{\ell_{\max}}(E) & P^T M_a^{\ell_{\max}} \\ M_b^{\ell_{\max}} P & M_c^{\ell_{\max}} \end{pmatrix}, \tag{5.125}$$

where I represents a unit matrix, and $\alpha^{\ell_{\max}}(E)$ is a diagonal matrix containing $\alpha_i^{\ell_{\max}}(E)\delta_{i,j}$.

As is shown in Appendix I, for sufficiently large ℓ the inverse t-matrix, m_ℓ, can be written as a product of two factors, one of which is strongly dependent on ℓ and on E, but is independent of the potential (except for its radius) and a second factor which depends on the details of the potential but is only weakly dependent on ℓ and E (see also Eq.(C.4)),

$$m_\ell(E) \approx \frac{[(2\ell+1)!!]^2}{(\kappa r^0)^{2\ell+1} P(\ell, E)}. \tag{5.126}$$

For a spherical-well potential with depth V and radius r^0, $P(\ell, E)$ is given by the expression,

$$P(\ell, E) = \frac{V r^{0^2}}{2\ell+3} + O[\ell^{-2}]. \tag{5.127}$$

Similarly, it is shown in Appendix J that when ℓ and ℓ' are both large, the structure factors, $G_{\ell,\ell'}(E,S)$ may be written as

$$G^m_{\ell,\ell'}(E,S) \approx (-1)^{\ell+m} 2\sqrt{\frac{\ell+\ell'}{\pi}} \frac{(2\ell+2\ell'-1)!!}{(\kappa S)^{\ell+\ell'+1}}. \tag{5.128}$$

Clearly, for large ℓ, we have

$$|G^m_{\ell,\ell'}(E,S)|^2 << |m_\ell(E) m_{\ell'}(E)|. \tag{5.129}$$

In other words, the matrix $M(E,S)$ is nearly diagonal for high angular-momentum states (as we might expect, because for large values of L the solution approaches that for free space and becomes insensitive to the details of the potential). This allows us to use perturbation theory to evaluate $\alpha_0^\infty(E)$. We note that $\alpha_0^{\ell_{\max}}(E_{\ell_{\max}}) = 0$, and use Eq.(5.125), to obtain,

$$\begin{aligned} \alpha_0^\infty(E_{\ell_{\max}}) &= -\sum_{\ell>\ell_{\max}} \frac{(P^T M_a^{\ell_{\max}})_{0,\ell} (M_b^{\ell_{\max}} P)_{\ell,0}}{M_{\ell,\ell}(E,S)} \\ &= -\sum_{\ell>\ell_{\max}} \frac{\left(\sum_{\ell'=0}^{\ell_{\max}} (-1)^{\ell'} G^m_{\ell,\ell'}(E,S) b^A_{\ell'}(\ell_{\max})\right)^2}{M_{\ell,\ell}(E,S)} \\ &= -\sum_{\ell>\ell_{\max}} m^A_\ell [b^A_\ell(\ell_{\max})]^2. \end{aligned} \tag{5.130}$$

In order to estimate α_0, we need an estimate for $b^A_\ell(\ell_{\max})$. Equation (5.108), which may be written as

$$\frac{b^A_\ell (2\ell+1)!!}{(\kappa r^0)^\ell} = \frac{(\kappa r^0)^{\ell+1}}{(2\ell+1)!!} P(\ell) \sum_{\ell'=0}^{\ell_{\max}} G^m_{\ell,\ell'}(E,S)(-1)^{\ell'} b^A_{\ell'}, \tag{5.131}$$

suggests an *ansatz* of the form,

$$b_\ell^A(\ell_{\max}) \approx \frac{x^{\ell-\ell_0} Y_{\ell,0}(-\hat{z})}{(2\ell - 2\ell_0 + 1)!!}$$
$$\approx 2^{\ell_0} \frac{d^{\ell_0}}{d\, x^{\ell_0}} j_\ell(x) Y_{\ell,0}(-\hat{z}), \tag{5.132}$$

where x is to be determined, ℓ_0 is the angular momentum of the corresponding single-scatterer state, and $b_{\ell_0}^A(\ell_{\max}) = Y_{\ell_0,0}(-\hat{z})$. For the moment we consider only σ, $(m = 0)$, states.

Substituting Eq.(5.132) into (5.131), we obtain,

$$\frac{x^{\ell-\ell_0}}{(2\ell - 2\ell_0 + 1)!!} \approx \frac{(\kappa r^0)^{2\ell+1} P(\ell, E)}{[(2\ell+1)!!]^2} 2^{\ell_0} \frac{d^{\ell_0}}{d\, x^{\ell_0}} h_\ell(\kappa S - x)$$
$$\approx \frac{(\kappa r^0)^{2\ell+1} P(\ell, E)}{[(2\ell+1)!!]^2} \frac{(2\ell + 2\ell_0 - 1)!!}{(\kappa S - x)^{\ell+\ell_0+1}}, \tag{5.133}$$

where we have used the expansion property of $h_\ell Y_L$, Eq.(D.9). We have also omitted terms containing $b_{\ell_i}^A(\ell_{\max})$ for $\ell_i < \ell_0$, because these terms scale as $r^{0\,2\ell+1}/S^{\ell+1}$, a much smaller ratio than that on the right-hand side of Eq.(5.133). If we denote the ratio $x/(\kappa r^0)$ by ζ, we can write this approximate equality as

$$\zeta^{\ell-\ell_0} \left(\frac{S}{r^0} - \zeta \right)^{\ell+\ell_0+1} \approx \frac{P(\ell, E)}{2\ell + 1}. \tag{5.134}$$

Now the right-hand side of this expression vanishes as $\ell \to \infty$ as a low power of $1/\ell$. The left-hand side, however, either grows or vanishes exponentially depending on whether the quantity $\zeta(S/r^0 - \zeta)$ is smaller or larger than unity. This inconsistency can be resolved through a weak dependence of ζ on ℓ,

$$\zeta \approx \frac{S}{2r^0} - \sqrt{\left(\frac{S}{2r^0} \right)^2 - \ell^{-2/\ell}}. \tag{5.135}$$

In solving Eq.(5.134) for ζ, it was assumed that $P(\ell)$ is proportional to $1/\ell$ in the limit of large ℓ, and that a constant, c, raised to the power $1/\ell$ can be approximated by unity. (It was also assumed that the expansion property of $h_\ell Y_L$ used in Eq.(5.133) remains valid despite the very weak ℓ dependence of x.)

Now, if $S/2r^0$ is greater than unity, Eq.(5.132) with $x = \kappa r^0 \zeta$ obtained from Eq.(5.135) provides an accurate representation of the leading terms describing the ℓ-dependence of b_ℓ^A. If $S/2r^0$ is greater than unity and ℓ is very large, the term $\ell^{-2/\ell}$ under the radical may be approximated by unity, making ζ independent of ℓ. The ℓ dependence of ζ is also described well for sufficiently large values of ℓ. These results are illustrated in Fig. 5.9 which shows $\zeta(\ell)$ as calculated, that is, $b_\ell^A(2\ell+1)!!/(\kappa r^0)^\ell$, and as predicted, for the $1\sigma_g$ state of the spherical wells.

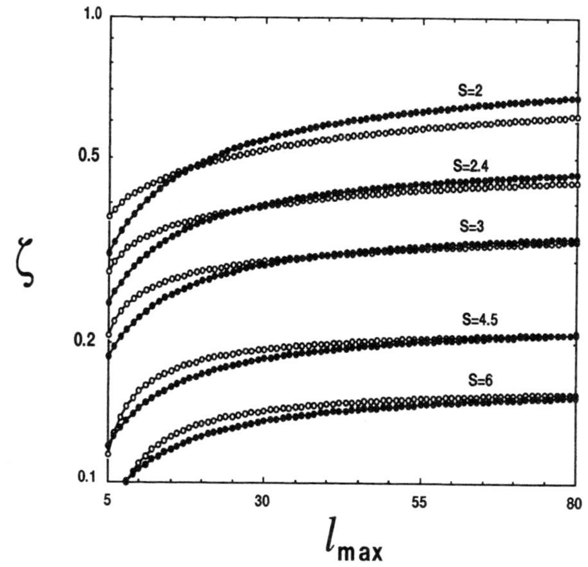

FIGURE 5.9. Predicted (filled circles) and calculated (open circles) values of ζ, the parameter that describes the rate of convergence with $\ell_{\max}$ in MST. Calculated values are for the $1\sigma_g$ state.

If $S/2r^0$ is equal to unity, the ℓ dependence of ζ becomes significant. However, in this case, even though ζ always varies slowly with ℓ, Eq.(5.134) would still hold approximately if ζ were approximately equal to unity because the summation of Eq.(5.131) into a Hankel function has its major contributions from the terms near $\ell' \approx \ell$ when $\zeta = 1$. However, because ζ is smaller than unity, Eq.(5.135) overestimates its size. This can be seen in the uppermost curves of Fig. 5.9 where the calculated and predicted values of ζ are shown for the $1\sigma_g$ state calculated using touching ($S = 2$) spherical-well potentials. For $\ell_{\max}$ between 80 and 180 Eq.(5.135) overestimates ζ by an essentially constant factor of approximately 1.1.

Combining Eqs. (5.130),(5.132), and (5.135) we have

$$\alpha_0^{\infty}(E_{\ell_{\max}}) \sim \frac{-1}{4\pi\kappa r^0} \sum_{\ell>\ell_{\max}} \frac{\zeta^{2\ell}(2\ell+1)}{P(\ell)}, \tag{5.136}$$

which, as is shown below, can be used to estimate the error in the energy.

If $S > 2r^0$, ζ becomes independent of ℓ for sufficiently large $\ell_{\max}$. In this case Eq.(5.136) can be summed with the result that the error in the energy is predicted to decrease exponentially with $\ell_{\max}$,

$$E_{\ell_{\max}} - E_{\infty} \propto \zeta^{2\ell_{\max}}, \qquad \ell_{\max} \to \infty. \tag{5.137}$$

In the special case of $S = 2r^0$, Eq. (5.135) reduces to

$$\zeta = 1 - \sqrt{1 - \ell^{-2/\ell}} \approx e^{-\sqrt{2\ln\ell/\ell}} \qquad (5.138)$$

although, as mentioned above, this result overestimates ζ and understates the convergence rate. Using this result in Eq. (5.136) yields

$$\alpha_0^\infty(E_{\ell_{\max}}) \sim \frac{-1}{4\pi\kappa r^0} \sum_{\ell>\ell_{\max}} \frac{e^{-2\sqrt{2\ell\ln\ell}}(2\ell+1)}{P(\ell)}, \qquad (5.139)$$

which indicates a slower than exponential convergence. The series *does* converge, however, because it is easily bounded by the series

$$\sum_{\ell>\ell_{\max}} \ell^2 e^{-2\sqrt{2\ell}} \sim \sqrt{\frac{1}{2}} \ell_{\max}^{5/2} e^{-2\sqrt{2\ell_{\max}}}. \qquad (5.140)$$

However, this result should not be used to estimate the convergence rate because the true convergence rate is much faster than that indicated by the last expression.

In the case of antisymmetric states, b_ℓ^A, can change sign with increasing ℓ which complicates the expression for b_ℓ^A. However, calculations indicate [21] that Eq.(5.135) still gives a good estimate of the magnitude of b_ℓ^A in the large ℓ limit, and therefore Eq.(5.136) can also be used for the antisymmetric states. The effect of $m \neq 0$ is the introduction of a factor of $(-1)^m$ on the right-hand side of Eq.(5.133) whose effect would be to interchange the behavior of the "gerade" and "ungerade" states. Thus the $1\sigma_g$ and $1\pi_u$ states and the $1\sigma_u$ and $1\pi_g$ states behave similarly with respect to convergence.

5.10.2 Convergence of the Wave Function

The error in the wave function can be expressed in terms of the mismatch between Ψ_{IA}, Eq.(5.90) and Ψ_{II} Eq.(5.91), at the surface of muffin-tin A. Since Ψ_{IIA}, Eq.(5.93), matches Ψ_{IA} identically term by term in both value and derivative at the muffin-tin radius, we can write the error in the wave function in the form,

$$\begin{aligned} \Delta\Psi(\ell_{\max}) &= \Psi_{IIA}(\ell_{\max}) - \Psi_{II}(\ell_{\max}) \\ &= \sum_{\ell=0}^{\ell_{\max}} b_\ell^A m_\ell^A \tilde{j}_\ell(\kappa r_A) Y_L(\hat{r}_A) - \sum_{\ell'=0}^{\ell_{\max}} b_{\ell'}^B \tilde{h}_{\ell'}(\kappa r_B) Y_{L'}(\hat{r}_B) \end{aligned} \qquad (5.141)$$

After expanding the Hankel function using Eq.(5.99) we have

$$\Delta\Psi(\ell_{\max}) = \sum_{\ell=0}^{\ell_{\max}} m_\ell^A b_\ell^A \tilde{j}_\ell(\kappa r_A) Y_L(\hat{r}_A)$$

$$-\sum_{\ell=0}^{\infty}\tilde{j}_\ell(\kappa r_A)Y_L(\hat{r}_A)(\sum_{\ell'=0}^{\ell_{\max}}G^m_{\ell,\ell'}b^B_{\ell'}). \tag{5.142}$$

The MST secular equation, Eq.(5.104) (truncated at $\ell_{\max}$), can be used to eliminate the sum over ℓ' and allows the cancellation of the first $(\ell_{\max}+1)$ terms in the sum over ℓ, leading to the result

$$\Delta\Psi(\ell_{\max}) = \sum_{\ell>\ell_{\max}} \tilde{j}_\ell(\kappa r_A)Y_L(\hat{r}_A)m^A_L b^A_\ell(\ell_{\max}). \tag{5.143}$$

The average of the square of $\Delta\Psi(\ell_{\max})$ over the surface of muffin-tin A is

$$\left\langle[\Delta\Psi(\ell_{\max})]^2\right\rangle = \sum_{\ell>\ell_{\max}} [\tilde{j}_\ell(\kappa r_A)m^A_\ell b^A_\ell]^2, \tag{5.144}$$

which on substituting forms valid for large ℓ becomes

$$\left\langle[\Delta\Psi(\ell_{\max})]^2\right\rangle = \frac{1}{(4\pi\kappa r^0)^2}\sum_{\ell>\ell_{\max}}(2\ell+1)\frac{\zeta^{2\ell}}{P(\ell)^2}. \tag{5.145}$$

Thus the rate of convergence of the mean square error of the wave function to zero is quite similar to the convergence of the energy to its asymptotic value, being slightly slower because $Y_L(-\hat{z})^2 \sim \ell$ while $P(l)^{-2} \sim \ell^2$.

The rate of convergence of the normal derivative of the wave function can be obtained from Eq.(5.143)

$$\Delta\Psi'(\ell_{\max}) = \sum_{\ell>\ell_{\max}} \kappa\tilde{j}'_\ell(\kappa r_A)Y_L(\hat{r}_A)m^A_L b^A_\ell(\ell_{\max}), \tag{5.146}$$

so that in the limit of large $\ell_{\max}$ the mean square error in the normal derivative of the wave function is

$$\left\langle[\Delta\Psi'(\ell_{\max})]^2\right\rangle = \frac{1}{(4\pi r^0)}^2\sum_{\ell>\ell_{\max}}(2\ell+1)\frac{\ell^2\zeta^{2\ell}}{P(\ell)^2}. \tag{5.147}$$

Thus, the predicted convergence rates of the energy, the mean square error of the wave function, and the mean square error of the derivative of the wave function are all the same in the limit of $\ell_{\max} \to \infty$. These predictions are compared with the results of the numerical calculations in Fig. 5.10. If we define $R_p(\ell)$ to be the ratio of successive terms in the series defined by Eqs.(5.136), (5.145), and (5.147), we have (ignoring factors of $[\ell/(\ell+1)]^n$)

$$R_p(\ell) = [\zeta(\ell)]^{2\ell}/[\zeta(\ell+1)]^{2\ell+2}. \tag{5.148}$$

These predicted ratios are plotted as the solid curves without symbols in Fig. 5.10 and may be compared with the corresponding ratios calculated numerically and defined by

$$R^E_c(\ell) = \frac{E(\ell_{\max}) - E(\ell_{\max}-1)}{E(\ell_{\max}+1) - E(\ell_{\max})}, \tag{5.149}$$

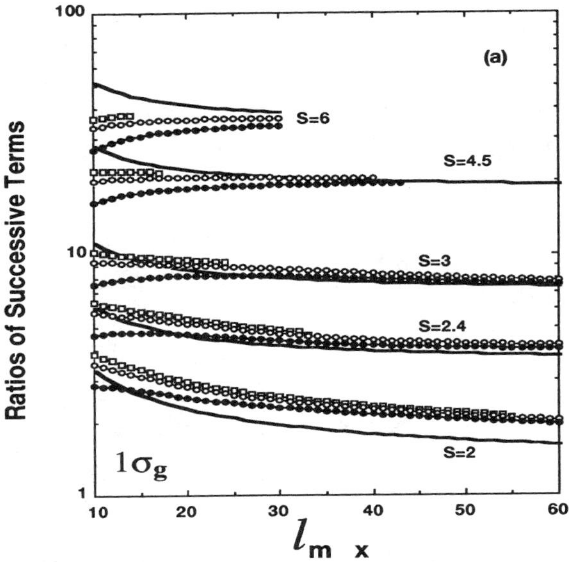

FIGURE 5.10. Predicted (solid lines) and calculated values for the ratios $f[(\ell_{\max}) - f(\ell_{\max} - 1)]/f[(\ell_{\max} + 1) - f(\ell_{\max}])$ where f represents the energy (squares), mean square error in the wave function (open circles), or mean square error in the normal derivative of the wave function (filled circles). Results are for the $1\sigma_g$ state calculated for two spherical-well potentials.

for the energy, by

$$R_c^{\Psi}(\ell) = \frac{\langle[\Delta\Psi(\ell_{\max})]^2\rangle - \langle[\Delta\Psi(\ell_{\max} - 1)]^2\rangle}{\langle[\Delta\Psi(\ell_{\max} + 1)]^2\rangle - \langle[\Delta\Psi(\ell_{\max})]^2\rangle}, \tag{5.150}$$

for the wave function, and by

$$R_c^{\Psi'}(\ell) = \frac{\langle[\Delta\Psi'(\ell_{\max})]^2\rangle - \langle[\Delta\Psi'(\ell_{\max} - 1)]^2\rangle}{\langle[\Delta\Psi'(\ell_{\max} + 1)]^2\rangle - \langle[\Delta\Psi'(\ell_{\max})]^2\rangle}, \tag{5.151}$$

for the normal derivative of the wave function. The results shown correspond to the $1\sigma_g$ state for spherical wells.

As is seen in these plots, the agreement between the analytical predictions of the convergence rates and those calculated is quite good. Generally, the observed convergence rates are slightly higher than those predicted. The predicted convergence rate is particularly conservative for the case of touching spheres, a result that may be traced back to the fact that Eq.(5.135) overestimates ζ for this case. Note that larger numbers in Fig. 5.10 mean faster convergence. A ratio of 10, for example, means that each succeeding term is decreasing by that factor. Constant ratios indicate exponential convergence which is observed asymptotically in all cases except for touching spheres.

The observation that exponential convergence (i.e. error$\sim e^{-a\ell_{\max}}$ where a is a constant) is not obtained for touching spheres raises the question of what happens to the convergence of the MST equations when the spheres overlap. A careful study of this case does not appear to have been made, but from Eq.(5.135) we expect that the MST bound-state energies will not converge with $\ell_{\max}$ for overlapping spheres. A limited number of calculations indicate that this is in fact the case. If the overlap is small, the asymptotic nature of the procedure only becomes apparent at fairly high values of $\ell_{\max}$. In any case, overlapping spheres can easily violate the conditions for expanding the free-particle propagator, and divergences associated with the amount of overlap can be expected.

5.10.3 Convergence of Single-Center Expansion of the Wave Function

Our analysis of the convergence of MT-MST has been based on the use of the single-center expansion, Eq.(5.90), to represent the wave function inside the muffin-tins and the multicenter expansion, Eq.(5.91), outside the muffin-tins. It is common practice, however, in MST calculations to use the single-center expansion within the interstitial region. The explicit expression that we have obtained for the asymptotic dependence of the terms in the expansion of the wave function on ℓ, allows us to investigate the validity of this procedure.

Outside the muffin-tin radius the single-center expansion has the form

$$\Psi_{In}(\mathbf{r}) = \sum_L a_L^n[J_L(\mathbf{r}) + t_\ell H_L(\mathbf{r})] = \sum_L b_L^n[m_\ell^n J_L(\mathbf{r}) + H_L(\mathbf{r})]. \quad (5.152)$$

For the case of two spherical square wells, and considering σ states for simplicity, we have the following asymptotic forms for large values of ℓ:

$$b_\ell \to \frac{(\zeta\kappa r^0)^\ell}{(2\ell+1)!!} \quad (5.153)$$

$$m_\ell \to \frac{[(2\ell+1)!!]^2}{(\kappa r^0)^{2\ell+1}P(\ell,E)} \quad (5.154)$$

$$\tilde{j}_\ell \to \frac{(\kappa r)^\ell}{(2\ell+1)!!} \quad (5.155)$$

$$\tilde{n}_\ell \to \frac{(2\ell-1)!!}{(\kappa r)^{(}\ell+1)}. \quad (5.156)$$

The term involving $\sum_\ell b_\ell m_\ell \tilde{j}_\ell$ is the only one whose convergence is in question. Asymptotically, the terms in this expansion are proportional to

$(\zeta(\ell)r/r^0)^\ell$. From Figure 5.9, we can see that ζ is a weakly increasing function of ℓ reaching approximately 0.61 for $\ell = 80$. This means that successive terms in the expansion are decreasing for ℓ as large as 80, for values of r/r^0 as large as $1/0.61 \approx 1.64$. Thus for values of r that exceed the muffin-tin radius, the single-center expansion should be considered formally asymptotic but practically quite useful because the divergence of the expansion only becomes apparent at a value of $\ell_{\max}$ which depends on r/r^0 and which may be quite large. It should be noted that the estimates of the ratios of successive terms in the single-center expansion obtained here are for square-well potentials. The t-matrix for an atomic potential will decrease much faster with ℓ, at least for small values of ℓ.

5.10.4 Summary

Let us now summarize the results of the numerical and analytic investigations presented above. The numerical calculations indicate that MST can be used to obtain an extremely accurate solution to the Schrödinger equation for the case of muffin-tin potentials. Both the numerical and analytical results indicate that the errors in the wave function and its derivative decrease exponentially with $\ell_{\max}$ in the limit in which $\ell_{\max}$ is large provided there is a finite separation between the muffin-tin potentials. According to these analytic results the error in the wave function or its derivative evaluated at the muffin-tin radius decreases by the factor ζ given by Eq.(5.134) when $\ell_{\max}$ increases by unity. Consistent with the variational nature of the solutions of the MST equations (at least in the case of MT potentials), the error in the energy decreases by the square of this factor when $\ell_{\max}$ increases by unity.

In the limit of large $\ell_{\max}$, the convergence factor ζ is a function of S/r^0, the ratio of the separation between the potentials to the muffin-tin radius, and thus depends only on geometry, not on the details of the potential. Geometry will also figure prominently in the next chapter where we begin our discussion of the application of MST to nonMT, space- filling potential cells. The details of the potential do, however, affect the rate at which the actual convergence rate approaches the asymptotic rate,

$$\zeta_\infty = \frac{S}{2r^0} - \sqrt{\left(\frac{S}{2r^0}\right)^2 - 1}. \tag{5.157}$$

The *approach* to the asymptotic convergence rate is also strongly dependent on the value S/r^0. As $S/2r^0$ approaches unity (the case of touching muffin-tins) higher values of $\ell_{\max}$ are required to attain the asymptotic regime. For the case of $S/2r^0 = 1$, the asymptotic regime characterized by exponential convergence is never attained. MST is, however, exact and convergent even in this case.

References

[1] J. Korringa, Physica **13**, 392 (1947).
[2] W. Kohn and N. Rostoker, Phys. Rev. **94**, 1111 (1954).
[3] W. H. Butler, Phys. Rev. B**14**, 468 (1976).
[4] B. L. Györffy and M. J. Stott, in *Band Structure Spectroscopy of Metals and Alloys*, edited by D. J. Fabian and L. M. Watson (Academic Press, London, 1972).
[5] F. S. Ham and B. Segal, Phys. Rev. **124**, 1786 (1961).
[6] V. L. Moruzzi, J. F. Janak, and A. R. Williams, *Calculated Electronic Properties of Metals* (Pergamon, New York, 1978).
[7] P. Soven, Phys. Rev., **156** 809 (1967).
[8] H. Shiba, Prog. Theor. Phys. **45**, 77 (1971).
[9] B. L. Gyorffy, Phys. Rev. B **5**, 2382 (1972).
[10] G. M. Stocks, W. M. Temmerman, and B. L. Gyorffy, Phys. Rev. Lett. **41** 339 (1978).
[11] A. Bansil, Phys. Rev. Lett. **41**, 1670 (1978).
[12] J. S. Faulkner, Prog. Mat. Sci., **27** 1 (1982).
[13] W. H. Butler, Phys. Rev. B**31** 3260 (1985).
[14] B. Ujfalussy, L. Szunyogh, P. Weinberger, and J. Kollar, in *Metallic Alloys: Experimental and Theoretical Perspectives*, edited by J. S. Faulkner and R. G. Jordan, Proceedings of NATO Advanced Research Workshop, Boca Raton, FL. July 16–21, 1994 (Kluwer Academic Publishers, Dordrecht, 1994), Vol. 256, p. 301.
[15] L. Szunyogh, B. Ujfalussy, P. Weinberger, and J. Kollar, Phys. Rev. B**49**, 2721 (1994) .
[16] P. J. Braspenning and A. Lodder, Phys. Rev. B**49**, 10222 (1994).
[17] R. Zeller, P. H. Dederichs, B. Újfalussy, L. Szunyogh, and P. Weinberger Phys. Rev. B**52**, 8807 (1995).
[18] N. Moghadam and G. M. Stocks *to be published.*
[19] O. K. Anderson, A. V. Postnikov, and S. Yu. Savrasov, in *Applications of Multiple Scattering Theory to Materials Science*, W. H. Butler, P. H. Dederichs, A. Gonis, and R. L. Weaver, eds., Proceedings of the Materials Research Society, Vol. 253, p. 37 (1992).
[20] W. H. Butler and X.-G. Zhang, Phys. Rev. **44**, 969 (1991).
[21] W. H. Butler, Phys. Rev. **41**, 2684, (1990).
[22] F. C. Smith, Jr., and K. H. Johnson, Phys. Rev. Lett. **22**, 1168 (1969).
[23] J. C. Slater, *Quantum Theory of Molecules and Solids, Volume 1, Electronic Structure of Molecules* (McGraw Hill, New York, 1963, pp. 5–7).
[24] This problem corresponds to an *elliptic* equation with *Dirichlet* boundary conditions for a *closed* surface. It is well known that such problems have a unique and well-defined solution. P. M. Morse and H. Feshbach, *Methods of Theoretical Physics* (McGraw-Hill, New York, 1953, p. 706).

6
MST for Space-Filling Cells

In this chapter, we discuss the generalization of MST to the case of non-MT, space-filling potential cells. As we did in the case of MT potentials, we develop the formalism within the angular-momentum representation. We show that MST is valid in the general case of nonoverlapping, space-filling scattering regions, retaining the basic features that characterize its application to MT potentials. In particular, it retains the separation of potential and structure from which MST derives much of its conceptual transparency and computational usefulness. We begin with a summary of the developments that led ultimately to the generalization of MST to space-filling potentials.

6.1 Historical Development of Full-Cell MST

The need to study the electronic structure of solids using non-MT, space-filling potential cells becomes evident when one considers the cases of open (nonclose-packed) structures, of surfaces, interfaces, and other defects. In an attempt to fill this need, a number of methods that allowed the treatment of the entire potential inside the WS cell, without approximation to its shape, were developed using various computational and formal approaches. Perhaps the most successful one of these is the full-potential linearized augmented plane wave (FLAPW) method [1]. Full-potential linear muffin-tin orbital (LMTO) [2] methods have also been introduced in recent years. However, none of these methods can match the power and versatility of a full-potential method based on the formalism of multiple scattering theory, either in terms of providing a complete treatment of the Schrödinger equation in solids, or of the spectrum of different physical systems that could be treated. In particular, none of these methods leads easily to the construction of the Green function which, as already mentioned, is indispensable in the study of many physical systems and properties. And yet, a generalization of the KKR method to full cell potentials has been very slow in developing. It is only very recently that calculations based on such generalized Green function, MST computer codes have begun to appear.

The reasons for the reluctance to develop such techniques are not altogether clear, but some educated guesses can be made. Chief among these was the perception that MST was valid only in the case of MT potentials. The argument was made [3–5] that the consecutive scatterings from two potential cells close enough together that their bounding spheres overlapped were so inextricably coupled together that they could not be described in terms of the scattering properties of the individual cells. More precisely, it was argued that a spherical wave emanating from the center of a given scatterer would begin to scatter from the potential in an adjacent cell before the scattering from the initial cell had been completed. Thus, one would need to take account of the so-called "near-field corrections" (NFCs) in applications of MST to non-MT, space-filling cells. The need to calculate NFCs, coupled with the need to solve a fairly complicated system of coupled differential equations to determine the local (cell) solutions has contributed greatly to the slow development of a full-potential methods based on MST.

The recent history of the application of multiple scattering theory in determining the electronic structure of solid materials began in 1974 with the publication of a paper by Williams and Morgan [6] (WM) which, contrary to accepted wisdom, suggested that MST in its well-known muffin-tin form remained valid in certain cases of space-filling cells. There were some restrictions placed on the shape of the cells, most notably that any vector, $\mathbf{r}$, in the cell be smaller than any intercell vector, $\mathbf{R}$, but the suggestion was provocative enough to elicit a storm of criticism as well as many attempts to generalize MST to space-filling cell potentials [3–12]. There were also works [13–17] that attempted to justify the conjecture of WM, and some [13,18] which provided mathematical justification for the validity of multiple scattering theory to non-MT potentials.

As the dust began to settle, a rather surprising picture was revealed. The conjecture of WM turned out in fact to be correct, and the muffin-tin form of the secular equation of multiple scattering theory does indeed remain valid in the case of space-filling cells. In deriving their theory, WM were guided by physical principles that indicate that MST is a valid construct in all cases, not just in that of muffin-tin potentials (recall the arguments based on Huygens' principle in Chapter 2). The cause of the criticism elicited by the work of WM is a technical error, having to do with the expansion of the free-particle propagator in terms of the regular and irregular solutions of the Helmholtz equation.

In their derivation, Williams and Morgan [6] relied on a procedure that violates the conditions for the validity of this expansion. In fact, numerical calculations can easily confirm that this procedure leads to divergent results, manifestly invalidating the derivation of the secular equation given by WM (without necessarily showing that the conclusions derived are incorrect.) Indeed, for very elongated cells, a situation not covered by the conditions placed on the geometry by WM, divergent behavior sets in from the very beginning in quite a pronounced fashion. Thus the criticism was well taken. At the same time, given this state of affairs it is quite remarkable that multiple scattering theory remains a valid construct retaining the separability of structure and potential in the case of space-filling cells, even of interpenetrating or elongated shapes that violate *all* of the conditions for the expansion of the free-particle propagator. That this is the case is shown in detail and in a number of ways in this chapter.

At this point, it may be worth emphasizing the logical content of the controversy that erupted with the publication of the paper by WM. As follows from the discussion in previous chapters, the MT geometry constitutes a *sufficient* condition for the validity of MST. As it turns out, it is not a necessary condition. Thus, it should not be thought unlikely that MST is indeed valid in the case of space-filling cells, even though it cannot be derived along lines identical to those used in the MT case. In fact, this turns out to be the case, provided that one modifies the usual derivations so as to account for the absence of the MT conditions. Therefore, the mathematical justification provided by WM for the general validity of multiple scattering theory is incorrect, but the claim itself is valid.

The work of WM was followed by a number of other works [3–5,8,11,12] in which the issue of NFCs and the form assumed by MST in the case of space-filling cells were studied. As already mentioned, it appeared that the failure of multiple scattering theory was due to underlying physical reasons: Namely, that a spherical wave emanating from the center of a nonspherical cell would begin scattering from an adjacent cell before the scattering inside the original cell was completed. It was then argued [3,5] that the scattering events emanating from those cells must be coupled in some very complicated way, making it impossible to describe this scattering in terms of the individual cell t-matrices. Thus the separation of the potential and structural aspects of a collection of scattering centers (potentials), which was found to be of such great computational utility in connection with MT potentials, was lost. One was now faced with the seemingly arduous and ill-defined task of calculating near- field corrections when setting up the secular equation for the band structure.

Fortunately, both the physical and the mathematical arguments given for the existence of NFCs can be bypassed on the basis of physical principle and analytic procedures. Let us first consider the physical aspect of the problem. As we saw in our discussion of Huygens' principle in Chapter 2, wave scattering (and indeed quantum mechanics in the regime where that principle holds) is local in nature. This implies that one could collect together all scattering events that emanate from points inside any one given cell and rearrange the scattering series in terms of cell-scattering matrices. This is what leads to the equations of multiple scattering theory in abstract operator space. This principle would have to fail in the angular-momentum representation to justify the existence of NFCs. Taken at face value, such a disparity between the nature of the physical process of scattering and its expression in an angular-momentum representation would imply that the latter is of very limited usefulness within scattering theory.

The study of and the search for near-field corrections reveals that the problems encountered in generalizations of multiple scattering theory are purely of geometric origin. It also suggest strongly that the form assumed by the t-matrix in the MT case remains valid in all cases, although it does not necessarily imply any particular rate of convergence that should be expected in numerical applications. Both of these aspects, the formal validity and the numerical characteristics of generalized multiple scattering theory are discussed in the following sections.

Computational studies of the effects of NFCs have also been carried out. Faulkner [19] has attempted a numerical investigation of the magnitude of NFCs for the case of the two-dimensional square lattice with a constant potential (the *empty lattice test*). His calculations indicated that NFCs are small, at least in the case of the two-dimensional empty lattice, amounting to about 2.5% of the $l = 0$ eigenvalue of the secular equation. Later work [17] demonstrated that MST yields indeed the exact solution of the empty lattice problem, provided that all internal sums over angular momentum states are properly converged.

In attempting to avoid the difficulties of near-field effects, Brown and Ciftan [7](BC) arrived at a secular equation determining the band structure of a material that is of the MT form *but* with the role of the cell t-matrix being played by a quantity defined in a rather different way. In BC's formalism the scattering properties of a cell depend on the entire medium surrounding the cell which, it was claimed, was equivalent to a dependence in that part of that medium included inside a sphere circumscribing the cell, the so-called moon region. They suggested that in determining the scattering properties of a cell it is necessary to consider explicitly the potential in the moon region. Thus, one first determines the wave function associated with the potential inside the bounding sphere and subsequently constructs the phase functions of the cell by means of integrals extending only over the domains of the cells.

As we have seen in Chapter 3, one of the surprising results to emerge from recent studies [20] of MST is that the potential in the moon region does not affect the calculation of the cell t-matrix. *Thus the t-matrices obtained in the BC and WM procedures are identical.* However, the rate of convergence of the solution of the associated Schrödinger equation for a non-spherical potential does depend on the value chosen for the potential in the moon region. By a proper selection of that potential (not necessarily equal to the crystal potential) many internal sums that arise naturally when the angular-momentum representation is used in connection with nonspherical potentials (or cell shapes) can be brought rapidly to convergence. Thus, the original suggestion by BC that one should include the crystal potential in the moons, although unduly restrictive, contains the germ of a result of considerable computational power.

The resolution of the various difficulties associated with the potential in the moon region is both surprising and reassuring. As was originally pointed out by Nesbet [16, 20] and as discussed at length in Chapter 3, in the converged limit the procedures of BC and of WM yield identical results for the cell t-matrix [21]. Potentially divergent expressions in the latter can be eliminated by means of convergent double sums (when carried out in the proper order), leading to the conclusion that the integrals that include the potential in the moon region do indeed allow the determination of the cell-scattering matrices [21].

6.2 Derivations of MST for Space-Filling Cells

As part of our formal development, we derive expressions for the total t-matrix of an assembly of scatterers, the corresponding Green function, and the wave function of a periodic material. We also derive the secular equation for the eigenenergies of a system of scattering potentials and for the band structure of a periodic system. These equations exhibit clearly the separation of the potential aspects, embodied in the single-cell phase functions and t-matrices, from the structure of the system, represented by the KKR structure constants (or modified structure constants as discussed below). In these various derivations, we attempt to treat explicitly the geometric subtleties associated with the application of MST to non-MT, space-filling cells.

As we pointed out in our discussion of MT potentials, in deriving MST one can take either a scattered-wave approach, or one can view the problem as that of obtaining the solution of a second-order partial differential equation. Within the scattering formalism, introduced for electronic structure calculations by Korringa [22], one considers the wave function of a collection of potentials as being built up from contributions arising from the multiple scattering of waves from the potential cells comprising the system. The condition that the disturbance incident upon a given cell (scatterer) consists of the sum of the outgoing waves scattered by all cells, including the one in question, leads to the secular equation that determines the eigenvalues and eigenenergies of the system of scatterers. This formal approach to MST is particularly attractive as it endows the theory with a pictorial quality that greatly clarifies its content.

In the differential equation approach, one attempts to construct the global solution for a collection of potential cells from solutions obtained for each of the cells separately. Thus, the individual cell solutions are combined so that the final wave function is continuous with continuous derivative across all cell boundaries. This matching of the solutions across cell boundaries can be effected either directly or within a variational formalism. Kohn and Rostoker [23] used both approaches in their original derivation of MST for spherically symmetric MT potentials. It is a tribute to the power and generality of MST that all of these formalisms give equivalent expressions for the secular equation, the wave function, and other quantities.

Attempts to generalize MST to non-MT, space-filling cells have traditionally been plagued by two interrelated difficulties. First, the loss of spherical symmetry implies that angular momentum is no longer a good quantum number so that one must work with matrices that are not diagonal in the L indices, and with wave functions, for example, Eq.(3.99), which include internal sums that, in principle, should be carried to convergence. By contrast, in the case of spherically symmetric MT potentials one need only consider matrices for the cell phase functions that are diagonal in L, and no internal sums arise in the wave function in Eq.(3.16). The need to consider infinite summations over angular momentum exacerbates the second and more serious difficulty, that of expanding the cell off-diagonal part of the free-particle propagator in terms of the KKR structure constants. As is pointed out in the following discussion, in the case of space- filling cells the expansion in Eq.(5.13) is in general not justified. The sums over L and L' become conditionally convergent or may even diverge. This purely *geometric* difficulty has seriously hindered attempts to generalize MST beyond the case of MT potentials.

These difficulties were originally construed [24–27] to imply that MST is not rigorously valid for space-filling cells. Potential-dependent modifications to the MST structure constants called near-field corrections (NFCs) were proposed [24] to deal with this apparent difficulty. It was conjectured that because of NFCs it would no longer be possible to describe the multiple scattering process solely in terms of the t-matrices of individual cells. However, subsequent work [28–35] has shown that these fears are not justified. We may recall our discussion in Chapter 3 where it was shown that the t-matrix indeed contains all of the information necessary to reconstruct the wave function anywhere outside the boundary of a cell, even in the case of cells of concave shape. This is accomplished by calculating the scattered wave outside the bounding sphere and analytically continuing the solution inside the moon region using the cell t-matrix. This feature of the t-matrices is retained in the case of a collection of cells, allowing the description of the t-matrix and the system wave function in terms of the t-matrices of the individual cells. We shall show that there are no NFCs, and that the MT form of MST remains valid in the general, non-MT case, with a possible difference in the form in which the structure constants, which continue to depend only on geometry, may have to be constructed to guarantee convergence. The only fundamental difference with the case of MT potentials is that now it is often necessary to carry certain internal sums to convergence before reliable results are obtained. At the same time, different forms may exhibit different mathematical properties from those of the corresponding MT case. For example, although all forms of the MT secular equation are variational, that is, yield variations in energy that are of the second order in the change of the wave function, only selected forms of the secular equation for space-filling cells have this property.

6.3 Full-Cell MST

In this section we derive the secular equation of multiple scattering theory for space-filling potentials generalizing one of the derivations of MT-MST used by Kohn and Rostoker [23]. The development given here also closely parallels that of Section 5.6.

We begin with the Lippmann–Schwinger equation,

$$\psi(\mathbf{r}) = \chi(\mathbf{r}) + \int G_0(\mathbf{r}-\mathbf{r}')V(\mathbf{r}')\psi(\mathbf{r}')\mathrm{d}^3r'. \tag{6.1}$$

Because we are looking for the bound states of the system, that is, those with finite amplitude even in the absence of an incident wave, we shall make Eq.(6.1) homogeneous by setting the incident wave, $\chi(\mathbf{r})$, to zero. By use of the Schrödinger equation, this Lippmann–Schwinger equation can be written as

$$\psi(\mathbf{r}) = \int G_0(\mathbf{r}-\mathbf{r}')\left(\nabla'^2 + E\right)\psi(\mathbf{r}')\mathrm{d}^3r'. \tag{6.2}$$

Using Eq.(2.40), satisfied by the free-particle propagator, we also have

$$\psi(\mathbf{r}) = \int \left[\left(\nabla'^2 + E\right)G_0(\mathbf{r}-\mathbf{r}')\right]\psi(\mathbf{r}')\mathrm{d}^3r', \tag{6.3}$$

which can be subtracted from Eq.(6.2) to yield the result

$$\begin{aligned} 0 &= \int \left[G_0(\mathbf{r}-\mathbf{r}')\nabla'^2\psi(\mathbf{r}') - \nabla'^2 G_0(\mathbf{r}-\mathbf{r}')\psi(\mathbf{r}')\right]\mathrm{d}^3r' \\ &= \sum_n \int_{\Omega_n} \left[G_0(\mathbf{r}-\mathbf{r}')\nabla'^2 - \nabla'^2 G_0(\mathbf{r}-\mathbf{r}')\right]\psi(\mathbf{r}')\mathrm{d}^3r'. \end{aligned} \tag{6.4}$$

The second line of this equation follows upon partitioning the volume of integration into the volumes of individual cells, with a subsequent summation over cells. Through the use of Green's theorem, this equation can be converted to a sum of surface integrals,

$$\sum_n \int_{S_n} [G_0(\mathbf{r}-\mathbf{r}')\nabla'\psi(\mathbf{r}') - \nabla' G_0(\mathbf{r}-\mathbf{r}')\psi(\mathbf{r}')]\, \mathrm{d}^3 r' = 0, \tag{6.5}$$

where now the symbol ∇' denotes the outward derivative over the surface, S_n, of the cell at n. Upon substituting Eq.(3.66) for $\psi(\mathbf{r})$ in Eq.(6.5), and writing,

$$G_0(\mathbf{r}-\mathbf{r}') = \sum_L G_L(\mathbf{r},\mathbf{r}'), \tag{6.6}$$

we can write Eq.(6.5) in the form

$$0 = \sum_n \int_{S_n} \mathrm{d}^2 r' \sum_{LL'} [G_L(\mathbf{r}-\mathbf{r}')\nabla' - \nabla' G_L(\mathbf{r}-\mathbf{r}')]\, \psi^n_{L'}(\mathbf{r}') A^n_{L'}. \tag{6.7}$$

Now, consider $G_0(\mathbf{r}-\mathbf{r}')$ with $\mathbf{r}$ in cell p and $\mathbf{r}'$ on the *surface* of cell n. We now expand G_0 about the center of cell p using standing-wave boundary conditions[1]

$$G_L(\mathbf{r}_p, \mathbf{r}'_p) = k J_L(\mathbf{r}_p) N_L(\mathbf{r}'_p), \tag{6.8}$$

and restricting $\mathbf{r}$ to lie inside a spherical region surrounding the center of cell p, such as the MT sphere, Eq.(6.7) takes the form

$$0 = k \sum_L J_L(\mathbf{r}_p) \sum_{L'n} \int_{S_n} \mathrm{d}^2 r' [N_L(\mathbf{r}'_p)\nabla' - \nabla' N_L(\mathbf{r}'_p)] \psi^n_{L'}(\mathbf{r}'_n) A^n_{L'}. \tag{6.9}$$

Finally, defining the expression

$$C^{pn}_{LL'} = -k \int_{S_n} \mathrm{d}^2 r' [N_L(\mathbf{r}'_p)\nabla' - \nabla' N_L(\mathbf{r}'_p)] \psi^n_{L'}(\mathbf{r}'_n), \tag{6.10}$$

we obtain

$$\sum_L J_L(\mathbf{r}_p) \sum_{L'n} C^{pn}_{LL'} A^n_{L'} = 0. \tag{6.11}$$

The cell off-diagonal quantities, $C^{pn}_{LL'}$, are the generalizations of the corresponding single-cell quantities that were defined in Chapter 3. In fact, we have (assuming the ψ_L has the form of Eq.(3.70) $\sum_{L'}[C_{LL'}J_{L'} - S_{LL'}N_{L'}]$),

$$C^{nn}_{LL'} = C^n_{LL'}. \tag{6.12}$$

[1]Outgoing-wave boundary conditions could also be used and would lead to a different secular equation as discussed in Sections 5.6 and 6.3.1. Generally, the choice of outgoing- or standing-wave boundary conditions is determined by the physical situation under study.

The secular equation determining the coefficients A_L^n, and hence the wave function for the system, now follows from Eq.(6.11). Because $J_L(\mathbf{r}_p)$ does not vanish identically, we must have

$$\sum_{L'n} C_{LL'}^{pn} A_{L'}^n = 0, \tag{6.13}$$

which has nontrivial solutions for the A_L^n only if the determinant of the coefficients vanishes,

$$\det C_{LL'}^{pn} = 0. \tag{6.14}$$

To see that Eq.(6.14) is equivalent to the usual MST equations, we use the condition $|\mathbf{R}_{pn}| > |\mathbf{r}_n|$ (which now does become a restriction on the derivation) and expand $N_{L'}(\mathbf{r}'_p)$ in the form

$$N_{L'}(\mathbf{r}'_p) = \sum_L B_{L'L}(\mathbf{R}_{pn}) J_L(\mathbf{r}'_n), \tag{6.15}$$

and use Eq.(6.5) to write

$$C_{LL'}^{pn} = C_{LL'}^n \delta_{np} + \sum_{L''} B_{LL''}(\mathbf{R}_{pn}) S_{L''L'}^n (1 - \delta_{np}). \tag{6.16}$$

Thus, the MST equations become

$$\sum_{nL'} \left[C_{LL'}^n \delta_{np} + \sum_{L''} B_{LL''}(\mathbf{R}_{pn}) S_{L''L'}^n (1 - \delta_{np}) \right] A_{L'}^n = 0. \tag{6.17}$$

6.3.1 Outgoing-Wave Boundary Conditions

Equation (6.17) is the MST secular equation for standing-wave boundary conditions. In an infinite system, boundary conditions that assume outgoing waves at infinity are usually more appropriate. They are also more appropriate at negative energy because the analytic continuation of outgoing spherical waves to negative energy yields functions that decay exponentially at large distances.

The outgoing wave analog to Eq.(6.17) can be derived by replacing the standing-wave Green function expansion, Eq.(6.8), by the outgoing-wave Green function expansion

$$G_L(\mathbf{r}_p, \mathbf{r}'_p) = -ik J_L(\mathbf{r}_p) H_L(\mathbf{r}'_p). \tag{6.18}$$

With this replacement, Eqs. (6.10) and (6.11) become, respectively,

$$E_{LL'}^{pn} = C_{LL'}^{pn} - iS_{LL'}^{pn} = ik \int_{S_n} \mathrm{d}^2 r' [H_L(\mathbf{r}'_p)\nabla' - \nabla' H_L(\mathbf{r}'_p)] \psi_{L'}^n(\mathbf{r}'_n), \tag{6.19}$$

and

$$\sum_L J_L(\mathbf{r}_p) \sum_{L'n} E_{LL'}^{pn} A_{L'}^n = 0. \tag{6.20}$$

Thus, an expansion of the Hankel function, $H_L(\mathbf{r}_p)$ about the centers of cells n when $n \neq p$ using Eqs.(D.9) and (D.10) yields the full potential MST secular equation for outgoing wave boundary conditions,

$$\sum_{nL'}[E^p_{LL'}\delta_{pn} - i\sum_{L''} G_{LL''}(\mathbf{R}_{pn})S^n_{L''L'}(1-\delta_{pn})]A^n_{L'} = 0. \qquad (6.21)$$

6.3.2 Empty-Lattice Test

One simple but important test of multiple scattering theory for space filling cells is the empty lattice test. In this test the cells are arranged on a periodic lattice and the potential within each cell is uniform. Thus $V(\mathbf{r}) = \sum_n v_0\Theta_n(\mathbf{r})$, where $\theta_n(\mathbf{r})$ is unity for $\mathbf{r}$ in cell n and vanishes otherwise. Because $V(\mathbf{r})$ is a constant, v_0, it is trivial to obtain the exact solution to the wave equation that can be compared with the results obtained from multiple scattering theory.

For a periodic system, the wave functions that result from solving Eq.(6.21) must satisfy Bloch's theorem so $A^n_L = A_L(\mathbf{k})e^{i\mathbf{k}\cdot\mathbf{R}_n}$ which allows one to perform the sum on n to obtain

$$\sum_{L'}\left[E_{LL'} - i\sum_{L''} G_{LL''}(\mathbf{k})S_{L''L'}\right]A_{L'}(\mathbf{k}) = 0, \qquad (6.22)$$

where $G_{LL'}(\mathbf{k}) = \sum_{n\neq 0} G_{LL''}(\mathbf{R}_n)e^{i\mathbf{k}\cdot\mathbf{R}_n}$.

Figure 6.1 shows the convergence of the energy eigenvalue at a particular value of $\mathbf{k}$ as the maximum number of angular momentum states is varied in evaluating Eq.(6.22). The exact value of $E(\mathbf{k})$ at $\mathbf{k} = (0.1.0.2.0.3)$ equals 0.4, as given by the relation $E = k^2 - \nu_0$, with $\nu_0 = 0.1$. It is clear from the figure that it is crucial to converge the internal angular-momentum sum over L''.

6.3.3 Note on Convergence

We have now derived the secular equation for space-filling cell potentials, Eq.(6.21), and illustrated its convergence in the case of the empty lattice with the results of numerical calculations. However, in spite of the care that we took in this derivation, a subtle point was missed. To see this, consider the secular equation for the empty lattice, Eq.(6.22), described by elongated cells that deviate drastically from spherically symmetry. In that case, Eq.(6.22) diverges because it contains terms that correspond to the argument of the structure constants (the intercell vectors) being smaller than intracell vectors. In fact, the same difficulty would arise even in the case of cubic cells if the angular-momentum expansions were carried out to large enough values of L.

We now examine what went wrong in the derivation of the secular equation. There is no difficulty with the derivation up to Eq.(6.5). However,

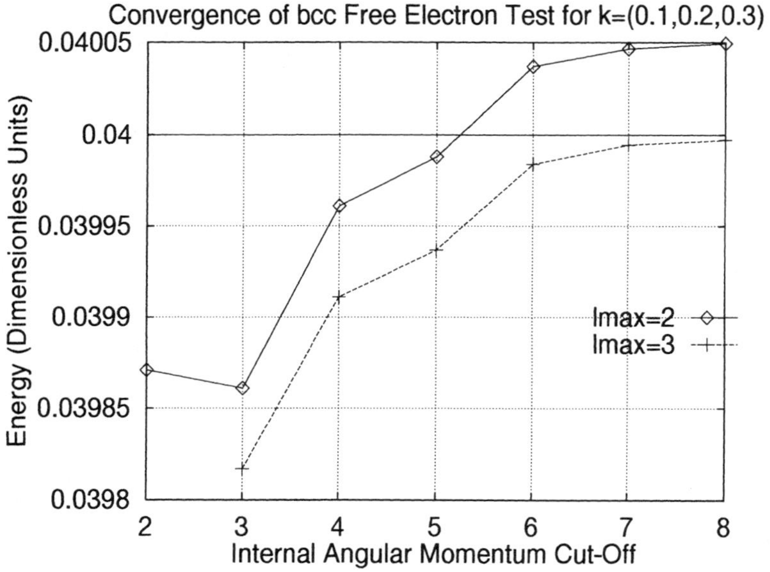

FIGURE 6.1. Empty lattice test for the lowest energy state at **k**=(0.1,0.2,0.3) of the bcc lattice as a function of the internal angular-momentum cut-off.

beginning with Eq.(6.8) further expansions were introduced and, most importantly, the order of summations over angular-momentum indices in different expansions were often interchanged. For example, this occurred in going from Eq.(6.10) to Eq.(6.15) (see the note of caution preceding the last equation) where we replace a sum over N_L by a double sum resulting from expanding N_L. This can be done provided it is known a priori that all sums taken in any order are convergent. Because this cannot be proven in general, that is to say, the condition on the derivation noted before Eq.(6.15) cannot be satisfied for space-filling cells, the derivation of the secular equation contains a flaw that prevents a formal proof of convergence.

This difficulty has been at the root of the discussions about NFCs. Fortunately, it can be easily bypassed by resorting to a different expansion of the irregular spherical functions. As is shown in Sections 6.5 and 6.7, as well as in the last chapter of this text, it is possible to write the equations of MST in terms of "modified" structure constants that can be shown formally to lead to converged results. This is done through the replacement of a single, potentially divergent sum by a conditionally convergent double sum over angular-momentum indices. These structure constants depend only on the geometry of the lattice and are independent of the potential inside the cells in a material. This point will hopefully become clearer in subsequent discussion when the explicit forms of these double expansions are given. For the moment, we proceed with our derivations as if all expansions converged secure in the knowledge that such expansions can be rendered convergent through the use of appropriately defined angular-momentum expansions of the spherical functions.

6.3.4 Full-Potential Wave Functions

Equations (6.17) or (6.21) determine the eigenstates of the system through $\det|C+BS|=0$ or $\det|E-iGS|=0$. In principle, they also determine the eigenfunctions of the system through the condition

$$\psi(\mathbf{r})=\sum_n A_L^n \psi_L^n(\mathbf{r})\Theta_n(\mathbf{r}), \tag{6.23}$$

where $\Theta_n(\mathbf{r})$ is unity for $\mathbf{r}$ in cell n and zero otherwise.

However, in the case of space-filling cells, this single-center expansion may diverge at points inside the cell but outside the MT sphere. In a latter section, we discuss techniques for deriving an expansion for the wave function that can be proved to converge throughout the cell.

The derivation of Eq.(6.17) implies that it represents a necessary condition on the wave function. That it is also a sufficient condition follows from the fact that under the normalization chosen, $\psi_L(\mathbf{r}\to 0)=J_L(\mathbf{r})$, the wave function inside the MT sphere suffices to determine the wave function throughout the cell, as was discussed in Chapter 3.

6.4 The Green Function and Bloch Function

The discussion in the previous section has shown that the secular equation for an assembly of space-filling potential cells takes the same form as in the case of muffin-tin potentials. In this section, we show that the same is true for the Green function of a collection of space-filling cells.

6.4.1 The Green Function

In the previous chapter, we derived expressions for the Green function associated with an assembly of scattering centers using a nonvariational approach. In keeping with the nature of the approach followed in this chapter, we provide a derivation of the Green function of MST based on a functional derivative formalism.

The formalism presented below is based on two observations. First, we note that from the Lippmann–Schwinger Eq.(6.1) we obtain the expression

$$\frac{\delta\psi(\mathbf{r})}{\delta V(\mathbf{r}')}=G_0(\mathbf{r},\mathbf{r}')\psi(\mathbf{r},\mathbf{r}')+\int \mathrm{d}\mathbf{r}'' G_0(\mathbf{r},\mathbf{r}'')V(\mathbf{r}'')\frac{\delta\psi(\mathbf{r}'')}{\delta V(\mathbf{r}')}, \tag{6.24}$$

which, when iterated and resummed, yields

$$\frac{\delta\psi(\mathbf{r})}{\delta V(\mathbf{r}')}=G(\mathbf{r},\mathbf{r}')\psi(\mathbf{r}'). \tag{6.25}$$

In particular, we have

$$\frac{\delta|\psi^m\rangle}{\delta V} = G_m|\psi^m\rangle, \tag{6.26}$$

where G_m is the Green function associated with the potential V_m and which, according to Eq.(4.85), can be written in the form

$$G_m(\mathbf{r},\mathbf{r}') = \sum_L F_L^m(\mathbf{r})\psi_L^m(\mathbf{r}') \equiv \langle F^m(\mathbf{r})|\psi^m(\mathbf{r}')\rangle, \quad \text{for} \quad r > r', \tag{6.27}$$

in terms of the regular, $|\psi^m\rangle$, and irregular, $|F^m\rangle$, solutions of the Schrödinger equation for the potential in cell m. Using the expansion of ψ in terms of basis functions, Eq.(6.91), we have

$$\begin{aligned} \frac{\delta\psi(\mathbf{r})}{\delta V_m(\mathbf{r}')} &= \sum_{n,L} \frac{\delta\psi_L^m(\mathbf{r})}{\delta V_m(\mathbf{r}')} A_L^m + \sum_L \frac{\delta A_L^m}{\delta V_m(\mathbf{r}')}\psi_L^m(\mathbf{r}) \\ &= \sum_n \left\langle \frac{\delta A^n}{\delta V_m(\mathbf{r}')} \middle| \psi^n(\mathbf{r}') \right\rangle \\ &\quad + \langle A^m|\psi^m(\mathbf{r}')\rangle\langle F^m(br)|\psi^m(\mathbf{r}')\rangle, \end{aligned} \tag{6.28}$$

for $r > r'$, and with both vectors $\mathbf{r}$ and $\mathbf{r}'$ in cell m.

The second relevant observation is that the matrix $\underline{M}^{nm}$, Eq.(6.102), can be written in the form

$$\underline{M}^{nm} = [|H^n\rangle, \langle\psi^m|]_m, \tag{6.29}$$

where the brackets denote an integral of the Wronskian of the two functions, $H_L^n(\mathbf{r})$ and $\psi_L^m(\mathbf{r})$, centered about the origins in cells n and m, respectively, over the surface of cell m. The bracket notation to represent surface integrals of Wronskians will be used often in the discussion that follows. That the last equation provides a proper representation of the matrix $\underline{M}^{nm}$ follows from Eqs.(3.81) and (3.82). Now, we recall that the matrix $\underline{M}^{nm}$ and the vectors $|A^n\rangle$ satisfy the set of homogeneous linear equations,

$$\sum_m \underline{M}^{nm}|A^m\rangle = 0. \tag{6.30}$$

Varying Eq.(6.30) with respect to $V(\mathbf{r}')$ we obtain the relation

$$\frac{\delta\underline{M}^{nm}}{\delta V_m(\mathbf{r}')}|A^m\rangle + \sum_j \underline{M}^{nj}\frac{\delta|A^j\rangle}{\delta V_m(\mathbf{r}')} = 0, \tag{6.31}$$

so that

$$\frac{\delta|A^j\rangle}{\delta V_m(\mathbf{r}')} = \sum_j [\underline{M}^{-1}]^{nj}\frac{\delta\underline{M}^{jm}}{\delta V_m(\mathbf{r}')}|A^m\rangle. \tag{6.32}$$

But from Eq.(6.29) we have (for $\mathbf{r}'$ in cell m)

$$\frac{\delta\underline{M}^{nm}}{\delta V_m(\mathbf{r}')} = [|H^n\rangle, \frac{\delta}{\delta V_m(\mathbf{r}')}\langle\psi^m(\mathbf{r}')|]_m, \tag{6.33}$$

so that Eq.(6.32) can be written in the form,

$$\frac{\delta|A^n\rangle}{\delta V_m(\mathbf{r}')} = \sum_j [\underline{M}^{-1}]^{nj} \underline{N}^{jm} |\psi^m(\mathbf{r}')\rangle\langle\psi^m(\mathbf{r}')|A^m\rangle. \tag{6.34}$$

In the last expression the matrix $\underline{N}^{im}$ is defined by its elements,

$$\underline{N}^{nm} = [|H^n\rangle, \langle F^m|]_m, \tag{6.35}$$

and is strictly site off-diagonal. This last property follows from the form of the irregular solution,

$$|F^m\rangle = [\underline{C}^m]^{-1}|H^m\rangle. \tag{6.36}$$

In fact, assuming that intercell vectors are longer than intracell vectors, we can write

$$\underline{N}^{nm} = \underline{G}(\mathbf{R}_{nm})[\underline{C}^m]^{-1}, \tag{6.37}$$

an expression that displays clearly the off-diagonal nature of the matrix $\underline{N}^{nm}$.

Now, using the expansion of the total wave function in terms of cell basis functions, Eq.(6.91), we can write Eq.(6.34) in the form

$$\frac{\delta|A^n\rangle}{\delta V_m(\mathbf{r}')} = \sum_j [\underline{M}^{-1}]^{nj} \underline{N}^{jm} |\psi^m(\mathbf{r}')\rangle\psi(\mathbf{r}'). \tag{6.38}$$

An expression for the variation of the total wave function with respect to the potential in cell m can now be obtained through the use of Eq.(6.28),

$$\begin{aligned}\frac{\delta\Psi(\mathbf{r})}{\delta V_m(\mathbf{r}')} = & \sum_{nj} \langle\psi^n(\mathbf{r})|[\underline{M}^{-1}]^{nj} \underline{N}^{jm} |\psi^m(\mathbf{r}')\rangle\Psi(\mathbf{r}') \\ & + \langle F^m(\mathbf{r})|\psi^m(\mathbf{r}')\rangle\Psi(\mathbf{r}') \quad \text{for } r > r'.\end{aligned} \tag{6.39}$$

Finally, comparing the last equation with Eq.(6.25) we obtain the expression for the Green function,

$$G(\mathbf{r},\mathbf{r}') = \langle\psi^n(\mathbf{r})|\tilde{\underline{\tau}}^{nm}|\psi^m(\mathbf{r}')\rangle + \langle\psi^m(\mathbf{r})|F^m(\mathbf{r}')\rangle\delta_{nm}, \quad \text{for } r' > r, \tag{6.40}$$

where the vectors $\mathbf{r}$ and $\mathbf{r}'$ are confined in cells n and m, respectively, and the matrix $\tilde{\underline{\tau}}^{nm}$ is defined by the expression

$$\tilde{\underline{\tau}}^{nm} = \sum_j [\underline{M}^{-1}]^{nj} \underline{N}^{jm}. \tag{6.41}$$

The expression for the Green function in Eq.(6.40) has a familiar form, consisting of singular term and and multiscatterer contributions. In the case of crystal lattices and for a choice of cell centers in which the intercell vectors are larger than intracell vectors, Eq.(6.40) can be readily converted to the form,

$$G(\mathbf{r},\mathbf{r}') = \langle\psi^n(\mathbf{r})|\hat{\underline{\tau}}^{nm}|\psi^m(\mathbf{r}')\rangle + \langle\psi^m(\mathbf{r})|F^m(\mathbf{r}')\rangle\delta_{nm}, \tag{6.42}$$

where $\hat{\underline{\tau}}^{nm}$ is the inverse of the matrix,

$$\hat{\underline{M}}^{nm} = \underline{S}^{n\dagger}\underline{C}^{n}\delta_{nm} - \underline{S}^{n\dagger}[\underline{G}(\mathbf{R}_{nm})\underline{S}^{m}](1-\delta_{nm}). \tag{6.43}$$

In the form of the last expression the Green function agrees with Eq.(6.45). Even though Eqs.(6.40) and (6.42) do not involve individual t-matrices, and consequently they tend to obscure somewhat the scattering aspects of an ensemble of cells, they also do *not* require the inversion of any phase functions – a feature that can be of great computational value.

6.4.2 Alternative Expressions for the Green Function

For some applications, it is convenient to have expressions for the single particle Green function explicitly in terms of the basis states $|\phi(\mathbf{r})\rangle$, Eq.(4.90). It is a matter of some straightforward algebra to derive the expressions,

$$\begin{aligned}\underline{G}(\mathbf{r},\mathbf{r}') = \; & \langle\psi^{i}(\mathbf{r})\,|\tilde{\underline{\tau}}^{ij}|\,\psi^{j}(\mathbf{r}')\rangle \\ & -[\,\Theta(r-r')\langle Z^{i}(\mathbf{r}')\,|\,S^{i}(\mathbf{r})\rangle \\ & +\Theta(r'-r)\langle Z^{i}(\mathbf{r})\,|\,S^{i}(\mathbf{r}')\rangle\,]\,\delta_{ij},\end{aligned} \tag{6.44}$$

or

$$\begin{aligned}\underline{G}(\mathbf{r},\mathbf{r}') = \; & \langle\psi^{i}(\mathbf{r})\,|\hat{\underline{\tau}}^{ij}|\,\psi^{j}(\mathbf{r}')\rangle \\ & -[\,\Theta(r-r')\langle\psi^{i}(\mathbf{r}')\,|\,F^{i}(\mathbf{r})\rangle \\ & +\Theta(r'-r)\langle\psi^{i}(\mathbf{r})\,|\,F^{i}(\mathbf{r}')\rangle\,]\,\delta_{ij}.\end{aligned} \tag{6.45}$$

In these expressions, we have defined the quantities

$$\tilde{\underline{\tau}} = [\underline{C} - \underline{GS}]^{-1}\underline{S}^{-1} \tag{6.46}$$

and

$$\hat{\underline{\tau}} = [\underline{S}^{\dagger}\underline{C} - \underline{S}^{\dagger}\underline{GS}]^{-1}. \tag{6.47}$$

The functions $|\psi^{i}\rangle$ and $|F^{i}\rangle$ are those regular and irregular solutions of the Schrödinger equation associated with the potential in cell i which at the surface of the sphere bounding the cell join smoothly to the functions $\underline{C}^{i}|J\rangle + \underline{S}^{i}|H\rangle$ and $\underline{S}^{-1}|J\rangle$, respectively. Some of these alternative forms may be more convenient than others in certain applications.

6.4.3 Bloch Functions for Periodic, Space-Filling Cells

It is now fairly straightforward to derive an expression for the wave function of a scattering assembly through a diagonalization of the Green function. This can be easily demonstrated for the case of lattice periodic systems. Neglecting the (real) single-scatterer term for real values of the energy, we obtain after a Fourier transformation of Eq.(6.44),

$$G_{\mathbf{k}}(\mathbf{r},\mathbf{r}') = \langle Z(\mathbf{r})\,|\underline{\tau}(\mathbf{k})|\,Z(-\mathbf{r}')\rangle, \tag{6.48}$$

where $\mathbf{r}$ and $\mathbf{r}'$ are cell vectors. Near a singularity of the matrix $\underline{\tau}(\mathbf{k})$, for example, where the secular Eq.(5.26) is satisfied, we can write

$$G_{\mathbf{k}}(\mathbf{r},\mathbf{r}') = \left\langle \phi_{\mathbf{k}}(\mathbf{r}) \left| [E - E(\mathbf{k})]^{-1} \right| \phi_{\mathbf{k}}(\mathbf{r}') \right\rangle . \tag{6.49}$$

By definition, the $\langle \phi_{\mathbf{k}}(\mathbf{r}) \mid$ are the eigenstates of the system, which in this case are the Bloch functions of the material, and are given by the relation

$$|\phi_{\mathbf{k}}(\mathbf{r})\rangle = \underline{C}(\mathbf{k}) |\, Z(\mathbf{r})\rangle, \tag{6.50}$$

where $\underline{C}(\mathbf{k})$ is the matrix that diagonalizes $\underline{\tau}(\mathbf{k})$. Again we note that this equation has precisely the same form as in the case of MT potentials.

Now, an expression for the cell charge density can be obtained. By definition, at energy E, we have,

$$\begin{aligned} \rho_E(\mathbf{r}) &= -\frac{1}{\pi} \text{Im Tr } G(\mathbf{r},\mathbf{r}') \\ &= -\frac{1}{\pi}\frac{1}{\Omega_{\mathrm{BZ}}} \int \mathrm{d}^3k \langle Z(\mathbf{r}) \,|\mathrm{Im}\underline{\tau}(\mathbf{k})|\, Z(\mathbf{r})\rangle. \end{aligned} \tag{6.51}$$

Upon using Eq.(6.50) we obtain the well-known and general expression

$$\rho_E(\mathbf{r}) = \frac{1}{\Omega_{\mathrm{BZ}}} \int_{\mathrm{BZ}} \mathrm{d}^3k \langle \phi_{\mathbf{k}}(\mathbf{r}) \mid \phi_{\mathbf{k}}(\mathbf{r})\rangle, \tag{6.52}$$

where the integration extends over the volume, Ω_{BZ}, of the first Brillouin zone (BZ) of the reciprocal lattice. The charge density follows from Eq.(6.52) upon integration over the energy, and the density of states is given by an integral over the coordinates, $\mathbf{r}$.

6.5 Variational Formalisms

In Section 6.3, we presented a simple derivation of MST for the case of non-MT, space-filling cells. This derivation shows that MST in principle can provide an exact solution of the Schrödinger equation for any system consisting of nonoverlapping potential cells.[2] In this section, we investigate the variational properties of the energy and wave function obtained within MST. That is, given an error of order $\delta\psi$ in the wave function, what is the order of the error in the energy obtained from the MST equations? It is often desirable that the error in the energy be of higher order than the error in the wave functions. In this chapter we also derive an expression for the Green function for MST using a variational derivative technique.

---

[2]The explicit construction of the wave function is discussed in the next chapter.

6.5.1 Variational Derivation of MST

A variational derivation of MST for the case of spherically symmetric MT potentials was originally given by Kohn and Rostoker [36]. The final expressions obtained within that formalism are identical to those obtained by Korringa [37] who followed a scattering approach. In this chapter, we generalize the variational formalism of Kohn and Rostoker to nonspherical MT potentials as well as to space-filling cells. In contrast with the spherical MT case, for which all forms of MST yield a variational energy, we will see that only those forms of MST derived variationally are guaranteed to satisfy the following stationarity property of the energy with respect to the wave function: If δE is the error in the energy associated with the variation $\delta\psi$ in the wave function, then

$$\delta E \approx (\delta\psi)^2. \tag{6.53}$$

In the case of spherically symmetric potentials the relation in Eq.(6.53) is satisfied by *all* forms of the secular equation of MST, regardless of whether or not they were derived within a variational formalism. However, this is *not* the case when MST is applied to nonspherical potentials. The reason that all forms of MST remain variational in the case of spherical potentials is that for such potentials the phase functions and scattering matrices are diagonal L-space, and converting one form to another involves multiplication by diagonal, and hence square, matrices. The invariance of the stationarity property of the secular equation in this case follows from the fact that the determinant of a product of square matrices is equal to the product of the individual determinants. In the case of nonspherical potentials, the expansion of the scattering amplitude in angular-momentum eigenstates requires the convergence of internal summations in products of matrices indexed by L. The conversion of one form of the secular equation to another requires multiplication by rectangular matrices (because of the presence of internal sums) with the result that the new form may not be variational.

6.5.2 A Variational Principle for MST

In our development, we follow essentially the formalism of Kohn and Rostoker's original argument [36], which was presented for the case of spherically symmetric muffin-tin potentials. The major differences between the formalisms of this section and theirs are technical ones: The radial derivatives that entered the discussion in the MT case must be replaced by full gradients across a surface that encloses a given potential cell, and it must be shown how convergent expansions for G_0 are obtained in the general non-MT case. The variational principle of Kohn and Rostoker is a close variant of Schwinger's variational principle, discussions of which can be found in works cited in the references [38].

The integral Lippmann–Schwinger equation, Eq.(6.1), is equivalent to the variational principle,

$$\delta\Lambda = 0, \tag{6.54}$$

where

$$\Lambda = \int \psi^*(\mathbf{r})V(\mathbf{r})\psi(\mathbf{r})\mathrm{d}^3r - \int\int \psi^*(\mathbf{r})V(\mathbf{r})G_0(\mathbf{r}-\mathbf{r}')V(\mathbf{r})\psi(\mathbf{r})\mathrm{d}^3r\mathrm{d}^3r'. \tag{6.55}$$

The existence of this variational principle implies that the error in the energy is of the second order compared to that of a trial wave function. To show this, we consider the value of Λ for a trial wave function ψ_t and trial energy E_t. We can expand $\Lambda(\psi_t, E_t)$ about its value at the exact eigenfunction and energy, (ψ, E) to obtain,

$$\Lambda(\psi_t, E_t) = \Lambda(\psi, E) + \frac{\delta\Lambda}{\delta\psi}(\psi_t - \psi) + \frac{\delta^2\Lambda}{\delta\psi^2}(\psi_t - \psi)^2 + \frac{\delta\Lambda}{deltaE}(E_t - E) + \cdots. \tag{6.56}$$

Using the fact that Λ vanishes and the condition that it be stationary with respect to variations in the wave function, we obtain the relation

$$\delta E \cong \left(\frac{\delta\Lambda}{\delta E}\right)^{-1}\left(\frac{\delta^2\Lambda}{(\delta\psi)^2}\right)(\delta\psi)^2, \tag{6.57}$$

showing that the error in the energy is of second order in the error in the wave function. This variational principle can also be used to obtain a secular equation determining the wave function through the following procedure: Upon using a trial function of the form

$$\psi = \sum_{j=0}^{n} a_j\phi_j, \tag{6.58}$$

with a_j a complex coefficient and ϕ_j an element of some basis set, and substituting into Eq.(6.54) we obtain,

$$\Lambda = \sum_{i,j=0}^{n} a_i^*\Lambda_{ij}a_j, \tag{6.59}$$

where

$$\Lambda_{ij} = \int \phi_i^*(\mathbf{r})V(\mathbf{r})\phi_j(\mathbf{r})d^3r - \int\int \phi_i^*(\mathbf{r})V(\mathbf{r})G_0(\mathbf{r}-\mathbf{r}')V(\mathbf{r})\phi_j(\mathbf{r})\mathrm{d}^3r\mathrm{d}^3r', \tag{6.60}$$

is a Hermitian matrix. Now, the conditions,

$$\frac{\delta\Lambda}{\delta a_i^*} = 0 \quad i = 1, 2, ..., n, \tag{6.61}$$

which follow from Eq.(6.54), yield the set of homogeneous linear equations,

$$\sum_{j=1}^{n} \Lambda_{ij} a_j = 0, \quad i = 1, 2, ..., n. \tag{6.62}$$

These equations have non-trivial solutions only if the determinant of the coefficients vanishes, which leads to the equation,

$$\det |\Lambda_{ij}| = 0. \tag{6.63}$$

This is the general form of the secular equation of MST. In the angular-momentum representation, and in the case of translationally invariant materials described by spherically symmetric potentials, Eq.(6.63) becomes the well-known secular equation of Korringa, Kohn, and Rostoker (KKR) [36, 37]. Our aim is to show that the MT form of the secular equation holds in the general case of space-filling cells.

6.5.3 *First Variational Derivation of MST*

For convenience of exposition we treat the case in which the total potential $V(\mathbf{r})$ is confined to a finite region of space. Our final formulae will, however, be applicable to infinite systems as well. We divide this region into non-overlapping, but otherwise arbitrarily shaped cells, denoting by Ω_n the volume occupied by cell n, and by $v_n(\mathbf{r})$ the potential within that cell.

We begin with an expression for Λ which can be written as $\Lambda = \sum_n \Lambda^n$, where

$$\Lambda^n = \int_{\Omega_n} \mathrm{d}r^3 \psi^*(\mathbf{r}) v_n(\mathbf{r}) B(\mathbf{r}), \tag{6.64}$$

and where $B(\mathbf{r})$ is given by the expression

$$B(\mathbf{r}) = \psi(\mathbf{r}) - \int \mathrm{d}r'^3 G_0(\mathbf{r}, \mathbf{r}') V(\mathbf{r}') \psi(\mathbf{r}'). \tag{6.65}$$

By use of the identities,

$$V(\mathbf{r}')\psi(\mathbf{r}') = (\nabla'^2 + E)\psi(\mathbf{r}') \tag{6.66}$$

and

$$\psi(\mathbf{r}) = \int \mathrm{d}r'^3 (\nabla'^2 + E) G_0(\mathbf{r}, \mathbf{r}') \psi(\mathbf{r}'), \tag{6.67}$$

together with Green's theorem, Eq.(6.65) can be converted to a surface integral,

$$B(\mathbf{r}) = \int_{S'} \mathrm{d}S' \hat{n}' \cdot [\nabla' G_0(\mathbf{r}, \mathbf{r}') - G_0(\mathbf{r}, \mathbf{r}')\nabla'] \psi(\mathbf{r}'). \tag{6.68}$$

In these expressions, ∇' signifies that the operator ∇ is applied to the primed variable. Similarly, the identities,

$$(\nabla^2 + E)B(\mathbf{r}) = 0 \tag{6.69}$$

and

$$\psi^*(\mathbf{r})v_n(\mathbf{r}) = (\nabla^2 + E)\psi^*(\mathbf{r}) \tag{6.70}$$

can be used with Green's theorem to write Λ^n as

$$\Lambda^n = \int_{S_n} \mathrm{d}S\hat{n} \cdot [\nabla\psi^*(\mathbf{r}) - \psi^*(\mathbf{r})\nabla]B(\mathbf{r}) \tag{6.71}$$

$$= \int_{S_n} \mathrm{d}S\hat{n} \cdot [\nabla\psi^*(\mathbf{r}) - \psi^*(\mathbf{r})\nabla] \times \int_{S'} \mathrm{d}S'\hat{n}' \cdot [\nabla' G_0(\mathbf{r},\mathbf{r}') - G_0(\mathbf{r},\mathbf{r}')\nabla']\psi(\mathbf{r}'). \tag{6.72}$$

We now assume that the integral over S can be written as the sum over integrals over the surfaces of the cells in the assembly and we write,

$$\Lambda^n = \int_{S_n} \mathrm{d}S\,\hat{n} \cdot [\nabla\psi^*(\mathbf{r}) - \psi^*(\mathbf{r})\nabla] \times \sum_m \int_{S_m} \mathrm{d}S'\,\hat{n}' \cdot [\nabla' G_0(\mathbf{r},\mathbf{r}') - G_0(\mathbf{r},\mathbf{r}')\nabla']\psi(\mathbf{r}') \tag{6.73}$$

This expression is valid provided that the contributions to the integrals from adjacent cell boundaries cancel out. This cancellation is guaranteed if the cell boundaries do not contain singularities or discontinuities of $\psi(\mathbf{r})$. It can be expected that this condition will be easily fulfilled because the functions $\psi(\mathbf{r})$ are regular solutions of the Schrödinger equation for the cell potentials which themselves are taken to have the proper analytic properties. We can now use the representability theorem and its analog for $\psi^*(\mathbf{r})$ to write

$$\Lambda^n = \int_{S_n} \mathrm{d}S\,\hat{n} \cdot [\nabla \sum_{L_1} a_{L_1}^{n*}\psi_{L_1}^{n*}(\mathbf{r}_n) - \sum_{L_1} a_{L_1}^{n*}\psi_{L_1}^{n*}(\mathbf{r}_n)\nabla] \times \sum_m \int_{S_m} \mathrm{d}S'\,\hat{n}' \cdot [\nabla' G_0(\mathbf{r},\mathbf{r}') - G_0(\mathbf{r},\mathbf{r}')\nabla'] \sum_{L_2} a_{L_2}^{m}\psi_{L_1}^{m*}(\mathbf{r}'_m). \tag{6.74}$$

Finally, the functional $\Lambda = \sum_n \Lambda^n$ takes the form,

$$\Lambda = \sum_{nm} \sum_{L_1 L_2} a_{L_1}^{n*}\lambda_{L_1 L_2}^{nm} a_{L_2}^{m}, \tag{6.75}$$

where

$$\lambda_{L_1 L_2}^{nm} = \int_{S_n} \mathrm{d}S\,\hat{n} \cdot [\nabla\psi_{L_1}^{n*}(\mathbf{r}_n) - \psi_{L_1}^{n*}(\mathbf{r}_n)\nabla]$$

$$\times \int_{S_m} dS' \, \hat{n}' \cdot [\nabla' G_0(\mathbf{r},\mathbf{r}') - G_0(\mathbf{r},\mathbf{r}')\nabla']\psi^m_{L_1}(\mathbf{r}'_m). \quad (6.76)$$

Demanding that Λ be stationary with respect to variations in $a^n_{L_1}$ leads to the set of linear homogeneous equations,

$$\sum_{mL_2} \lambda^{nm}_{L_1L_2} a^m_{L_2} = 0. \quad (6.77)$$

A similar set involving $a^{n*}_{L_1}$ is obtained upon variation with respect to $a^m_{L_2}$.

The set of equations in the last expression possesses nontrivial solutions provided that the determinant of the coefficients vanishes. This leads to the secular equation,

$$\det\left|\lambda^{nm}_{L_1L_2}\right| = 0. \quad (6.78)$$

It is clear that $\underline{\lambda}^{nm}$ as defined in Eq.(6.76), preserves the fundamental property of MST, namely, the separation of the structure of the material, embodied in the free-particle propagator, from the potential, contained in the local (cell) solutions, $\psi^n_L(\mathbf{r})$. It is useful in computational applications of MST to make this feature more explicit by writing $\underline{\lambda}^{nm}$ in a form involving the structure constants of the lattice and the cell functions (or t-matrices).

We begin with the single-site term, $\underline{\lambda}^{nn}$. In this case, the argument $\mathbf{r}'$ in $G_0(\mathbf{r},\mathbf{r}')$ in Eq.(6.76) can be taken to lie on a sphere circumscribing cell n so that we can write,

$$G_0(\mathbf{r},\mathbf{r}') = \sum_L J_L(\mathbf{r}_n) H_L(\mathbf{r}'_n). \quad (6.79)$$

Inserting this expression into Eq.(6.76), carrying out the indicated surface integrals, and using the definitions of the generalized sine and cosine functions, we can write

$$\underline{\lambda}^{nm} = \tilde{\underline{S}}^n \underline{C}^n, \quad (6.80)$$

which involves explicitly the cell functions $\tilde{\underline{S}}^n$ and $\underline{C}^n$, and a tilde denotes the transpose of a matrix. This equation also involves an internal summation (in the product of the matrices on the right) which should be carried to convergence when solving the secular equation. The definition of convergence can be taken to mean that the change in the secular matrix with increasing values of L become smaller than some predetermined range. In any case, the sums can be carried to full convergence through the use of Eq.(6.76) involving G_0 rather than its expansion.

When the cells n and m are sufficiently far apart so that a cell vector in cell n lies outside a sphere bounding cell m, and vice versa, we can again expand the free-particle propagator as in Eq.(6.79) and obtain the expression,

$$\begin{aligned}\underline{\lambda}^{nm} &= \tilde{\underline{S}}^n \underline{C}^{nm} \\ &= \tilde{\underline{S}}^n \underline{G}(\mathbf{R}_{nm}) \underline{S}^m. \end{aligned} \quad (6.81)$$

Here, we have defined the expression,

$$C_{LL'}^{nm} = \int_{S_m} \mathrm{d}S' \, \hat{n}' \cdot [\nabla' H_L(\mathbf{r}'_n) - H_L(\mathbf{r}'_n)\nabla']\psi_{L'}^m(\mathbf{r}'_m), \tag{6.82}$$

and $\underline{G}(\mathbf{R}_{nm})$ denotes the structure constants associated with the intercell vectors, $\mathbf{R}_{nm}$. We note that $\underline{C}^{nm}$ involves an integral over the surface of cell m of the Wronskian between the cell basis functions for cell m, $\psi_L^m(\mathbf{r}_m)$, and the Hankel functions, $H_L(\mathbf{r}_n)$, expanded about the center of a neighboring cell n. The use of $\underline{C}^{nm}$ effectively completes the sum (matrix product) between $\underline{G}$ and $\underline{S}^m$ in Eq.(6.81). The remaining product, $\underline{\tilde{S}}^n \underline{C}^{nm}$, can also be converged by using Eq.(6.76) directly in terms of G_0 rather than expanding it in terms of free-particle solutions.

We now consider the case in which the cells n and m are nearest neighbors. (It is assumed that if the cells are farther apart than nearest neighbors, then each lies outside a bounding sphere bounding the other. If this is not the case and the bounding sphere of a cell intersects the potential regions of the other cell, then the cells should be treated in the manner of nearest neighbor pairs described here.) In this case, there are two distinct ways for obtaining $\underline{\lambda}^{nm}$. First, one may use Eq.(6.76) as it stands, carrying out the surface integrals in terms of G_0 without expanding in terms of free-particle solutions. This procedure leads to fully converged results, but obscures the fact that the quantity $\underline{\lambda}^{nm}$ can be written in terms of scattering matrices or cell functions connected by free-particle structure constants. To see that such an expression can be obtained even in this case, we recall the shifted-cell construction used to obtain the wave function in the moon region of a convex cell in terms of its t-matrix. Given the form of the wave function in Eq.(3.121), and realizing that if $\mathbf{r}_0$ lies in an adjacent cell it has the form $\mathbf{r}_0 = \mathbf{R}_{nm} + \mathbf{r}'$, we can write

$$\underline{\lambda}^{nm} = \underline{\tilde{S}}^n \underline{\hat{C}}^{nm}, \tag{6.83}$$

where the matrix $\underline{\hat{C}}^{nm}$ is given explicitly by the expression,

$$\begin{aligned} \hat{C}_{LL'}^{nm} = & \sum_{L_2} J_{L_2}(\mathbf{b}) \int_{S_m} \mathrm{d}S' \, \hat{n}' \cdot [\nabla' G_{L_2 L}(\mathbf{r}'_n + \mathbf{R}_{nm} + \mathbf{b}) \\ & - G_{L_2 L}(\mathbf{r}'_n + \mathbf{R}_{nm} + \mathbf{b})\nabla']\psi_{L'}^m(\mathbf{r}'_m). \end{aligned} \tag{6.84}$$

When using this expression, the sum over L_2 must *follow* any sums over L' to lead to converged results. Reversing the sums effectively eliminates the vecor $\mathbf{b}$ from the expressions, and the resulting expressions can no longer be shown to converge.

It is also possible to expand the quantity $G_{LL'}(\mathbf{r}'_n + \mathbf{R}_{nm} + \mathbf{b})$ so that $\mathbf{r}_n$ appears as an argument of a Bessel function and transform Eq.(6.83) to the form,

$$\underline{\lambda}^{nm} = \underline{\tilde{S}}^n \underline{\hat{G}}(\mathbf{R}_{nm}) \underline{S}^m, \tag{6.85}$$

where the *modified structure constants* $\hat{\underline{G}}(\mathbf{R}_{nm})$ are given by the expression,

$$\hat{G}_{L_1L_2}(\mathbf{R}_{nm}) = \sum_{L'} J_{L'}(\mathbf{b}) \left[\sum_{L_3} \mathrm{i}^{\ell'-\ell_3+\ell_1} C(L'L_2L_1) G_{L_3L_2}(\mathbf{b}+\mathbf{R}_{nm}) \right]. \tag{6.86}$$

It is important to point out again the significance of the modified structure constants in the last expression. Adding the vector **b** in the argument of $\hat{\underline{G}}$ effectively moves every point in the moon region of convex cell outside the bounding sphere. Now, the sum over L_2 with a regular solution, J_{L_2}, involved in the definition of the cell t-matrices converges absolutely. Once this sum is converged, the sumation over $J_(\mathbf{b})\mathbf{L}'$ effectively removes the vector **b**. For this reason it is necessary for the outer sum (over L') to follow that over L_2. Reversing the order of these summations results in the ordinary expansion of the irregular solutions (H_L or N_L) that may involve intercell vectors that are smaller than intracell vectors. Such expansions cannot be shown to converge.

Although the last two expressions are revealing in a formal sense, they are cumbersome for computational purposes. For practical applications, Eq.(6.84) provides a more convenient choice than Eq.(6.85). It involves essentially a linear combination of matrices $\underline{C}^{nm}$ folded over Gaunt coefficients. As before the sum over L_1 in this expression can be summed to full convergence through the use of Eq.(6.76) without expanding G_0.

The previous discussion shows that the secular equation resulting from a variational derivation of MST for space-filling cells can be written in the general form,

$$\det |\tilde{\underline{S}}^n \hat{\underline{C}}^{nm}| = 0, \tag{6.87}$$

where $\hat{\underline{C}}^{nm} = \underline{C}^{nm}$ for n and m farther apart than nearest neighbors. In the discussion that follows, we will often neglect the distinction between hatted and plain quantities, but the proper quantities must be used as required. Note also that it is only in a variational derivation where products of the form $\underline{\tilde{S}GS}$ come into play that it is necessary to use modified structure constants. Other, nonvariational forms of the secular matrix will not in general require the use of such structure constants unless intracell vectors exceed in length the distance between adjacent cells.

If the shape of the cell is such that the intercell vectors that connect the expansion centers of each cell are larger than all of the intracell vectors between a cell center and its boundary, the generalized cosine matrix, $C_{LL'}^{nn'}$, can be be expanded using the addition theorem for the irregular solid harmonics, allowing the generalized cosine matrix to be written, for $n \neq n'$, as

$$C_{LL'}^{nn'} = -\sum_{L''} G_{LL''}(\mathbf{R}_{nn'}) S_{L''L}^{n'}. \tag{6.88}$$

For periodic materials, the Fourier transform of $G_{LL'}(\mathbf{R}_{nn'})$ yields the well-known structure constants of the KKR method. With the use of Eq.(6.88), the secular equation that determines the allowed energies can be written in the form,

$$\det \left| \underline{\tilde{S}}^{n} \underline{C}^{n} - \underline{\tilde{S}}^{n} \{ \underline{G}^{nm} \underline{S}^{m} \} \right| = 0. \tag{6.89}$$

Here, underlines denote matrices in L-space, and the braces indicate that for nearby cells, for example, nearest neighbors, the product of the (possibly modified) structure constants and one of the sine-function matrices must be carried to convergence before the other sine-function matrix is multiplied by the resulting product. At the same time, it should be mentioned that for many systems of practical interest, for example, monatomic solids, and for some model systems, for example, the empty lattice in two dimensions, the modified structure constants can be replaced with the ordinary structure constants of the KKR formalism, provided that the proper order of summations is maintained.

The last expression shows that the secular equation for MST for space filling, even nonconvex, but nonoverlapping cells has the same form as the secular equation for MT potentials. In particular the separation of structure and potential, that is, the structure constants and the sine and cosine matrices, that characterizes MST in the MT case remains a feature of the non-MT case subject to the constraint mentioned above, namely, that modified structure constants must in general be used and that sums over angular-momentum indices must be carried out in a particular order.

It is important to realize moreover that the rate of convergence of the expansion in the last equation depends on L as well as on the ratio $r_{n'}/R_{nn'}$ with larger values of L requiring that the summation over L' be truncated at higher values. For muffin-tin potentials of equal radii, the ratio $r_{n'}/R_{nn'}$ never exceeds 0.5 and the convergence of this sum is seldom a critical consideration, but as this ratio approaches unity great care must be exercised to maintain convergence. The usual practice of treating all matrices as square and truncating them at a common cut-off ($\ell_{\max}$) can easily lead to a sequence of solutions that as a function of $\ell_{\max}$ appears to converge for low values of $\ell_{\max}$ but eventually diverges.

Formally, the secular equation can be written in a number of equivalent forms which, however, exhibit different convergence characteristics. As is shown in the next section it is possible, for example, to derive rigorously a version of the MST equations that omits the $\tilde{S}$, that is,

$$\sum_{n'L'} C^{nn'} a_{L'}^{n'} = 0, \tag{6.90}$$

but the energies for which this equation is satisfied will not, in general, be variational with respect to the wave function as are those obtained from Eq.(6.89). The result that some forms of the KKR secular equation are variational, while others are not, is a new feature that distinguishes non–

muffin-tin MST from the muffin-tin limit and is due to the conditional convergences associated with non–muffin-tin MST and the consequent necessity of converging internal angular-momentum sums. The practical consequence of this is that it is not correct to view Eq.(6.89) or Eq.(6.83) as consisting of the products of *square* matrices $\tilde{S}^n$ and $C^{nn'}$.

It is instructive to point out some important differences between the secular Eq.(6.14), which is not variational, and its variational counterpart, Eq.(6.87). In the nonvariational form, it is possible to replace the modified cosine functions by the ordinary ones which leads to a secular equation identical to that of the MT case but with the replacement of L-diagonal sine and cosine functions with their nondiagonal, full-cell counterparts. However, even though all MT expressions for the secular matrix can be derived within a variational formalism and lead to a variational secular equation, in the full cell case only those forms of the matrix derived within a variational principle are variational.

6.6 Second Variational Derivation (*)

In this section we present a variational derivation of the secular equation of MST along somewhat different lines from the one given in the previous section. This derivation confronts directly the problems arising with respect to expansions of the free-particle propagator in the case of space-filling cells, and suggests an additional formal method by which these problems can be resolved. Thus, it highlights the geometric nature of the obstacles encountered in applying MST to space-filling potentials. For the sake of clarity of presentation, we first consider the case of nonspherical muffin-tin potentials, that is, nonspherical potentials that can be separated by nonoverlapping spheres.

6.6.1 The Secular Equation for Nonspherical MT Potentials

We begin with the solutions, $\psi_L^n(\mathbf{r}_n)$, associated with the potential in cell Ω_n in an assembly of scatterers. Again, based on the representability theorem, we attempt to construct a solution of the Schrödinger equation for a collection of potentials in terms of a linear combination of ψ_L^n's, the solutions for individual cells,

$$\psi(\mathbf{r}) = \sum_L \psi_L^n(\mathbf{r}_n) A_L^n, \tag{6.91}$$

for $\mathbf{r}$ in cell Ω_n. Following essentially the steps that lead to Eq.(6.72) we write,

$$\Lambda = \int \mathrm{d}S \int \mathrm{d}S' \left[\nabla\psi^*(\mathbf{r}) - \psi*(\mathbf{r})\nabla\right]$$

$$\times \left[\psi(\mathbf{r}'_n)\nabla' G_0(\mathbf{r}-\mathbf{r}'_n) - G_0(\mathbf{r}-\mathbf{r}'_n)\nabla'\psi(\mathbf{r}'_n)\right], \qquad (6.92)$$

where the prime on ∇' indicates that the gradient is to be applied to functions of $\mathbf{r}'$, and the surfaces of integration are those used in Eq.(6.72). From now on we suppress the conditions on r and r' in the surface integrals. In order to make further progress, we recall that if the vectors $\mathbf{r}$ and $\mathbf{r}'$ are confined inside nonoverlapping spheres around cells Ω_n and Ω_m, we can express the free-particle propagator in the "factored" form,

$$\begin{aligned} G_0(-\mathbf{r}_n + \mathbf{R}_{nm} + \mathbf{r}_m) = \quad & -ik\sum_L J_L(\mathbf{r}_n) H_L(\mathbf{r}_m)\delta_{nm} \\ & + \sum_{LL'} J_L(\mathbf{r}_n) G_{LL'}(\mathbf{R}_{nm}) \\ & \times J_{L''}(\mathbf{r}_m)\,(1-\delta_{nm}), \end{aligned} \qquad (6.93)$$

where we have set $r_m > r_n$, and in all expansions of the form $\langle J|H\rangle$ the argument of the Hankel function is the larger of r and r'. The expression in Eq.(6.93) can now be inserted into Eq.(6.92), and the integrals in that equation can be broken up over the surfaces of spheres bounding individual cells to lead to the expression,

$$\begin{aligned} \Lambda = \quad & \sum_n\sum_m \int_{S_n}\int_{S_m} \{\langle A^n\,|\nabla\psi^{n*}\rangle\,[\langle\nabla J\,|H\rangle\delta_{nm} \\ & + \langle J\,|G_{nm}|\,\nabla J\rangle\,(1-\delta_{nm})]\,\langle\psi^m\,|A^m\rangle \\ & - \langle A^n\,|\nabla\psi^{n*}\rangle\,[\langle J\,|H\rangle\delta_{nm} \\ & + \langle J\,|G_{nm}|\,J\rangle\,(1-\delta_{nm})]\,\langle\nabla\psi^m\,|\,A^m\rangle \\ & - \langle A^n\,|\psi^{n*}\rangle\,[\langle\nabla J\,|\nabla H\rangle\delta_{nm} \\ & + \langle\nabla J\,|G_{nm}|\,\nabla J\rangle\,(1-\delta_{nm})]\,\langle\psi^m\,|A^m\rangle \\ & + \langle A^n\,|\psi^{n*}\rangle\,[\langle\nabla J\,|H\rangle\delta_{nm} \\ & + \langle\nabla J\,|G_{nm}|\,J\rangle\,(1-\delta_{nm})]\,\langle\nabla\psi^m\,|A^m\rangle\}, \end{aligned} \qquad (6.94)$$

where $\underline{G}_{nm} \equiv \underline{G}(\mathbf{R}_{nm})$. In this expression, we have suppressed integration variables and have made use of the bra and ket notation to denote various linear combinations and summations over L. In fact, we can condense the notation even further through the introduction of the matrices $\underline{\mathbf{G}}$ with elements $[\underline{\mathbf{G}}]^{nm}_{LL'} = G_{LL'}(\mathbf{R}_{nm})$, and the "vectors" $\langle A\,|\,\psi\rangle$ with elements, $[\langle A\,|\,\psi\rangle]_n = \langle A^n\,|\,\psi^n\rangle$, and obtain the expression,

$$\begin{aligned} \Lambda = \quad & \int\!\!\int \{\,\langle A|\,\nabla\psi^*\rangle\,[\langle\nabla J\,|\,H\rangle + \langle J\,|\underline{\mathbf{G}}|\,\nabla J\rangle]\,\langle\psi\,|\,A\rangle \\ & - \langle A|\,\nabla\psi^*\rangle\,[\langle J\,|\,H\rangle + \langle J\,|\underline{\mathbf{G}}|\,J\rangle]\,\langle\nabla\psi\,|\,A\rangle \\ & - \langle A|\,\psi^*\rangle\,[\langle\nabla J\,|\,\nabla H\rangle + \langle\nabla J\,|\underline{\mathbf{G}}|\,\nabla J\rangle]\,\langle\psi\,|\,A\rangle \\ & + \langle A|\,\psi^*\rangle\,[\langle\nabla J\,|\,H\rangle + \langle\nabla J\,|\underline{\mathbf{G}}|\,J\rangle]\,\langle\nabla\psi\,|\,A\rangle\,\} \\ = & \int\!\!\int\langle A|\,\{\,[|\,\nabla\psi^*\rangle\langle J\,| - |\,\psi^*\rangle\langle\nabla J\,|] \end{aligned}$$

$$\times \, [\, \underline{\mathbf{G}} \, (| \, \nabla J \rangle \langle \psi \, | - | \, J \rangle \langle \nabla \psi \, |)$$
$$- \, (| \, \nabla H \rangle \langle \psi \, | - | \, H \rangle \langle \nabla \psi \, |) \,] \, \} \, | \, A \rangle. \tag{6.95}$$

Upon using the definitions of the phase functions in terms of the usual surface integrals, we can now write,

$$\Lambda = \langle A \, | \, \underline{\tilde{\mathbf{S}}\mathbf{G}\mathbf{S}} - \underline{\tilde{\mathbf{S}}\mathbf{C}} \, | \, A \rangle, \tag{6.96}$$

where underlined boldface symbols denote matrices in both angular-momentum and site space. Note that $\underline{\mathbf{S}}$ and $\underline{\mathbf{C}}$ are cell-diagonal matrices. Note also that the last expression for Λ is made possible by the MT geometry which allows G_0 to be written in the form of Eq.(6.93), and consequently allows the independent integration of the cell variables $\mathbf{r}_n$ and $\mathbf{r}'_m$. It is precisely at this stage that difficulties may arise in attempts to apply the present formalism to space-filling cells. Two different ways of handling the non-MT geometry are discussed in the following sections.

Taking the variation of Λ with respect to $\langle A^n |$ equal to zero leads to the set of linear homogeneous equations,

$$\left[\underline{\tilde{\mathbf{S}}\mathbf{G}\mathbf{S}} - \underline{\tilde{\mathbf{S}}\mathbf{C}} \right] | \, A \rangle = 0, \tag{6.97}$$

which may have nontrivial solutions only when the determinant of the coefficients vanishes,

$$\det \left| \underline{\tilde{\mathbf{S}}\mathbf{G}\mathbf{S}} - \underline{\tilde{\mathbf{S}}\mathbf{C}} \right| = 0, \tag{6.98}$$

which is identical to that derived in the previous section. Provided that we can treat the $\underline{\mathbf{S}}$ matrices as being invertible we can also write

$$\det \left[\underline{\mathbf{G}\mathbf{S}} - \underline{\mathbf{C}} \right] = 0, \tag{6.99}$$

or

$$\det \left[\underline{\mathbf{G}} - \underline{\mathbf{C}\mathbf{S}}^{-1} \right] = 0. \tag{6.100}$$

In the last expression, we can identify the quantity $\underline{\mathbf{C}\mathbf{S}}^{-1}$ with the inverse of the cell t-matrix, provided the proper choice of normalization of the cell wave function has been made. Then we can write the secular equation in the familiar form,

$$\det \left[\underline{\mathbf{M}} \right] = 0, \tag{6.101}$$

where

$$\underline{M}^{ij} = \underline{m}^{i} \delta_{ij} - \underline{G}(\mathbf{R}_{ij})(1 - \delta_{ij}). \tag{6.102}$$

Not unexpectedly, this has the well-known form of the KKR secular equation for MT potentials.

6.6.2 Space-Filling Cells of Convex Shape

We now extend the variational formalism of the previous section to space-filling cells. In order to illustrate as clearly as possible the underlying concepts, we confine our discussion to convex-shaped cells.

We recall that an important feature of the derivation of Eq.(6.102) was that the free-particle propagator couples one pair of cells at a time. Thus, it is convenient to begin by treating explicitly the case of two adjacent cells of convex, polyhedral shape. The incorporation of the resulting argument into the discussion of an arbitrary number of cells can be effected in a straightforward way. In the following discussion, it is formally useful to consider the cells as being separated by a thin strip of zero potential and infinitesimal width, η. The width of the strip will be allowed to go to zero at the end of formal arguments.

Now, for $\mathbf{r}_n$ and $\mathbf{r}_m$ cell vectors in cells Ω_n and Ω_m, respectively, an expansion of $G_0(-\mathbf{r}_n + \mathbf{R}_{nm} + \mathbf{r}_m)$ in the factored form, Eq.(6.93), can be obtained. We take the intercell vector to be perpendicular to the common face of the cells, and imagine the centers of the cells as being displaced along $\mathbf{R}_{nm}$ and away from this face. As long as η is finite, albeit infinitesimal, there exist vectors, $\mathbf{a}_n$ and $\mathbf{a}_m$, such that spheres centered at the displaced centers, $\mathbf{R}_n + \mathbf{a}_n$ and $\mathbf{R}_m + \mathbf{a}_m$, and covering the cells, Ω_n and Ω_m, respectively, do not overlap. (See Fig 6.2 for an illustration.)

Therefore, in terms of vectors measured with respect to the shifted centers, we recover the MT case. With the definitions,

$$\bar{\mathbf{r}}_n = \mathbf{r}_n - \mathbf{a}_n, \quad \bar{\mathbf{r}}'_n = \mathbf{r}'_n - \mathbf{a}_n, \quad \bar{\mathbf{R}}_{nm} = \mathbf{R}_{nm} - \mathbf{a}_n + \mathbf{a}_m, \tag{6.103}$$

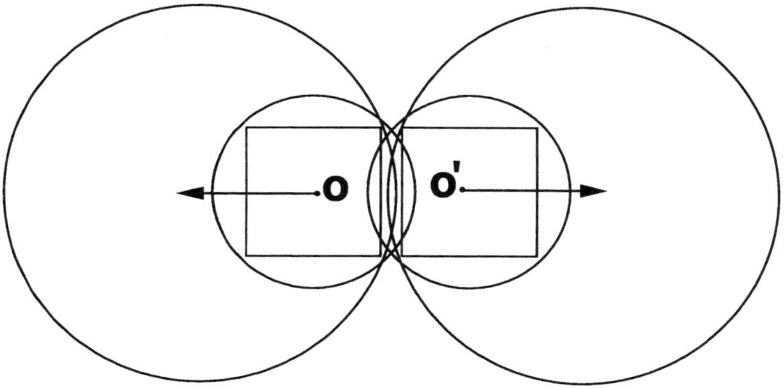

FIGURE 6.2. Two adjacent nonspherical cells separated by a narrow strip of zero potential. The spheres bounding the cells overlap the adjacent potential when centered at positions O and O', but not when the centers are displaced in the manner indicated.

we have the inequalities,

$$\begin{aligned} \mid \bar{\mathbf{r}}_m + \bar{\mathbf{R}}_{nm} \mid > \bar{r}_n, \quad \mid \bar{\mathbf{R}}_{nm} - \bar{\mathbf{r}}_m \mid > \bar{r}_m \\ \bar{R}_{nm} > \bar{r}_n, \quad \bar{R}_{nm} > \bar{r}_m, \end{aligned} \tag{6.104}$$

and we can write,

$$\begin{aligned} G_0\,(\,-\,\mathbf{r}_n + \mathbf{R}_{nm} + \mathbf{r}'_m\,) &= G_0\,(\,-\,\bar{\mathbf{r}}_n + \bar{\mathbf{R}}_{nm} + \bar{\mathbf{r}}'_m\,) \\ &= \sum_{LL'} J_L(\bar{\mathbf{r}}_n) G_{LL'}(\bar{\mathbf{R}}_{nm}) J_{L'}(\bar{\mathbf{r}}_m) \\ &= \langle J(\bar{\mathbf{r}}_n) \mid G(\bar{\mathbf{R}}_{nm}) \mid J(\bar{\mathbf{r}}_m)\rangle. \end{aligned} \tag{6.105}$$

Furthermore, the functions $\langle J(\bar{\mathbf{r}}_n) \mid$ can be expanded (see Appendix D) in the form,[3]

$$\begin{aligned} \langle J(\bar{\mathbf{r}}_n) \mid &= \langle J(\mathbf{r}_n - \mathbf{a}_n) \mid \\ &= \langle J(\mathbf{r}_n) \mid \underline{g}(\mathbf{a}_n), \end{aligned} \tag{6.106}$$

for all values of $\mathbf{r}_n$ and $\mathbf{a}_n$. A similar expression holds for $\mid J(\bar{\mathbf{r}}'_m)\rangle$. We now substitute these expansions into Eq.(6.105), and obtain the expression,

$$\begin{aligned} &G_0\left(-\bar{\mathbf{r}}_n + \bar{\mathbf{R}}_{nm} + \bar{\mathbf{r}}'_m\right) = \\ &\quad \langle J(\mathbf{r}_n) \left|\underline{g}(-\mathbf{a}_n) \,||\, \underline{G}(\mathbf{R}_{nm} + \mathbf{a}_n + \mathbf{a}_m) \,||\, \underline{g}(\mathbf{a}_m)\right| J(\mathbf{r}'_m)\rangle. \end{aligned} \tag{6.107}$$

Here, the double bars indicate the order of operations: They are a reminder that the multiplications of the $\underline{g}$'s by the structure constant matrices, $\underline{G}$, must *follow* those between the $\underline{g}$'s and the J's. Otherwise, one obtains formally the expansion in Eq.(6.93) in terms of the vectors defined with respect to undisplaced centers which, in this case, may diverge for certain values of the cell vectors. With the proper order of the sums retained, Eq.(6.107) can be substituted into Eq.(6.92), and the integrals over the cell vectors $\mathbf{r}_n$ and $\mathbf{r}'_m$ can be performed in straightforward fashion, leading to the expression,

$$\Lambda = \langle A \mid \underline{\tilde{\mathbf{S}}\mathbf{g}} \,||\, \underline{\mathbf{G}} \,||\, \underline{\mathbf{g}\mathbf{S}} - \underline{\tilde{\mathbf{S}}\mathbf{C}} \mid A\rangle. \tag{6.108}$$

Note, that because the integrals over *both* cell vectors were confined to the surfaces of the cells, this equation reflects the direct, point-by-point matching of the wave function across cell boundaries. Once again, using the variational condition, Eq.(6.106), with respect to A_L^n, we obtain the secular equation,

$$\det\left[\underline{\tilde{\mathbf{S}}\mathbf{g}} \,||\, \underline{\mathbf{G}} \,||\, \underline{\mathbf{g}\mathbf{S}} - \underline{\tilde{\mathbf{S}}\mathbf{C}}\right] = 0. \tag{6.109}$$

---

[3]For the sake of simplifying the presentation, we do not distinguish between the $\underline{g}$ and $\underline{g}^\dagger$, or between $\underline{g}$'s that multiply a structure constant matrix on the left or on the right. The proper forms of $\underline{g}$ to be used in each case can be inferred from the material in Appendix D.

We point out that this equation is meaningful under the condition that the order of operations is indicated by the double bars. However, we know from our discussion in Section 6.3 that the products denoted by the double bars can be performed sequentially as long as either one is carried to convergence before the other. This leads directly to the secular equation in the form of Eq.(6.89). Moreover, provided that the matrix $\underline{\mathbf{S}}$ and its transpose are invertible, their determinants can be factored out and the last equation takes the form,

$$\det\left[\underline{\mathbf{g}} \,||\, \underline{\mathbf{G}} \,||\, \underline{\mathbf{g}} - \underline{\mathbf{C}}\underline{\mathbf{S}}^{-1}\right] = 0. \tag{6.110}$$

But at this point the products of the $\underline{g}'s$ with the structure constants *can* be carried out as the presence of the $\underline{\mathbf{S}}$ matrices which imposed the restriction on the summations has been removed. Using the property, $\underline{g}(-\mathbf{a}_n)\underline{G}(-\mathbf{a}_n + \mathbf{R}_{nm} + \mathbf{a}_m)\underline{g}(\mathbf{a}_m) = \underline{G}(\mathbf{R}_{nm})$, we obtain the secular equation,

$$\det\left[\underline{\mathbf{G}} - \underline{\mathbf{C}}\underline{\mathbf{S}}^{-1}\right] = 0, \tag{6.111}$$

which is precisely Eq.(6.100), and also (6.102). In the case in which the intracell vectors are smaller than any intercell vector, Eq.(6.110) can also be converted to Eq.(6.99).

At this point, we can generalize these considerations to the case of an arbitrary number of cells. Following the lines of the discussion in the previous subsection, we treat each pair of scatterers in the manner indicated above and arrive at Eq.(6.99) for an assembly of scattering potential cells.

It is important to realize that the passage from Eq.(6.109) to Eq.(6.110) does not involve an invalid exchange of summation indices. Note that for the proper order of operations, the (LL') element of the matrix in Eq.(6.109),

$$\sum_{L_1 L_2}\sum_{L_3 L_4}\left\{S^{\mathrm{T}}_{LL_1} g_{L_1 L_2}\right\} G_{L_2 L_3}\left\{g_{L_3 L_4} S_{L_4 L'}\right\} - \sum_{L_1} S^{\mathrm{T}}_{LL_1} C_{L_1 L'}, \tag{6.112}$$

is well defined for *all* values of L and L'. Therefore, it is allowed to multiply this expression on the left by ${S^{\mathrm{T}}}^{-1}_{L_0 L_1}$ and sum over L_1. In fact, multiplication by *any* matrix is allowed provided that the product can be shown independently to converge. The multiplication by ${S^{\mathrm{T}}}^{-1}_{L_0 L_1}$ and the subsequent summation has the effect of replacing the quantity $S^{\mathrm{T}}_{LL_1}$ by $\delta_{L_0 L_1}$. The matrix $\underline{\mathbf{S}}$ can also be removed in a similar manner. As the resulting expressions are manifestly convergent, the sums can be performed to yield Eq.(6.112). The procedure just described, when satisfied in detail, can be used to justify multiplication by rectangular matrices and the conversion of one form of the secular equation to another. However, in contrast the case of square matrices, multiplications by rectangular matrices in general alter the convergence and variational properties of the equations. In particular, even though Eq.(6.109) yields energy eigenvalues that are variational with respect to the wave function, Eq.(6.111) does not have this property.

It cannot be stressed too strongly that preserving the order of conditionally converging sums is crucial in the derivation of Eq.(6.111). This feature has lasting effects that are present in Eq.(6.111), even though this equation no longer involves double summations. For example, Eq.(6.111) cannot be converted to Eq.(6.96), in spite of the formal analogies between the two. In the case of space-filling cells, Eq.(6.96) diverges. Nor can the matrix in Eq.(6.111) be expanded in terms of the cell t-matrices in a "power series" like manner, as that expansion may also diverge.[4] What we have shown amounts to the statement that the *summed* form of the inverse of the total t-matrix is the matrix in Eq.(6.111). Thus, we have succeeded in proving the formal existence of the matrix $\underline{\mathbf{M}}$, and to show that it takes the MT form even in the case of space- filling potential cells. The numerical aspects of the various forms of the secular equation will be discussed in Section 7.7.

The remainder of this section contains derivations of MST that emphasize the role of the cell shape and thus may illuminate further the importance of geometry in determining the validity of MST. At the same time, we have seen before that such "geometric" proofs can be rather involved mathematically. The reader not particularly interested in such mathematical detail may skip to the next section.

6.6.3 Displaced-Cell Approach: Convex Cells (*)

In this section, we present a variational derivation of the secular equation of MST using the displaced-cell formalism of Chapter 3. The advantage of this approach is that it can be applied to the case of concave cells, as is indicated in the next subsection. We examine in some detail the case of convex cells.

We start again with Eq.(6.72) and the observation that we need consider only two cells at a time. In view of Eq.(3.118), the free-particle propagator, $G_0(-\mathbf{r}_n+\mathbf{R}_{nm}+\mathbf{r}_m)$, connecting two points in adjacent cells, can be written in the form,

$$
\begin{aligned}
&G_0(-\mathbf{r}_n + \mathbf{R}_{nm} + \mathbf{r}'_m) \\
&\quad = \sum_{L_1} J_{L_1}(\mathbf{b}_{nm})\Big[\sum_{L_2} G_{L_1L_2}(-\mathbf{r}_n + \mathbf{R}_{nm} + \mathbf{b}_{nm})J_{L_2}(\mathbf{r}'_m)\Big] \\
&\quad = \sum_{L_1} J_{L_1}(\mathbf{b}_{nm})\left[\sum_{L_2L_3} g_{L_1L_3}(-\mathbf{r}_n)G_{L_3L_2}(\mathbf{R}_{nm} + \mathbf{b}_{nm})J_{L_2}(\mathbf{r}'_m)\right] \\
&\quad = \sum_{L_1} J_{L_1}(\mathbf{b}_{nm})
\end{aligned}
$$

---

[4]Divergences can be avoided in the case in which intercell vectors are larger than intracell vector, provided that each sum over L is carried fully to convergence before a subsequent sum is evaluated. As is discussed in Section 7.8, even though the product $\underline{t}^i\underline{G}^{ij}\underline{t}^j$ may diverge if the internal sums in it are carried out in arbitrary fashion, the conditionally evaluated product, $[\underline{t}^i\underline{G}^{ij}]\underline{t}^j$ converges.

$$\times \left[\sum_{L_2 L_3 L_4} \mathrm{i}^{\ell_1 - \ell_3 + \ell_4} C\left(L_1 L_3 L_4\right) J_{L_4}(\mathbf{r}_n) G_{L_3 L_2}(\mathbf{R}_{nm} + \mathbf{b}_{nm}) J_{L_2}(\mathbf{r}'_m) \right]. \tag{6.113}$$

This expression can be substituted into Eq.(6.72) and the integrals over the cell vectors can be performed, leading to the expression,

$$\begin{aligned}\Lambda^{nm} = &\sum_{LL'} A_L^n \Bigg\{ \sum_{L_1} J_{L_1}(\mathbf{b}_{nm}) \\ &\times \left[\sum_{L_2 L_3 L_4} \mathrm{i}^{\ell_1 - \ell_3 + \ell_4} C(L_1 L_3 L_4) S_{LL_4}^{n\mathrm{T}} G_{L_3 L_2}(\mathbf{R}_{nm} + \mathbf{b}_{nm}) S_{L_2 L'}^{m} (1 - \delta_{nm}) \right] \\ &- \delta_{nm} \sum_{L_1} S_{LL_1}^{n\mathrm{T}} C_{L_1 L'}^{n} \Bigg\} A_{L'}^{m}\end{aligned} \tag{6.114}$$

Now, the variational condition,

$$\frac{\delta \Lambda^{nm}}{\delta A_L^n} = 0, \tag{6.115}$$

leads to

$$\begin{aligned}&\sum_{L'} \Bigg\{ \sum_{L_1} J_{L_1}(\mathbf{b}_{nm}) \\ &\times \left[\sum_{L_2 L_3 L_4} \mathrm{i}^{\ell_1 - \ell_3 + \ell_4} C(L_1 L_3 L_4) S_{LL_4}^{n\mathrm{T}} G_{L_3 L_2}(\mathbf{R}_{nm} + \mathbf{b}_{nm}) S_{L_2 L'}^{m} (1 - \delta_{nm}) \right] \\ &- \delta_{nm} \sum_{L_1} S_{LL_1}^{n\mathrm{T}} C_{L_1 L'}^{n} \Bigg\} A_{L'}^{m} = 0.\end{aligned} \tag{6.116}$$

This last expression yields the secular equation

$$\det \underline{\mathbf{M}} = 0, \tag{6.117}$$

where,

$$\begin{aligned}M_{LL'}^{nm} = &\sum_{L_1} J_{L_1}(\mathbf{b}_{nm}) \\ &\times \left[\sum_{L_2 L_3 L_4} \mathrm{i}^{\ell_1 - \ell_3 + \ell_4} C(L_1 L_3 L_4) S_{LL_4}^{n\mathrm{T}} G_{L_3 L_2}(\mathbf{R}_{nm} + \mathbf{b}_{nm}) S_{L_2 L'}^{m} (1 - \delta_{nm}) \right] \\ &- \delta_{nm} \sum_{L_1} S_{LL_1}^{n\mathrm{T}} C_{L_1 L'}^{n}.\end{aligned} \tag{6.118}$$

As was the case with Eq.(6.109), we note that the last expression is valid for all values of the *outside* indices, LL'. Once again, multiplication by the matrix elements ${S^{n\mathrm{T}}}^{-1}_{L_0L}$ on the left, and summation over L yields the expression,

$$M^{nm}_{L_0L'} = \sum_{L_1} J_{L_1}(\mathbf{b}_{nm}) \left[\sum_{L_2} \sum_{L_3} \sum_{L4} \mathrm{i}^{\ell_1-\ell_3+\ell_4} C\left(L_1L_3L_4\right) \delta_{L_0L_4} \right.$$
$$\times\, G_{L_3L_2}(\mathbf{R}_{nm}+\mathbf{b}_{nm}) S^{m}_{L_2L'}\left(1-\delta_{nm}\right) - \delta_{nm} C^{n}_{L_0L'}. \quad (6.119)$$

Finally, the matrix elements of $\underline{\mathbf{S}}$ can be removed by a multiplication on the right, the sums over L_2 and L_1 can be performed, leading to $g_{L_0L_3}$, and we obtain the result,

$$\begin{aligned} M^{nm}_{L_0L_5} &= \sum_{L_3} g_{L_0L_3}\left(\mathbf{b}_{nm}\right) G_{L_3L_5}\left(\mathbf{R}_{nm}+\mathbf{b}_{nm}\right)\left(1-\delta_{nm}\right) \\ &\quad - \sum_{L1} C^{n}_{L_0L_1} S^{-1}_{L_1L_5} \delta_{nm} \\ &= G_{L_0L_5}\left(\,\mathbf{R}_{nm}\,\right)\left(1-\delta_{nm}\right) - m^{n}_{L_0L_5}\delta_{nm}. \end{aligned} \quad (6.120)$$

This is clearly a secular equation of the muffin-tin form. It should be pointed out once again that all summations have been performed in a manner that ensures convergence. Thus, the secular equation was derived in a way that explicitly matches the wave function across cell boundaries in the case of arbitrarily shaped, convex cells.

6.6.4 Displaced-Cell Approach: Concave Cells(*)

The application of the displaced-cell approach to concave cells is straightforward, although somewhat tedious. For the reasons discussed in Chapter 3 with respect to the single scatterer, it is now necessary to replace the single vector $\mathbf{b}_{nm}$ by *two* series of vectors, each allowing the analytic continuation of the wave function from outside the sphere bounding each of a pair of adjacent cells to points inside their respective moon regions. As the intermediate expressions leading to the secular equation, Eq.(6.120), in this case are rather complicated, and as the reader has probably acquired the flavor of the process, we forego displaying these expressions here.

6.7 Construction of the Wave Function

Our discussion up to this point has established that the secular equation describing the scattering of an electron (wave) by an assembly of space-filling cells can be written in forms that are used to describe scattering in the case of muffin-tin potentials. We have also seen that in order to obtain converged results, it may be necessary to use modified structure constants in place of their familiar KKR counterparts. This will guarantee convergence in obtaining the band structure of a periodic system. Therefore, as far as the secular equation is concerned, the muffin-tin form of MST applies essentially intact.

However, there remains the question of constructing the wave function (or the Green function) of the scattering assembly as a linear combination of the products of local (cell) solutions and the eigenvectors of the secular matrix. It should be kept in mind that in such a construction one should use the coefficients obtained from a fully converged secular matrix, one which may require the use of modified structure constants. To be more explicit, we consider a secular equation written in terms of the modified structure constants,

$$\tilde{G}_{L'L_1}(\mathbf{R}) = \sum_{L} g_{L'L}(\mathbf{b}) G_{LL_1}(\mathbf{R}+\mathbf{b}), \tag{6.121}$$

with L, the summation index being carried to values "smaller than L' or L_1." In this form, the eigenvectors of the secular equation are L dependent, and the corresponding wave function acquires a similar dependence. In this construction, converged values of the wave function are obtained as L increases, always following sufficiently behind the cut-off values of L' and L_1. Also, the use of such structure constants leads to convergent expressions for $\underline{\tau}$ in terms of cell t-matrices of the form,

$$\underline{\tau} = t + t[\tilde{G}t] + \cdots. \tag{6.122}$$

An example of the use of this form of the structure constants is given in Eq.(6.109). In addition to the derivation of that equation, we can also consider the same expression in terms of structure constants of the form given in the last expression, in which a displacement vector, $\mathbf{b}$, appears. In this case, the sum over L must follow any sums or expansions in terms of the outside indices, L' and L_1.

This point is particularly worthy of note because the wave function (and the Green function) have been found in numerical work to have considerably more sensitive convergence properties than the secular equation. Thus, it is quite possible to arrive at an apparently convergent secular matrix, written in terms of the canonical structure constants of KKR, only to find that the wave function exhibits divergent behavior. This is an indication that indeed the secular matrix *has not* converged except in possibly asymptotic form. These difficulties can be remedied by reconstructing the secular matrix in terms of modified structure constants keeping the various sums over angular-momentum states in the proper order. Although this procedure can be fairly difficult computationally in the general case, requiring fairly large cut-off values of L in various expansions, for example, for a two-dimensional empty lattice, for three-dimensional systems it can be expected to be only moderately more demanding than the canonical approach to computing the secular equation.

6.8 The Closure of Internal Sums (*)

The implementation of MST with respect to space-filling cell potentials hinges crucially on the convergence of various internal summations. Such sums can either be carried to convergence directly, that is, through usually laborious matrix multiplication and the use of a possibly large cutoff for the corresponding internal index or, at least formally, can be evaluated in exact fashion in terms of surface or volume integrals. An example of this procedure is afforded by the expression for $\underline{C}^{nn'}$ in Eq.(3.81). In fact, it was only when the surface integrals in that equation were used that the calculations reported below and in Appendix E with respect to the two-dimensional empty lattice were found to converge.

It is possible to derive a formally exact expression for the closure of internal sums in the secular equation of MST in terms of volume integrals. To see this, we use the representability theorem to write the variational functional defined in Eq.(3.128) in the form

$$\sum_{nm} \langle A^n | \lambda^{nm} | A^m \rangle = 0 \tag{6.123}$$

where

$$\begin{aligned} \lambda^{nm} &\equiv \underline{S}^{n\dagger} \underline{C}^{nm} \\ &= \int_{\Omega_n} \mathrm{d}r^3 |\psi^n(\mathbf{r})\rangle v_n(\mathbf{r}) \{ \langle \psi^n(\mathbf{r})| \\ &- \int_{\Omega_m} \mathrm{d}r'^3 G_0(\mathbf{r},\mathbf{r}') v_m(\mathbf{r}') \langle \psi^{n'}(\mathbf{r}')| \}, \end{aligned} \tag{6.124}$$

which is just the functional of Kohn and Rostoker. This expression provides at least a formal way of completing the sums in the secular equation of MST. However, the practical implementation of the volume integrations in the last expression does pose computational problems of its own, particularly in the treatment of the poles of G_0. Unlike the case with Eq.(3.81) where the effects of the poles were bypassed through the use of surface integrals, these effects must be addressed directly in the performance of volume integrations. Therefore, whether the evaluation of volume integrals is in fact more expedient than the direct use of matrix products is a question that should be addressed numerically on a case-by-case basis.

6.9 Numerical Results

In order to demonstrate the validity of Eq.(6.89) and the variational nature of the energy for its solutions, we have calculated the energy and the wave function for certain states of a two-dimensional square lattice. In this test the individual potentials $v_n(\mathbf{r})$ were taken as a constant V_0 within cell n

$\ell_{\max}$	E_{exact}	E_{calc}	MSWFE	$\frac{E_{\text{calc}}-E_{\text{exact}}}{MSWFE}$	E_{var}
0	-5.0	-5.9291847	6.91×10^{-3}	-3.58	-4.906345
4	-5.0	-4.9999550	7.97×10^{-6}	5.65	-4.999792
4	-1.0	-0.994071	1.15×10^{-3}	5.15	-0.978936
8	-1.0	-0.99999989	1.55×10^{-8}	7.02	-0.9999989

TABLE 6.1. Calculated energies and wave function errors for the second and third $k = 0$ states of a square lattice. The depth of the potential in this example is taken as $V_0 = -9$, and the side of the square is π. The energy is measured in units of l^{-2} where l is the unit of length. The maximum angular momentum used in the calculation is denoted by $\ell_{\max}$, while E_{exact} and E_{calc} denote the exact and calculated energies and MSWFE denotes the mean square error in the wave function. E_{var} denotes the energy calculated by using the calculated MST wave function in the Rayleigh–Ritz variational expression for the energy. These calculations employed the variational version of MST.

and zero outside. The cells were squares arranged so as to completely fill the plane. Thus, although the total crystal potential, V_0, was a constant which allowed us to calculate trivially the exact wave functions[5] and eigenenergies of the Schrödinger equation, MST was faced with the formidable task of representing these functions using the free-space Green function. Details of the two-dimensional "empty lattice test" such as the two-dimensional versions of $H_L(\mathbf{r})$, $J_L(\mathbf{r})$, and $G_{LL'}(\mathbf{R})$ can be found in papers by Butler and Nesbet [39] and Faulkner [40].

Results for the empty lattice test are shown in Table 6.1. The column denoted by $\ell_{\max}$ shows the maximum value of the orbital angular momentum used in the expansion of the wave function. We emphasize, however, that it is necessary to converge all internal sums if one is to obtain meaningful results. Thus, internal sums were not truncated at $\ell_{\max}$, but carried to full convergence by means of calculating the quantities $C_{LL'}^{nn'}$ through direct integration over the surfaces of adjacent cells, for the first two nearest-neighbor shells of a given site. Because the state being investigated is a state with full square symmetry, only values of orbital angular momentum evenly divisible by four enter the calculation. The column denoted by E_{calc} contains the energies for which the secular Eq.(6.89) is satisfied. We also show the ratio of the error in the energy ($E_{\text{calc}} - E_{\text{exact}}$) to mean square error in the wave function (MSWFE) defined as

$$\int dr^3 |\psi_{\text{calc}}(\mathbf{r}) - \psi_{\text{exact}}(\mathbf{r})|^2 / \int dr^3 |\psi_{\text{exact}}|^2, \tag{6.125}$$

where the calculated value of the wave function, ψ_{calc}, was obtained from Eq.(3.66) using the coefficients A_L^n obtained form Eq.(6.89). Finally, we

[5] These are the Bessel functions of argument κr with $\kappa = \sqrt{E + V_0}$.

ℓ_{max}	E_{exact}	E_{calc}	MSWFE	$\frac{E_{calc}-E_{exact}}{MSWFE}$	E_{var}
0	–5.0	–4.0386580	1.88×10^{-1}	5.12	–4.2449788
4	–5.0	–5.0125155	3.39×10^{-5}	–369.22	–4.9996807
4	–1.0	–0.8783834	1.87×10^{-3}	65.10	–0.9704981
8	–1.0	–0.9992260	1.33×10^{-7}	5811.48	–0.9999960

TABLE 6.2. Calculated energies and wave function errors for the second and third $k = 0$ states of a square lattice. The parameters are the same as for Table 6.1. These calculations employed a nonvariational version of MST, as discussed in the previous chapter.

show E_{var}, a variationally defined value of the energy calculated by using the Rayleigh–Ritz variation principle with the calculated wave function,

$$E_{var} = \frac{\int dr^3 \psi^*_{calc}(\mathbf{r}) H \psi_{calc}(\mathbf{r})}{\int dr^3 |\psi_{calc}(\mathbf{r})|^2}. \tag{6.126}$$

The important points to notice are that the error in the energy is of the same order as the mean square error of the wave function, even though both of these vary over eight orders of magnitude for the different values of ℓ_{max}. Furthermore, it is seen that the variational procedure does not improve the energy over the value obtained from the secular equation itself. These results may be contrasted with similar calculations that used the version of the MST equations without the $\tilde{S}$, Eq.(6.90), which are shown in Table 6.2. We recall that that is the form of the secular equation arrived at within the non-variational approach of the previous chapter. The energy for this version improves with increasing ℓ_{max} but not as fast as the mean square error of the wave function. For this version, however, the Rayleigh–Ritz refinement does greatly improve the energy. At the same time, we see that the wave functions are of comparable accuracy in the two versions of MST.

6.10 Square Versus Rectangular Matrices (*)

The need to converge internal sums when MST is used in conjunction with nonspherical cell potentials results in the use of rectangular matrices to represent such quantities as the single-scatterer $\underline{S}$ and $\underline{C}$ matrices. This is in contrast to the square matrices, indeed diagonal matrices, that represented these quantities in the case of spherically symmetric potentials. This in turn raises the issue of the existence of the inverses of such matrices whose use must be justified in each case, as was done in our discussion of non-MT MST. However, while the convergence of internal sums is crucial in the application of MST to nonspherical potentials, it is by no means exclusively a characteristic of MST applied to space-filling cells. The issue of convergence is important in a computational sense, but its presence does not change the fundamental character of MST.

Let us devote a little space in amplifying the remark made above. To begin with, internal sums must be converged in *any* application that uses a specific representation in terms of basis states. Internal sums would have to be converged in connection with nonspherical potentials even if these potentials were separated by nonoverlapping spheres. The important point to keep in mind is that the convergence rate of an expression *does* depend on the particular form employed, even if this form is *formally* equivalent to other expressions. For example, the rate at which the eigenvalues of the secular equation converge depends on the particular form employed, as do the properties of these solutions, for example, their variational nature. Thus, the secular equation expressed in terms of phase functions involves rectangular matrices while the secular equation expressed in terms of the inverse of the scattering matrix, as in the MT form, contains square matrices only. These two forms are formally equivalent, although they may exhibit very different numerical behavior.

References

[1] H. Krakauer, M. Posternak, and A. J. Freeman, Phys. Rev. B**19**, 1706 (1979).

[2] M. S. Methfessel, *Multipole Green Functions for Electronic Structure Calculations*, Thesis, Katholike Universiteit te Nijmegen (1986), unpublished.

[3] J. S. Faulkner, Phys. Rev. B**19**, 6186 (1979).

[4] P. Ziesche and G. Lehmann, *Ergebnisse in der Elektronentheorie der Metalle* (Akademie-Verlag, Berlin, 1983), p. 151.

[5] P. Ziesche, J. Phys. C**7**, 1085 (1974).

[6] A. R. Williams and J. van W. Morgan J. Phys. C**7**, 37 (1974).

[7] R. G. Brown and M. Ciftan, Phys. Rev. B**27**, 4564 (1983).

[8] W. John, G. Lehmann, and P. Ziesche, Phys. Statys. Solidi b**53**, 287 (1972).

[9] L. G. Ferreira, A. Agostinbo, and D. Lida, Phys. Rev. B**14**, 354 (1976).

[10] L. Scheire, Physica A**81**, 613 (1975).

[11] R. G. Brown and M. Ciftan, Phys. Rev. B**32**, 3454 (1985).

[12] E. Badralexe and A. J. Freeman, Phys. Rev. B**36**, 1378 (1987); **36**, 1389 (1987); **36**, 1401 (1987); **38** 10469 (1988).

[13] A. Gonis, Phys. Rev. B**33**, 5914 (1986).

[14] R. Zeller, J. Phys. C**4**, 3155 (1971).

[15] J. Molenaar, J. Phys. C**21**, 1455 (1988).

[16] R. K. Nesbet, Phys. Rev. B**30**, 4230, (1984); **33**, 8027 (1986).

[17] R. Zeller, Phys. Rev. B**38**, 5993 (1988).

[18] A. Gonis, X.-G. Zhang, and D. M. Nicholson, Phys. Rev. B**38**, 3564 (1988).

[19] J. S. Faulkner, Phys. Rev. B**32**, 1339 (1985); **38**, 1686 (1988).

[20] W. H. Butler, R. G. Brown, and R. K. Nesbet, Mat. Res. Soc. Symp. Proc. Vol. **193**, 27 (1190).

[21] W. H. Butler, A. Gonis, and X.-G. Zhang, Phys. Rev. B**48**, 2118 (1993).

[22] J. Korringa, Physica **13**, 392 (1947)
[23] W. Kohn and N. Rostoker, Phys. Rev. **94**, 1111 (1954).
[24] J. S. Faulkner, Phys. Rev. B**19**, 6186 (1979).
[25] W. John, G. Lehmann, and P. Ziesche, Phys. Stat. Sol. b**53**, 287 (1972).
[26] P. Ziesche and G. Lehmann, *Ergebnisse in der Elektronentheorie der Metalle* (Akademic-Verlag, Berlin, 1983), p. 151.
[27] P. Ziesche, J. Phys. C**7**, 1085 (1974).
[28] A. Gonis, Phys. Rev. B**34**, 5914 (1986).
[29] R. Zeller, J. Phys. C**20**, 2347 (1987).
[30] R. Zeller, Phys. Rev. B**30**, 4230 (1984).
[31] J. Molenaar, J. Phys. C**21**, 1455 (1988).
[32] R. K. Nesbet, Phys. Rev. B**30**, 4230 (1984).
[33] R. K. Nesbet, Phys. Rev. B**33**, 8027 (1986).
[34] A. Gonis, X.-G. Zhang, and D. M. Nicholson, Phys. Rev. B**38**, 3564 (1988).
[35] A. Gonis, X.-G. Zhang, and D. M. Nicholson, Phys. Rev. B**40**, 947 (1988).
[36] W. Kohn and N. Rostoker, Phys. Rev. **94**, 1111 (1954).
[37] J. Korringa, Physica **13**, 392 (1947).
[38] B. L. Moiseiwitsch, *Variational Principles* (Interscience Publishers, London, 1966).
[39] W. H. Butler and R. K. Nesbet, Phys. Rev. B**42**, 1518 (1990).
[40] J. S. Faulkner, Phys. Rev. B**32**, 1339 (1985); **38**, 1686 (1988).

7

Augmented MST(*)

7.1 General Comments

This chapter is devoted to the study of the wave functions and Green functions of MST, with emphasis on the effects of the truncation of angular-momentum expansions. We will see that these effects can be largely alleviated through augmentation procedures that satisfy the proper normalization requirements on the wave function of an assembly of scattering cells. At the same time, although this chapter contains material of practical importance, it also involves concepts that are not essential for a basic understanding of MST and could be skipped in a first reading.

One of the great advantages afforded by MST is the use of a very efficient basis set. This efficiency is quite remarkable for a first-principles method. Acceptable band energies can be obtained with only four states per atom ($\ell = 0, 1$) for simple metals and semiconductors and only nine states ($\ell = 0, 1, 2$) for transition metals. The reason for this very fast convergence is somewhat subtle, however, and it is quite easy to obtain inconsistent results if the truncation of the basis set is not performed carefully.

One type of error that arises when the MST basis set is truncated is associated with the normalization of the wave functions or the density of states. The density of states is usually calculated from the Green function using the relation $\rho(\mathbf{r}, E) = -\frac{1}{\pi}\mathrm{Im}[G(\mathbf{r}, \mathbf{r}; E)]$, Eq.(5.43). For discrete states or for states of a given wave vector in a periodic system, $\int \mathrm{d}\mathbf{r}\rho(\mathbf{r}, E)$ should be unity. However, if the expansion for the wave function in a given basis set is truncated, the normalization to unity is only approximate.

A very nice feature of MST and KKR theory is the existence of a formula due to Lloyd [1] which allows the integrated density of states to be calculated directly at any energy through the relation $N(E) = -\frac{1}{\pi}\text{Im}[\ln \det M(E)]$ where $M(E)$ is the KKR secular matrix. However, the density of states as usually calculated from the truncated Green function is not consistent with that obtained from the Lloyd formula. The reason for this is that the Lloyd formula increments the integrated density of states by exactly unity, every time the determinant of the secular matrix goes through zero *regardless of the truncation in angular momentum.* However, the density of states, as usually calculated from the Green function, is not normalized, so that $\int |\psi(\mathbf{r})|^2 d^3r$ is not exactly unity unless an inconveniently large number of angular momenta are included in the expansion of the wave function. The electron density may be reasonably accurate near the center of the cell, but it will be incorrect near the cell boundary because it is there that the neglect of higher angular-momentum contributions is most significant.

The loss of electrons associated with high ℓ's can significantly affect the results of total energy and electronic structure calculations. In order to maintain charge neutrality, one normally adjusts the Fermi energy such that more electrons are put in at higher energies but with lower ℓ's. This effectively squeezes electrons into a smaller space near the nuclei. Therefore one often gets larger lattice constants with ℓ truncations at lower values. Calculations of quantities such as the bulk modulus that are sensitive to interstitial charge densities often have significant errors due to the ℓ truncations. Alternatively, to avoid this problem, one may require use of an inconveniently large secular matrix.

The truncation also introduces errors into calculations of quantities which involve dipole matrix elements, e.g. those involving the gradient of the potential or the gradient of the wave function. Matrix elements of this type appear in the electron-phonon interaction, in transport calculations, in calculations of the interaction of electrons with electromagnetic fields and photons, and in calculations of interatomic forces. In these calculations there will be coupling between states with angular momentum ℓ_{max} and $\ell_{\text{max}} + 1$ which generally cannot be neglected if the phase shift for the ℓ_{max} state is appreciable even if that for the $\ell_{\text{max}} + 1$ state is negligible.

The most straightforward remedy for the errors due to ℓ truncation is to use larger values of ℓ_{max}. However, this may increase the computation time by an order of magnitude. Other techniques, such as the matrix partition method [2], have also been used.

In this chapter we examine an alternative form for the MST wave function, which we will call the *augmented wave function.* We show that this augmented wave function, in contrast to the usual truncated one, is smooth and continuous everywhere in space, and normalizes properly. Consequently, the corresponding Green function yields a charge density that agrees with the Lloyd formula. Furthermore, using the augmented wave function in the calculation of dipole terms automatically includes the coupling between the coefficients of the $\ell_{\rm max}$ and $\ell_{\rm max}+1$ states. The continuity properties of the wave function are important in application of MST to calculations of the total energy using density functional theory.

7.2 MST with a Truncated Basis Set: MT Potentials

Applications of multiple scattering theory have traditionally been restricted [3, 4] to potentials of the muffin-tin form, that is, potentials that are spherically symmetric and bounded by nonoverlapping spheres. We recall our discussion of Section 5.7 where we derived the MST secular equation and the form of the wave function for an arbitrary number of muffin-tin potentials. The secular equation was derived through the condition of continuity between the single-center expansion and the multicenter expansion of the wave function in the vicinity of the nth MT. The underlying assumption in that derivation was that all expansions in angular momentum states were carried to convergence.

Now, when the angular momentum sums are truncated at $\ell_{\rm max}$, as must be done in practical applications of MST, one obtains the truncated wave functions,

$$\psi_t^n(\mathbf{r}) = \sum_L^{\ell_{\rm max}} a_L^n \Big\{ J_L(\mathbf{r}_n) + t_\ell^n H_L(\mathbf{r}_n) \Big\}, \tag{7.1}$$

$$\psi_t^o(\mathbf{r}) = \sum_{n'L'}^{\ell_{\rm max}} b_{L'}^{n'} H_{L'}(\mathbf{r}_{n'}), \tag{7.2}$$

and the secular equation

$$a_L^n = \sum_{n' \neq n, L'}^{\ell_{\rm max}} G_{LL'}(\mathbf{R}_{nn'}) t_{\ell'}^{n'} a_{L'}^{n'}. \tag{7.3}$$

The convergence of the energy and wave functions calculated this way for muffin-tin potentials has been studied previously [5], and some of the results were presented in Section 5.9.3. As was shown there, the discontinuity between ψ^o and ψ^n at the muffin-tin radius vanishes in the limit $\ell_{\rm max} \to \infty$.

The discontinuity in the wave function for a given value of $\ell_{\rm max}$ is

$$\Delta\psi_t = \psi_t^n - \psi_t^o = \sum_L^{\ell_{\rm max}} a_L^n J_L(\mathbf{r}_n) - \sum_{n' \neq n, L'}^{\ell_{\rm max}} a_{L'}^{n'} t_{L'}^{n'} H_{L'}(\mathbf{r}_{n'}). \tag{7.4}$$

Expanding the irregular solid harmonic we have

$$\Delta\psi_t = \sum_L^{\ell_{\max}} a_L^n J_L(\mathbf{r}_n) - \sum_{n'\neq n} \sum_{L'}^{\ell_{\max}} \sum_L^{\infty} a_{L'}^{n'} t_{L'}^{n'} J_L(\mathbf{r}_n) G_{LL'}(\mathbf{R}_{nn'}), \tag{7.5}$$

and using the secular equation, Eq.(7.3), we have

$$\Delta\psi_t = \sum_{L>\ell_{\max}} a_L^n J_L(\mathbf{r}_n). \tag{7.6}$$

The meaning of this result is that the discontinuity of the wave function (and of its derivative) at the muffin-tin radius can be eliminated by the use within the muffin-tins of the augmented wave function defined by

$$\psi_a^n(\mathbf{r}_n) = \psi_t^n(\mathbf{r}_n) + \sum_{L>\ell_{\max}} a_L^n J_L(\mathbf{r}_n). \tag{7.7}$$

The coefficients a_L^n for $\ell \geq \ell_{\max}$ are obtained from the secular equation truncated at $\ell_{\max}$, Eq.(7.3), so that very little additional effort is needed to generate the continuous wave function.

The augmented wave function has the additional advantage that it is normalized correctly. The physical reason for this is that the augmented wave function is the *exact* solution to the Schrödinger equation but for a modified Hamiltonian, one that leads to vanishing scattering amplitudes for partial waves that exceed $\ell_{\max}$. Thus, the secular equation is truncated at $\ell_{\max} < \ell_0$ by the approximation that the potential is such that its scattering amplitudes for $\ell > \ell_{\max}$ vanish. However, rather than truncating the wave function at the same number of partial waves as the secular equation, enough partial waves should be retained so that the truncation error of the wave function, which is of order $(\sqrt{E} r_{MT})^{\ell_0}/(2\ell_0 + 1)!!$ (where ℓ_0 is the cut-off), becomes negligible.

7.3 General Potentials

In this section we give a brief derivation of a nonvariational form of MST that is similar to that of Section 6.3. In this section, however, we will use outgoing-wave boundary conditions for the expansion of the Helmholtz Green function so that the formulas will look slightly different. As in Section 6.3 we base our derivation on the Lippmann–Schwinger equation,

$$\psi(\mathbf{r}) = \int \mathrm{d}^3 r' G_0(\mathbf{r}, \mathbf{r}') V(\mathbf{r}') \psi(\mathbf{r}'), \tag{7.8}$$

which by the use of Green's theorem can be converted into a sum of integrals over the surfaces of each of the cells of a scattering assembly,

$$\sum_n \int_{S_n} \mathrm{d}S\hat{n} \cdot [G_0(\mathbf{r}, \mathbf{r}')\nabla' - \nabla' G_0(\mathbf{r}, \mathbf{r}')]\psi(\mathbf{r}') = 0. \tag{7.9}$$

We expand the wave function within each cell in the same form as Eq.(5.73),

$$\psi^n(\mathbf{r}) = \sum_L a_L^n \psi_L^n(\mathbf{r}), \tag{7.10}$$

in terms of basis functions that satisfy the integral equation

$$\psi_L^n(\mathbf{r}) = J_L(\mathbf{r}) + \int \mathrm{d}^3r' G_0(\mathbf{r},\mathbf{r}')v_n(\mathbf{r}'), \psi_L^n(\mathbf{r}') \tag{7.11}$$

and are therefore locally exact solutions to the Schrödinger equation within cell n. Use of this expansion in Eq.(7.9) yields the equation,

$$\sum_{nL} \int_{S_n} \mathrm{d}S\hat{n} \cdot [G_0(\mathbf{r},\mathbf{r}')\nabla' - \nabla' G_0(\mathbf{r},\mathbf{r}')]\psi_L^n a_L^n = 0. \tag{7.12}$$

If $\mathbf{r}$ is confined inside a sphere inscribed within cell m in the above equation, the Green function may be expanded using regular and irregular solutions to the Helmholtz equation, Eq.(D.17), so that we obtain,

$$\sum_{L'Ln} J_{L'}(\mathbf{r}_m) \int_{S_n} \mathrm{d}S\hat{n} \cdot [H_{L'}(\mathbf{r}'_m)\nabla' - \nabla' H_{L'}(\mathbf{r}'_m)]\psi_L^n(\mathbf{r}'_n)a_L^n = 0. \tag{7.13}$$

Because $J_{L'}$ is not identically zero, we obtain the secular equation of MST for general potentials,[1]

$$\begin{aligned}\sum_{Ln} \int_{S_n} \mathrm{d}S\hat{n} \cdot [H_{L'}(\mathbf{r}'_m)\nabla' - \nabla' H_{L'}(\mathbf{r}'_m)]\psi_L^n(\mathbf{r}'_n)a_L^n \\ \equiv \sum_{nL} [H_{L'}^m, \psi_L^n]_{S_n} a_L^n = 0,\end{aligned} \tag{7.14}$$

where we use brackets to represent "Wronskian-like" surface integrals. The equivalence between this form of the secular equation and more familiar forms was established in Section 6.3.

In order to put the secular equation in a form that is more tractable for computations, we introduce a different set of basis functions very similar to those introduced in Section 3.4. These basis functions are closely related to basis functions suggested by Williams and Morgan [6] and have the advantage that they can be calculated by integrating outward from the origin as was done in Section 3.4. These basis functions satisfy the integral equation,

$$\phi_L^n(\mathbf{r}) = \sum_{L'} J_{L'}(\mathbf{r})E_{L'L}^n + \int \mathrm{d}^3r' G_0(\mathbf{r},\mathbf{r}')v_n(\mathbf{r}')\phi_L^n(\mathbf{r}'), \tag{7.15}$$

---

[1] For the sake of clarity of presentation, we will not indicate explicitly the possible need to use modified structure constants or other techniques for assuring convergence. However, previous discussions of this point should be kept in mind in the following presentation.

where $E_{L'L}$ is defined below. These basis functions may be obtained by solving a coupled set of integral equations,

$$\phi_L^n = \sum_{L'} \Big\{ J_{L'}(\mathbf{r}_n) e_{L'L}^n(r_n) + iH_{L'}(\mathbf{r}_n) s_{L'L}^n(r_n) \Big\},$$

$$e_{L'L}^n(r_n) = *\delta_{L'L} + ik \int_0^r \mathrm{d}r'_n H_{L'}(\mathbf{r}'_n) v_n(\mathbf{r}'_n) \phi_L^n(\mathbf{r}'_n),$$

$$s_{L'L}^n(r_n) = -k \int_0^r \mathrm{d}r'_n J_{L'}(\mathbf{r}'_n) v_n(\mathbf{r}'_n) \phi_L^n(\mathbf{r}'_n). \qquad (7.16)$$

Here $v_n(\mathbf{r}_n)$ is the potential within cell n. As usual, the "asymptotic" values of the phase functions, $e_{L'L}^n(r_n)$ and $s_{L'L}^n(r_n)$, reached for values of r_n that are greater than the radius of the sphere circumscribing cell n are represented by $E_{L'L}^n$ and $S_{L'L}^n$, respectively.

After $\phi_L^n(\mathbf{r})$ and the matrix $E_{L'L}^n$ have been obtained, the basis functions, $\psi_L^n(\mathbf{r})$, and the t-matrices, $t_{LL'}^n$, can be obtained through the relations,

$$\phi_L^n(\mathbf{r}) = \sum_{L'} \psi_{L'}^n(\mathbf{r}) E_{L'L}^n,$$

$$S_{LL'}^n = \sum_{L''} t_{LL''}^n E_{L''L'}^n. \qquad (7.17)$$

This last equation is a generalization of the familiar relation, $\sin\eta_\ell = t_\ell e^{-i\eta_\ell}$.

The expressions for the generalized exponential and sine matrices can be converted by use of Green's theorem into surface integrals,

$$S_{L'L}^n = \int_{S_n} \mathrm{d}S\hat{n} \cdot [J_{L'}(\mathbf{r}_n)\nabla - \nabla J_{L'}(\mathbf{r}_n)]\phi_L^n(\mathbf{r}_n)$$
$$\equiv [J_{L'}^n, \phi_L^n]_{S_n}, \qquad (7.18)$$

and

$$E_{L'L}^n = -ik \int_{S_n} \mathrm{d}S\hat{n} \cdot [H_{L'}(\mathbf{r}_n)\nabla - \nabla H_{L'}(\mathbf{r}_n)]\phi_L^n(\mathbf{r}_n)$$
$$\equiv -ik[H_{L'}^n, \phi_L^n]_{S_n}, \qquad (7.19)$$

where the surface integrals are over the surface of cell n. These results imply that

$$[J_{L'}^n, \psi_L^n(\mathbf{r}_n)]_{S_n} = t_{L'L}^n, \qquad (7.20)$$

and

$$-[H_{L'}^n, \psi_L^n(\mathbf{r}_n)]_{S_n} = \delta_{L'L}, \qquad (7.21)$$

which can be used to obtain a form of the secular equation similar to Eq.(7.3). Expanding the irregular solid harmonic in Eq.(7.14), using the addition theorem, Eq.(D.10), and using the above relations we have

$$a_L^n = \sum_{L'L'',n'\neq n} G_{LL'}^{nn'} t_{L'L''}^{n'} a_{L''}^{n'}, \tag{7.22}$$

which may be compared with Eq.(7.3).

Now, let us consider the effect of truncating the secular equation and the wave functions in ℓ as is necessary in calculations based on MST. The wave function as "normally" truncated is given by

$$\psi_t^n(\mathbf{r}) = \sum_L^{\ell_{\max}} a_L^n \psi_L^n(\mathbf{r}), \tag{7.23}$$

and the truncated secular equation is given by the expression,

$$a_L^n = \sum_{L''}^{\ell_{\max}} \sum_{L'}^{\ell_i} \sum_{n'\neq n} G_{LL'}^{nn'} t_{L'L''}^{n'} a_{L''}^{n'}. \tag{7.24}$$

The internal sum on L' *may* be truncated at $\ell_{\max}$ but to ensure convergence when n and n' are near neighbors it is best to retain the option of increasing the internal index so that $\ell_i > \ell_{\max}$. The wave functions and characteristic energies will not, in general, converge to the correct value as $\ell_{\max}$ increases if this internal sum is not converged. Use of ℓ_i large enough to converge the sum is equivalent to using the secular Eq.(7.14). Alternatively, and as is shown below, convergence in L *and* L' can be reached by replacing the product <u>Gt</u> by a surface integral.

For any ℓ truncation, ψ_t^n is an exact solution of the Schrödinger equation within cell n by construction since we assume (for the present argument) that the basis functions, ψ_L^n, can be obtained with negligible error. There is, however, a discontinuity at the boundary separating two cells because the wave functions and their derivatives will not match exactly except in the limit in which $\ell_{\max} \to \infty$. Consider a point $\mathbf{r}$ on the boundary between cells m and n and let $\mathbf{r}_m$ and $\mathbf{r}_n$ be the vectors to this point from the origins of cells m and n respectively. The discontinuity at $\mathbf{r}$ is given by

$$\Delta\psi_t(\mathbf{r}) = \sum_L^{\ell_{\max}} \left\{ a_L^m \psi_L^m(\mathbf{r}_m) - a_L^n \psi_L^n(\mathbf{r}_n) \right\}. \tag{7.25}$$

We can use Eq.(7.11) to write this discontinuity as

$$\Delta\psi_t(\mathbf{r}) = \sum_L^{\ell_{\max}} \Big\{ a_L^m J_L^m(\mathbf{r}_m) - a_L^n J_L^n(\mathbf{r}_n) + a_L^m [G_0, \psi_L^m]_{S_m}$$
$$-a_L^n [G_0, \psi_L^n]_{S_n} \Big\}. \tag{7.26}$$

In order to express $\int G_0 v_n \psi \mathrm{d}^3 r$ as a surface integral in the above equation it is necessary to assume that $\mathbf{r}_m$ and $\mathbf{r}_n$ are not exactly on the bounding

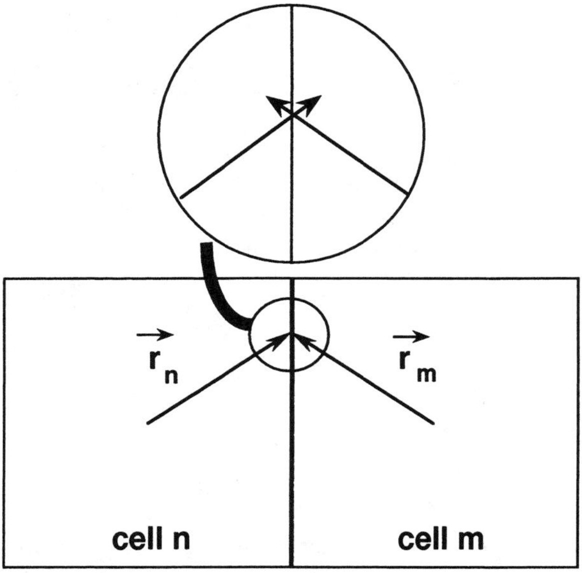

FIGURE 7.1. The vectors $\mathbf{r}_n$ and $\mathbf{r}_m$, where $\mathbf{r}_n$ is a vector from the center of cell n to the surface between cells n and m except that it extends slightly beyond the surface into cell m.

surface but are infinitesimally displaced outside their cells so that $\mathbf{r}_m$ is in cell n and $\mathbf{r}_n$ is in cell m, as is indicated in Fig.7.1.

Provided that $R_{nm} > r$, and that the sums are performed completely one at a time as is indicated by the brackets, $[G_0, \psi_L^n]_{S_n}$ can be written as[2]

$$[G_0, \psi_L^n] = \sum_{L'}^{\infty} J_{L'}^m(\mathbf{r}_m)[H_{L'}^m, \psi_L^n]_{S_n}, \tag{7.27}$$

where the Green function is expanded about the center of cell m. Similarly the surface integral over cell m, $[G_0, \psi_L^m]_{S_m}$, can be written as

$$\sum_{L'}^{\infty} J_{L'}^n(\mathbf{r}_n)[H_{L'}^n, \psi_L^m]_{S_m}. \tag{7.28}$$

If we use these in the expression for $\Delta\psi_t$, (Eq.7.26), and if we add and subtract the sum

$$\sum_{L}^{\ell_{\max}} \sum_{p \neq n,m} [G_0, \psi_L^p]_{S_p} a_L^p \tag{7.29}$$

---

[2]Again, it should be kept in mind that double expansions, in terms of modified structure constants, can be used to assure convergence.

in that equation we obtain

$$\begin{aligned}\Delta\psi_t(\mathbf{r}) = &\sum_L^{\ell_{\max}} \left\{ a_L^m J_L^m(\mathbf{r}_m) - a_L^n J_L^n(\mathbf{r}_n) \right\} \\ &+ \sum_L^{\ell_{\max}} \sum_{L'}^{\infty} \Big\{ \sum_{p \neq n} J_{L'}^n(\mathbf{r}_n)[H_{L'}^n, \psi_L^p]_{S_p} a_L^p \\ &- \sum_{p \neq m} J_{L'}^m(\mathbf{r}_m)[H_{L'}^m, \psi_L^p]_{S_p} a_L^p \Big\} \end{aligned} \tag{7.30}$$

Use of the secular equation (Eq.7.14) yields

$$\Delta\psi_t(\mathbf{r}) = \sum_{L>\ell_{\max}} \left\{ a_L^n J_L^n(\mathbf{r}_n) - a_L^m J_L^m(\mathbf{r}_m) \right\}. \tag{7.31}$$

This result is consistent with the result for muffin-tin potentials obtained in the previous section. In order to obtain a continuous wave function one needs to augment the wave functions within each cell by adding regular solid harmonics for $\ell > \ell_{\max}$ with coefficients that are determined by the secular equation truncated at $\ell_{\max}$. Looking more closely, we can see that Eq.(7.24) corresponds to a system whose t-matrix has zero elements for ℓ's above the truncated values. In fact, the unphysical situation caused by using t-matrices with finite dimensions can be removed by using t-matrices with infinite dimensions but with zero elements for ℓ's above $\ell_{\max}$. This interpretation of ℓ truncation leads to the same truncated secular equation, which is what we desire, but also gives an augmented wave function,

$$\psi_a(\mathbf{r}) = \psi_t(\mathbf{r}) + \sum_{\ell>\ell_{\max}} a_L^n J_L^n(\mathbf{r}), \tag{7.32}$$

where the coefficients a_L^n for $\ell > \ell_{\max}$ are defined as

$$a_L^n = \sum_{L''}^{\ell_{\max}} \sum_{L'}^{\ell_i} \sum_{n' \neq n} G_{LL'}^{nn'} t_{L'L''}^{n'} a_{L''}^{n'}. \tag{7.33}$$

and the second term in ψ_a cancels exactly the term from cell n contained in $\Delta\psi_t$.

7.4 Green Functions and the Lloyd Formula

7.4.1 Green Functions

One of the important features of MST is its facility for the direct calculation of the Green function using only quantities calculated at a single energy. The Green function can be expressed in terms of the t-matrices and the basis functions in the general form,

$$G(\mathbf{r},\mathbf{r}') = \sum_{L,L'} \psi_L^n(\mathbf{r}) D_{LL'}^{nn'} \psi_{L'}^{n'}(\mathbf{r}') + \sum_L F_L^n(\mathbf{r}) \psi_L^n(\mathbf{r}') \delta_{nn'}, \tag{7.34}$$

where $\mathbf{r}$ is in cell n, $\mathbf{r}'$ is in cell n', $r > r'$, and F_L^n is an irregular solution to the Schrödinger equation obtained by integrating the Lippmann–Schwinger equation inward from a sphere that circumscribes the cell starting with H_L^n. The second term in this equation is simply the Green function for a single potential.

In general, the matrix elements $D_{LL'}^{nn'}$ can be found as follows. We first use Eq.(7.34) to obtain the equation,

$$\sum_n [H_{L''}^k(\mathbf{r}), G(\mathbf{r},\mathbf{r}')]_{S_n} = \sum_{L'} \Big\{ \sum_{Ln} [H_{L''}^k(\mathbf{r}), \psi_L^n(\mathbf{r})]_{S_n} D_{LL'}^{nn'} + [H_{L''}^k(\mathbf{r}), F_{L'}^{n'}(\mathbf{r})]_{S_{n'}} \Big\} \psi_L^{n'}(\mathbf{r}'), \tag{7.35}$$

where $H_L^k(\mathbf{r}) = H_L(\mathbf{r} - \mathbf{R}_k)$ is the solid harmonic centered at the center of cell k. The Wronskian in the single-scatterer term involves only the irregular solution because one can enlarge the surface of the integral to enclose the circumscribing sphere of the cell without altering the result of the integral. The left hand side of the equation can be reduced to a single integral over a surface that encloses the *entire* potential. When $\mathbf{r}$ is on that surface the Green function can be written in the single-scatterer form,

$$G(\mathbf{r},\mathbf{r}') = \sum_L H_L(\mathbf{r}) \Phi_L(\mathbf{r}'),$$

where $\Phi_L(\mathbf{r}')$ is a wave function regular at the origin which matches to $\sum_{L'} T_{LL'} H_{L'}(\mathbf{r}') + J_L(\mathbf{r}')$ at the boundary of the system, with $T_{LL'}$ being the elements of the t-matrix of the entire potential. Because the center of $H_L^k(\mathbf{r})$ is always enclosed by the boundary of the entire potential, the Wronskian $\sum_n [H_{L''}^k(\mathbf{r}), G(\mathbf{r},\mathbf{r}')]_{S_n} = [H_{L''}^k(\mathbf{r}), G(\mathbf{r},\mathbf{r}')]_S$, where S denotes the surface over the entire potential, must vanish.

The Wronskians $[H_{L''}^k(\mathbf{r}), F^{n'}(\mathbf{r})]_{S_{n'}}$ and $[H_{L''}^k(\mathbf{r}), \psi_L^n(\mathbf{r})]_{S_n}$ can be expanded using Eqs.(7.20) and (7.21) and the expansion of the solid harmonics to obtain the expression,

$$[H_{L''}^k(\mathbf{r}), F^{n'}(\mathbf{r})]_{S_{n'}} = \sum_L G_{L''L}^{kn'} (1 - \delta_{kn'})$$

$$\sum_n [H_{L''}^k(\mathbf{r}), \psi_L^n(\mathbf{r})]_{S_n} = -\delta_{L''L} + \sum_{L'''} \sum_{n \neq k} G_{L''L'''}^{kn} t_{L'''L}^n.$$

These expressions substituted into Eq.(7.35) lead to the equation,

$$\sum_L \Big\{ G_{L''L}^{kn'} (1 - \delta_{kn'}) - D_{L''L}^{kn'} + \sum_{L'''L'} \sum_{n \neq k} G_{L''L'''}^{kn} t_{L'''L'}^n D_{L'L}^{nn'} \Big\} \psi_L^{n'}(\mathbf{r}') = 0. \tag{7.36}$$

Because $\psi_L^{n'}(\mathbf{r}')$ is generally nonzero, we obtain the equation for the matrix elements $D_{L'L}^{nn'}$,

$$D_{L'L}^{nn'} = G_{L'L}^{nn'}(1-\delta_{nn'}) + \sum_{L''L''',n''\neq n} G_{L'L''}^{nn''} t_{L''L'''}^{n''} D_{L'''L}^{n''n'}. \tag{7.37}$$

Equations (7.34) and (7.37) allow the calculation of the Green function directly without resort to a possibly poorly convergent spectral representation.

7.4.2 The Lloyd Formula

Another very important feature of multiple scattering theory is the existence of formulae that relate the integrated density of states to integrals over the Green function [1]. These formulae have the general form,

$$\frac{d}{dE}\left\{\mathrm{Tr}\ln M(E)\right\} = i\int \mathrm{d}^3r\, G(\mathbf{r},\mathbf{r}) + \text{other terms}, \tag{7.38}$$

where $M(E)$ is a secular matrix and the "other terms" arise because the MST secular equation generally has analytic structure in addition to that associated with the true roots. In fact, there is some arbitrariness in the MST secular matrix. If, for example, $\det M(E) = 0$ determines the roots then those, same roots may be found from $\det(AM) = 0$, where A is another matrix whose vanishing eigenvalues do not coincide with those of $\underline{M}$. Each version of the secular matrix has its own expression for the Green function and an associated Lloyd formula.

7.4.3 Single Scatterer

When a potential[3] is introduced into otherwise free space, the distribution of states in energy changes from that corresponding to the absence of the potential.[4] Lloyd [1] showed that this change is given by the expression

$$\Delta N(E) = \frac{1}{\pi}\mathrm{Im\ Tr\ }\ln t, \tag{7.39}$$

---

[3]For simplicity of presentation, we consider spatially bounded potentials.

[4]Strictly speaking, this change appears to be a contradiction in terms, because each free state evolves into a scattered state at the same energy under the influence of the potential, thus leaving the distribution of states in energy unaffected. Indeed, the density of states of the entire space does not change when a single scatterer is introduced into a spatially bounded region. Similar "difficulties" with interpretation arise when one considers the change in the density of states when an impurity is introduced into a pure host material. The density of states of the system as a whole *does not* change. However, changes do occur in the *local* density of states (and corresponding charge density), defined through integrals of the Green function only over the domain of the impurity cell.

where $N(E)$ is the integrated density of states up to energy E, and t is the t-matrix associated with the potential. For the case of spherically symmetric potentials, Friedel [7] showed that the change in the integrated density of states is given by the expression,

$$\Delta N(E) = \frac{1}{\pi}\sum_l (2l+1)\delta_l, \tag{7.40}$$

where δ_l is the phase shift for the lth partial wave. Friedel's result agrees with the Lloyd formula for the case of spherically symmetric potentials. In further work, Krein [8] showed that the change in the integrated density of states can be related to the S-matrix in the angular momentum representation.

In this subsection, we derive Lloyd's formula for a single scatterer entirely within the angular-momentum representation, and directly in terms of the t-matrix. The result obtained here will be useful in the next subsection where Lloyd's formula is extended to multiple scattering theory. For the sake of clarity of presentation, we begin with a consideration of spherically symmetric potentials, and generalize to nonspherical potentials subsequently.

It is convenient to work with real spherical harmonics,[5] so that the functions $J_L(\mathbf{r})$ and $N_L(\mathbf{r})$ are real. For spherically symmetric potentials, we saw in Chapter 3 that outside the bounding sphere the basis function corresponding to the Lth partial wave (the radial wave function) can be written in the form,

$$\begin{aligned} \psi_L(\mathbf{r}) &= J_L(\mathbf{r}) + t_l H_L(\mathbf{r}) \\ &= \mathrm{e}^{\mathrm{i}\delta_l}(J_L(\mathbf{r})\cos\delta_l - N_L(\mathbf{r})\sin\delta_l). \end{aligned} \tag{7.41}$$

Here, the t-matrix is given by the expression

$$t_l = \frac{1}{2k\mathrm{i}}(\mathrm{e}^{2\mathrm{i}\delta_l} - 1) = -\frac{1}{k}\sin\delta_l \mathrm{e}^{\mathrm{i}\delta_l}, \tag{7.42}$$

and, for the choice of basis functions used here, can be written as the surface integral of a Wronskian,

$$t_l = [J_L, \psi_L]. \tag{7.43}$$

Direct differentiation of the last expression with respect to energy yields

$$\begin{aligned} \dot{t}_l &= [\dot{J}_L, \psi_L] + [J_L, \dot{\psi}_L] \\ &= [\dot{J}_L, J_L] - [\dot{\psi}_L, \psi_L] + [\dot{J}_L, H_L]t_l + t_l[\dot{\psi}_L, H_L], \end{aligned} \tag{7.44}$$

where energy derivatives are denoted by dots over the symbols, and the second line of the last equation follows from the first upon use of Eq.(7.41).

---

[5]The use of complex spherical harmonics results in an equivalent but somewhat longer derivation.

Now, although J_L is real, ψ_L is a complex function, and use of Eq.(7.41) along with the fact that J_L and the corresponding ψ_L have the same normalization, Eq.(F.23), yields the result

$$[\dot{\psi}_L, \psi_L] = [\dot{\psi}_L, \psi_L^*]\mathrm{e}^{2\mathrm{i}\delta_l} = [\dot{J}_L, J_L]\mathrm{e}^{2\mathrm{i}\delta_l}. \tag{7.45}$$

It follows that Eq.(7.44) can be written in the form,

$$\begin{aligned}\dot{t}_l &= [\dot{J}_L, J_L](1 - \mathrm{e}^{2\mathrm{i}\delta_l}) + [\dot{J}_L, H_L]t_l + t_l[\dot{\psi}_L, H_L] \\ &= [\dot{J}_L, J_L]2\mathrm{i}kt_l + [\dot{J}_L, H_L]t_l + t_l[\dot{\psi}_L, H_L].\end{aligned} \tag{7.46}$$

¿From this expression, we obtain the result,

$$\mathrm{Tr}\ \dot{t}_l t_l^{-1} = \mathrm{Tr}\ 2\mathrm{i}k[\dot{J}_L, J_L] + \mathrm{Tr}[\dot{J}_L, H_L] + \mathrm{Tr}[\dot{\psi}_L, H_L]. \tag{7.47}$$

Using the expressions for volume integrals in terms of surface integrals of Wronskians involving energy derivatives given in Appendix K, and the expression for the Hankel functions in terms of Bessel and Neumann functions, Eq.(C.26), we obtain

$$\sum_L \int J_L(\mathbf{r})H_L(\mathbf{r})\mathrm{d}^3r = \mathrm{Tr}[\dot{J}_L, H_L] = \int \mathrm{d}^3r G_0(\mathbf{r}, \mathbf{r}), \tag{7.48}$$

$$\sum_L \int \psi_L(\mathbf{r})F_L(\mathbf{r})\mathrm{d}^3r = \mathrm{Tr}[\dot{\psi}_L, H_L] = \int \mathrm{d}^3r G_1(\mathbf{r}, \mathbf{r}), \tag{7.49}$$

where G_1 represents the Green function of a single-scatterer , and

$$2\mathrm{i}k\sum_L \int J_L(\mathbf{r})H_L(\mathbf{r})\mathrm{d}^3r = 2\mathrm{i}k\sum_L \int J_L(\mathbf{r})[J_L(\mathbf{r}) + \mathrm{i}N_L(\mathbf{r})]\mathrm{d}^3r, \tag{7.50}$$

so that Eq.(7.47) yields

$$\begin{aligned}\Im\ \mathrm{Tr}\dot{t}_l t_l^{-1} &= \Im\ \mathrm{Tr}\frac{\mathrm{d}}{dE}\ln t_l \\ &= \Im\int \mathrm{d}^3r[G_1(\mathbf{r}, \mathbf{r}) - G_0(\mathbf{r}, \mathbf{r})] \\ &= -\pi[n(E) - n_0(E)] \\ &= -\pi\Delta n(E).\end{aligned} \tag{7.51}$$

This is the Lloyd formula that we have been seeking for the case of a single spherically symmetric potential.

The result just obtained can be readily generalized to the case of non-spherical potentials. Beginning with the expression

$$\underline{t} = [|J\rangle, \langle\psi|], \tag{7.52}$$

and using the matrix forms of the cosine and sine functions, and the t-matrix, we obtain the generalized expression

$$\Delta n(E) = \frac{1}{\pi}\ \Im\ \mathrm{Tr}\frac{\mathrm{d}}{\mathrm{d}E}\ln \underline{t}. \tag{7.53}$$

7.4.4 A Collection of Scatterers

The single-scatterer results of the previous section can be readily generalized to a multicell scattering assembly. Such a generalization has been given by Faulkner [9] for the case of spherically symmetric MT potentials based on Krein's theorem [8]. In this subsection, we provide two different derivations of the Lloyd formula following somewhat different lines of argument. In the first derivation, we work explicitly and from the very beginning within the angular-momentum representation, making use of the formulae for the energy derivatives of the structure constants given in Appendix L. In the second derivation, we begin with expressions in operator space, and pass over to the angular-momentum representation at the end. Both derivations are shown in detail for the case of spherically symmetric potentials. The extension to nonspherical, space-filling cells is not particularly difficult and can be effected at the end.

Before proceeding with formal derivations, it might be helpful to point out that one can understand intuitively that the Lloyd formula must apply to a collection of cells, regardless of shape considerations. We recall Eq.(4.79) in which the quantity T^{ij} is given essentially by the MT form. Since the imaginary parts of the logarithm of the elements of the translation operator in that equation have opposite signs, direct substitution into the Lloyd formula for the single-scatterer yields the result

$$\Delta n(E) = -\frac{1}{\pi} \Im \, \mathrm{Tr} \frac{\mathrm{d}}{\mathrm{dE}} \ln \underline{M}. \tag{7.54}$$

7.4.5 First Derivation

It is convenient to work with the form $M(E) = 1 - Gt$ for which the Lloyd formula becomes relatively simple, as we now show. (A reader who is uninterested in the details of the demonstration should skip to Eq.(7.63).) Taking energy derivatives, we have,

$$\frac{d}{dE}\Big\{\mathrm{Tr}\ln(1 - Gt)\Big\} = \mathrm{Tr}\Big\{(1 - Gt)^{-1}\Big\}(-G\dot{t} - \dot{G}t). \tag{7.55}$$

The first term on the left-hand side of this equation can be evaluated quite simply because $(1 - Gt)^{-1}G$ is just D defined by Eq.(7.37). If we use X to represent $(1 - Gt)^{-1}$, we have

$$\frac{d}{dE}\Big\{\mathrm{Tr}\ln(1 - Gt)\Big\} = -\sum_i D^{ii}\dot{t}^i - \sum_{i \neq j} X^{ij}\dot{G}^{ji}t^i. \tag{7.56}$$

It is shown in Appendix L, and also in work by Kaprzyk and Bansil [10], that $\dot{G}$ is given by

$$-\dot{G}^{ij} = G^{ij}[\dot{J}^j, H^j]_j + [\dot{H}^i, J^i]_i G^{ij} + \sum_{k \neq i,j} G^{ik}[\dot{J}^k, J^k]_k G^{kj}, \tag{7.57}$$

which may be substituted into Eq.(7.56) to give

$$\begin{aligned}\frac{d}{dE}\Big\{\mathrm{Tr}\ln(1-Gt)\Big\} = &-\sum_i D^{ii}\dot{t}^i + \sum_{i\neq j} X^{ij}G^{ji}[\dot{J}^i, H^i]_i t^i \\ &+\sum_{i\neq j} X^{ij}[\dot{H}^i, J^i]_i G^{ji}t^i \\ &+\sum_{ij} X^{ij}\sum_{k\neq i,j} G^{jk}[\dot{J}^k, J^k]_k G^{ki}t^i \\ &-\sum_i X^{ii}\sum_{k\neq i} G^{ik}[\dot{J}^k, J^k]_k G^{ki}t^i. \end{aligned} \tag{7.58}$$

Then, using $D = G + GtD = G + DtG$, and $t^iX^{ii} = t^iD^{ii}t^i + t^i$, we obtain

$$\begin{aligned}\frac{d}{dE}\Big\{\mathrm{Tr}\ln(1-Gt)\Big\} = &-\sum_i D^{ii}\dot{t}^i + \sum_i D^{ii}[\dot{J}^i, H^i]_i t^i \\ &+\sum_i D^{ii}t^i[\dot{H}^i, J^i]_i \\ &+\sum_i D^{ij}\sum_{k\neq i}[\dot{J}^k, J^k]_k G^{ki}t^i \\ &-\sum_i (t^iD^{ii}t^i + t^i)\sum_{k\neq i} G^{ik}[\dot{J}^k, J^k]_k G^{ki}. \end{aligned} \tag{7.59}$$

Whence, using $D^{ii} = (GtD)^{ii}$ and $\dot{H}^i = \dot{G}^{ik}J^k + G^{ik}\dot{J}^k$, we obtain.

$$\begin{aligned}\frac{d}{dE}\Big\{\mathrm{Tr}\ln(1-Gt)\Big\} = &-\sum_i D^{ii}\dot{t}^i + \sum_i D^{ii}[\dot{J}^i, H^i]_i t^i \\ &+\sum_i D^{ii}t^i[\dot{H}^i, J^i]_i \\ &+\sum_i D^{ii}[\dot{J}^i, J^i]_i \\ &+\sum_i (t^iD^{ii}t^i + t^i)[\dot{H}^i, H^i]_i. \end{aligned} \tag{7.60}$$

Most of the surface integrals can be collected together using the expression

$$[\dot{\psi}^i, \psi^i]_i = [\dot{J}^i, J^i]_i - \dot{t}^i + t^i[\dot{H}^i, J^i]_i + [\dot{J}^i, H^i]_i t^i + t^i[\dot{H}^i, H^i]_i. \tag{7.61}$$

However, using the techniques of Appendix L, the surface integral involving ψ and $\dot{\psi}$ can be shown to be equal to the normalization integral over the cell

$$[\dot{\psi}^i_L, \psi^i_{L'}]_i = \int_{\Omega_i} \mathrm{d}^3r\, \psi^i_L(\mathbf{r})\psi^i_{L'}(\mathbf{r}) \tag{7.62}$$

Thus by comparison with Eq.(7.34) we have

$$\frac{\mathrm{d}}{\mathrm{dE}}\Big\{\mathrm{Tr}\ln(1-Gt)\Big\} = \sum_{iLL'}\Big\{D^{ii}_{LL'}\int_{\Omega_i}\mathrm{d}^3r\psi^i_L(\mathbf{r})\psi^i_{L'}(\mathbf{r})$$
$$+t^i_L[\dot{H}^i_L, H^i_{L'}]_i t^i_{L'}\Big\}$$
$$=\int \mathrm{d}r\Big\{G(\mathbf{r},\mathbf{r})-G_0(\mathbf{r},\mathbf{r})\}$$
$$-\sum_i\int \mathrm{d}^3r\{G_i(\mathbf{r},\mathbf{r})-G_0(\mathbf{r},\mathbf{r})\Big\}, \qquad (7.63)$$

which is the Lloyd formula we are seeking. Kaprzyk and Bansil [10] derived an analogous relation for the secular matrix $t^{-1} - G$.

Using Eq.(7.53) for the Lloyd formula in the limit of a single scatterer, we can express the last two terms in the previous equation in terms of the individual cell t-matrices. These can in turn be transposed to the left-hand side of the equation and combined with the expression there to yield the result

$$\frac{\mathrm{d}}{\mathrm{dE}}\left[\mathrm{Tr}\ln(1-Gt)^{-1}t\right] = \int \mathrm{d}^3r[G(\mathrm{r},\mathrm{r})-G_0(\mathrm{r},\mathrm{r})]. \qquad (7.64)$$

Taking the imaginary parts of both sides we obtain,

$$\Delta n(E) = -\frac{1}{\pi}\Im\,\mathrm{Tr}\frac{\mathrm{d}}{\mathrm{dE}}\ln\underline{M}$$
$$=\frac{1}{\pi}\Im\,\mathrm{Tr}\frac{\mathrm{d}}{\mathrm{dE}}\ln\tau, \qquad (7.65)$$

where $\underline{M}$ (and τ) is of the MT form.

For ease of presentation, in the steps connecting Eqs.(7.56) and (7.63) we used some expressions that are not valid for space-filling cell potentials. Consequently, our derivation is rigorously valid only for potentials that vanish outside of spheres inscribed within each cell. However, the final result is valid for full-cell non–muffin-tin potentials as well. To see this, recall that divergent sums can be converted into conditionally convergent sums, as was indicated in our discussion in the previou chapter. The details of applying this approach to the case of the Lloyd formula can possibly be filled in by the reader.

7.4.6 Second Derivation

Again we treat explicitly the case of spherical MT potentials and generalize to space-filling cells subsequently. Using the operator definition of the Green function, we define the "function"

$$F(z) = \mathrm{Tr}\ln(z-H)$$
$$= \mathrm{Tr;}\ln G^{-1} \qquad (7.66)$$

of the complex energy variable, z. Differentiating with respect to energy, we have

$$\dot{F}(z) = \text{Tr}(z - H)^{-1} = \text{Tr G}. \tag{7.67}$$

We also have

$$F(z) = \text{Tr} \ln \left[G_0^{-1}(1 - G_0 V)\right] = \text{Tr} \ln(1 - G_0 V) - \text{Tr} \ln G_0. \tag{7.68}$$

Now, using the expansion of the logarithm function, we obtain[6]

$$\text{Tr} \ln(1 - G_0 V) = \text{Tr}\, G_0 V + \frac{1}{2}\text{Tr}\, G_0 V G_0 V + \frac{1}{3}\text{Tr} G_0 V G_0 V G_0 V + \cdots. \tag{7.69}$$

Now, taking V to be the sum of non-overlapping potentials,

$$V = \sum_i v_i, \tag{7.70}$$

substituting in Eq.(7.69), and using the definition of the cell t-matrix,

$$\begin{aligned} t_i &= v_i(1 - G_0 v_i)^{-1} \\ &= (1 - tG_0)^{-1} v_i, \end{aligned} \tag{7.71}$$

we can resum the last expansion in terms of cell t-matrices in the form

$$\text{Tr} \ln(1 - G_0 V) = \sum_i \text{Tr} \ln(1 - G_0 v_i) + \text{Tr} \ln(1 - \hat{G}_0 t). \tag{7.72}$$

In this expression the symbol $\hat{G}_0$ denotes the strictly cell off-diagonal part of the free-particle propagator, so that

$$\hat{G}_0^{ii} = 0, \quad \text{and} \quad \text{Tr}\, \hat{G}_0 t = 0. \tag{7.73}$$

Thus, we have

$$F(z) = \sum_i \text{Tr} \ln[1 - G_0 v_i] - \text{Tr} \ln G_0 + \text{Tr} \ln \left[1 - \hat{G}_0 t\right]. \tag{7.74}$$

Furthermore, we note that

$$\begin{aligned} \frac{\mathrm{d}}{\mathrm{dE}} \text{Tr} \ln(1 - G_0 v_i) &= \text{Tr}(1 - G_0 v_i)^{-1} G_0 G_0 v_i \\ &= \text{Tr}\, G_0 t_i G_0. \end{aligned} \tag{7.75}$$

Using the last result, we can cast the right-hand side of Eq.(7.74) in the angular-momentum representation. To do this, we expand the logarithms, write the resulting expressions in the angular-momentum representation, and expand the free-particle propagator in terms of Bessel and Hankel

---

[6]We assume that the various expansions converge as this assumption does not affect the validity of the formal arguments.

functions, and real-space structure constants. Then the series, which is now in terms of matrices in L-space, can be resumed to yield the result

$$-\frac{1}{\pi}\mathrm{Im}\,\dot{F}(z) = \sum_i (n_i(z) - n_0(z)) - n_0(z) \\ - \frac{1}{\pi}\mathrm{Im}\frac{\mathrm{d}}{\mathrm{d}z}\mathrm{Tr}\,\ln[1 - \underline{G}_0\underline{t}]. \quad (7.76)$$

Using the single-scatterer result, Eq.(7.53), it is a simple matter to cast the last expression into the form of Eq.(7.65). Also, we note that all forms of the Lloyd formula can be written either in terms of the density of states (forms involving energy derivatives) or in terms of the integrated density of states (forms without energy derivatives). Finally, the derivation given above can be generalized to the case of space-filling cells through proper summation procedures, such as performing conditionally convergent sums in the proper manner, or by replacing divergent sums by conditionally convergent double sums.

Before leaving this section, we comment briefly on the choice of the form $(1 - G_0 t)$ used in our derivations of the Lloyd formula. In contrast to the more usual form, $(\underline{t}^{-1} - G_0)$, the former expression can be readily extended throughout the upper (lower) complex plane. One main disadvantage of forms based on $\underline{M}$ or $\underline{\tau}$ is that they contain the singularities of the single-cell t-matrices, leading often to difficulties in applications requiring the use of complex energies. The cell t-matrices do not appear explicitly in the form $(1 - G_0 t)$, having been converted into expressions in terms of the single cell Green functions. This feature often facilitates greatly the performance of numerical procedures.

7.4.7 The Effects of Truncation

So far in this section we have assumed that there are no restrictions on the angular-momentum sums. Practical calculations, of course, require that the sums be finite. The Green function as normally truncated, to be denoted by $G_t(\mathbf{r}, \mathbf{r}')$, is obtained by truncating both terms in Eq.(7.34) to $\ell_{\max}$, with the matrix $D^{nn'}_{LL',t}$ which is truncated to $\ell_{\max}$ in the place of $D^{nn'}_{LL'}$. The matrix $D^{nn'}_{LL',t}$ is the solution of

$$D^{nn'}_{LL',t} = G^{nn'}_{LL'}(1 - \delta_{nn'}) + \sum_{L'''}^{\ell_{\max}} \sum_{L''}^{\ell_i} \sum_{n'' \neq n} G^{nn''}_{LL''} t^{n''}_{L''L'''} D^{n''n'}_{L'''L',t}, \quad (7.77)$$

with $G^{nn'}_{LL'}$ being an element of the structure-constant matrix. This truncated Green function, however, is not properly normalized and gives a density of states that disagrees with the Lloyd formula.

Because the solution to the Schrödinger equation in MST is closely associated with the Green function, its normalization is not arbitrary but is determined by the secular matrix. Writing the Green function in the form,

$$G(\mathbf{r},\mathbf{r}') = \sum_i \frac{\psi_i(\mathbf{r})\psi_i(\mathbf{r}')}{E - E_i}, \tag{7.78}$$

and substituting the expansion for the wave functions, Eq. (7.10), one has,

$$G(\mathbf{r},\mathbf{r}') = N \sum_{inn'LL'} \frac{a_L^n \psi_L^n(\mathbf{r}) a_{L'}^{n'} \psi_{L'}^{n'}(\mathbf{r}')}{E - E_i}, \tag{7.79}$$

where N is a normalization factor. Here, it is understood that the coefficients a_L^n also depend on the state label i and that the basis functions $\psi_L^n(\mathbf{r})$ are evaluated at E_i. We compare this equation with $G(\mathbf{r},\mathbf{r}')$ from Eq.(7.34), which for $E \approx E_i$, can be written as

$$G(\mathbf{r},\mathbf{r}') \approx \sum_{nn'LL'} \frac{a_L^n \psi_L^n(\mathbf{r}) a_{L'}^{n'} \psi_{L'}^{n'}(\mathbf{r}')}{\alpha(E - E_i)}. \tag{7.80}$$

We have assumed that E_i is a discrete state that is not a bound state of a single potential. Here $\alpha(E - E_i)$ is the eigenvalue of $D_{LL'}^{nn'}$ that vanishes at E_i. One can see that in order to have a set of wave functions that is consistent with the Green function, the normalization of the wave function must be uniquely determined by the energy derivative of the eigenvalue of the secular matrix, that is, $N = \alpha^{-1}$. Because this derivative is almost independent of the truncation in ℓ, the normalization factor for the wave function is almost a constant with respect to ℓ_{max}. This means that the wave function cannot be normalized to 1 when ℓ is truncated. As a result, the truncated Green function usually yields fewer electron states within a given energy range. Normally one could compensate for this error by moving the Fermi energy. However, this corresponds to putting high ℓ electrons, which are mostly near the boundary of the cells, into low ℓ orbits closer to the nuclei at higher energies. This approximation effectively puts too many electrons close to the nuclei, resulting in an equilibrium lattice constant that is too large. It would also cause errors in forces and related quantities.

When the Lloyd formula is evaluated by taking the trace of the natural logarithm of the secular matrix or, equivalently, the logarithm of the determinant of the secular matrix truncated at ℓ_{max} one finds that the energy derivative of this quantity is not equal to the expected result using the truncated Green function. Thus,

$$\frac{\mathrm{d}}{\mathrm{dE}} \sum_i \sum_L^{\ell_{\text{max}}} \left\{ \ln(1 - Gt) \right\}_{LL}^{ii} \neq \int \mathrm{d}^3 r \{G_t(\mathbf{r},\mathbf{r}) - G_{0t}(\mathbf{r},\mathbf{r})\}; \tag{7.81}$$

rather, we have,

$$\frac{d}{dE} \sum_i \sum_L^{\ell_{\text{max}}} \left\{ \ln(1 - Gt) \right\}_{LL}^{ii} = \int \mathrm{d}^3 r \{G_a(\mathbf{r},\mathbf{r}) - G_0(\mathbf{r},\mathbf{r})\}. \tag{7.82}$$

where G_a is defined below.

Just as for the wave functions, we can define an augmented Green function, $G_a(\mathbf{r},\mathbf{r}')$,

$$G_a(\mathbf{r},\mathbf{r}') = \sum_{L,L'} \psi_L^n(\mathbf{r}) D_{LL',a}^{nn'} \psi_{L'}^{n'}(\mathbf{r}') + \sum_L F_L^n(\mathbf{r}) \psi_L^n(\mathbf{r}') \delta_{nn'}, \tag{7.83}$$

where the summation over ℓ is no longer truncated, and the elements of the augmented D-matrix, for $\ell, \ell' \leq \ell_{\max}$ are defined as $D_{LL',a}^{nn'} = D_{LL',t}^{nn'}$, and for $\ell > \ell_{\max}$, $\ell' \leq \ell_{\max}$ as

$$D_{LL',a}^{nn'} = G_{LL'}^{nn'}(1-\delta_{nn'}) + \sum_{L'''}^{\ell_{\max}} \sum_{L''}^{\ell_i} \sum_{n''\neq n} G_{LL''}^{nn''} t_{L''L'''}^{n''} D_{L'''L',t}^{n''n'},$$

$$D_{L'L,a}^{nn'} = D_{LL',a}^{n'n},$$

and, finally, for $\ell, \ell' > \ell_{\max}$,

$$D_{LL',a}^{nn'} = G_{LL'}^{nn'}(1-\delta_{nn'}) + \sum_{L'''}^{\ell_{\max}} \sum_{L''}^{\ell_i} \sum_{n''\neq n} G_{LL''}^{nn''} t_{L''L'''}^{n''} D_{L'''L',a}^{n''n'}.$$

Although the augmented Green function uses the truncated secular equation, it still needs an infinite matrix $D_{LL',a}^{nn'}$. In practice, one needs to truncate $D_{LL',a}^{nn'}$ at a higher value of ℓ than $\ell_{\max}$. However, the computational time of the above procedure is much less than that required to compute $D_{LL',t}^{nn'}$ to the same level of truncation.

7.5 Numerical Study of Two Muffin-Tin Potentials

In this section we use an example of two muffin-tin scatterers to demonstrate the difference between the truncated and the augmented wave functions. This case has been used in Ref. [5] to study the discontinuity of the truncated wave functions. Here we show numerically that the augmented wave function is continuous and smooth for any ℓ truncation, and that it normalizes exactly to unity within machine accuracy.

We study the bound states ($E < 0$) of this system. As we mentioned in the previous section, the normalization of the wave function in MST is uniquely determined by the Green function. In the case of a bound state with energy $E_i < 0$, the wave functions must satisfy the condition

$$\frac{1}{A}|\psi_i(\mathbf{r})|^2 = -\frac{1}{\pi}\mathrm{Im}\int_{E_i-\delta E}^{E_i+\delta E} \mathrm{d}E G(\mathbf{r},\mathbf{r},E), \tag{7.84}$$

where A is the normalization factor for ψ_i and E_i is the eigenenergy. We note that the secular matrix can be written in the form,

$$[M(E)^{-1}]_{LL'}^{nn'} = \left(\frac{dM_i}{dE}\right)^{-1} \frac{a_L^n a_{L'}^{n'}}{E - E_i}, \tag{7.85}$$

near $E \approx E_i$, where M_i is the eigenvalue of the secular matrix, and a_L^n are the components of the corresponding eigenvector. For the case of MT potentials, the secular matrix is $M_{LL'}^{nn'} = t_L^{n\,-1}\delta_{nn'}\delta_{LL'} - G_{LL'}^{nn'}$. Combining the above equations with Eq.(7.34), we can determine the normalization factor A,

$$A = -\frac{1}{\sqrt{-E_i}}\left(\frac{dM_i}{dE}\right). \tag{7.86}$$

If $\ell_{\max}$ is sufficiently large, the eigenvalues of the secular matrix and their derivatives will be approximately independent of $\ell_{\max}$ and A will be approximately a constant independent of ℓ truncation. Therefore at any finite ℓ truncation, the factor A does not give the proper normalization for a truncated wave function.

The following calculations were based on double precision (real*8) arithmetic on an IBM RISC 6000 work station. The secular equation was solved with $\ell_{\max}$ from 0 to 6. The integrations of the wave functions over three-dimensional space were converted to surface integrals over the muffin tins of the energy derivatives of the logarithm derivative of the wave functions. The augmented wave functions are calculated by using spherical waves up to $\ell = 60$. The results for the case of potentials with radius $r = 1$, depth $V = 5$, and separation $S = 2$ (touching spheres), are listed in Table 7.1. We see that the normalization of the truncated wave function has an error that is dependent on the ℓ truncation, while the normalization of the augmented wave function is exactly unity within machine accuracy. We list in Table 7.2 the discontinuity of the wave functions and their derivatives at the muffin tin boundary. The augmented wave function has essentially zero discontinuity in both its value and its derivative.

One should note that the augmented wave function is not only more accurate in terms of the normalization and continuity, it agrees with the exact solution more closely over the entire cell. In Fig.7.2 we show a comparison of the truncated wave function and the augmented wave function with the exact solution for the muffin-tin case discussed above. The wave functions are plotted along the axis that goes through the centers of the two muffin tins. The error of the truncated wave function and its discontinuity at the muffin tin boundary is apparent. On the other hand, the augmented wave function is continuous and smooth and agrees well with the exact solution. The main error in the augmented wave function is due to the approximate eigenenergy obtained from the truncated secular equation.

We note that in this example the truncated wave function overall is fairly accurate. This is due to the weak multiple scattering resulting from

$\ell_{\max}$	E	N_t	N_a
0	-1.28683287549593	0.99538989670782	1.00000000000000
1	-1.31812990252951	0.99867299791530	1.00000000000000
2	-1.32489887845514	0.99969260888189	1.00000000000000
3	-1.32629994207095	0.99993160043574	0.99999999999998
4	-1.32659348953932	0.99998461908932	1.00000000000000
5	-1.32665699438796	0.99999644921598	1.00000000000000
6	-1.32667124816526	0.99999915445965	1.00000000000000

TABLE 7.1. Wave function normalization for two muffin-tin scatterers. The parameters used are muffin-tin radius $r = 1$, depth of the potential $V = 5$, and the separation $S = 2$. N_t is the normalization of the truncated wave function and N_a is that of the augmented wave function.

$\ell_{\max}$	ψ_t	ψ'_t	ψ_a	ψ'_a
0	0.1697	0.1738	1.789×10^{-16}	2.358×10^{-16}
1	0.1030	0.1606	3.096×10^{-16}	1.436×10^{-16}
2	5.524×10^{-2}	0.1179	3.030×10^{-16}	3.048×10^{-16}
3	2.856×10^{-2}	7.775×10^{-2}	5.137×10^{-15}	4.376×10^{-15}
4	1.465×10^{-2}	4.861×10^{-2}	3.050×10^{-16}	4.929×10^{-16}
5	7.534×10^{-3}	2.952×10^{-2}	4.053×10^{-16}	8.589×10^{-16}
6	3.904×10^{-3}	1.766×10^{-2}	1.814×10^{-16}	1.503×10^{-15}

TABLE 7.2. Discontinuity of the wave functions at the muffin-tin boundary. The parameters used are the same as in preceding table. Columns ψ_t and ψ'_t list the root-mean-square difference across the muffin-tin boundary of the inside solution and the outside solution for the truncated wave function and its derivative, and columns ψ_a and ψ'_a list the same quantities for the augmented wave function.

the peculiar geometry of only two scatterers. One would expect the error of the truncated wave function to be much larger in a more realistic case where the multiple scattering effects are more dominant. In such a case the benefit of using the augmented wave function should be even greater.

7.6 Convergence of Electronic Structure Calculations

Perhaps the main point of the discussion in the previous sections is that it is important to view the convergence of MST not as a limiting process in which the expansion of the wave functions is truncated at $\ell_{\max}$, but one in which the wave function expansions are not truncated and the potential is approximated by setting its phase shifts equal to zero for $\ell > \ell_{\max}$. Failure to obey this rule may cause the Green function to be inconsistent with the

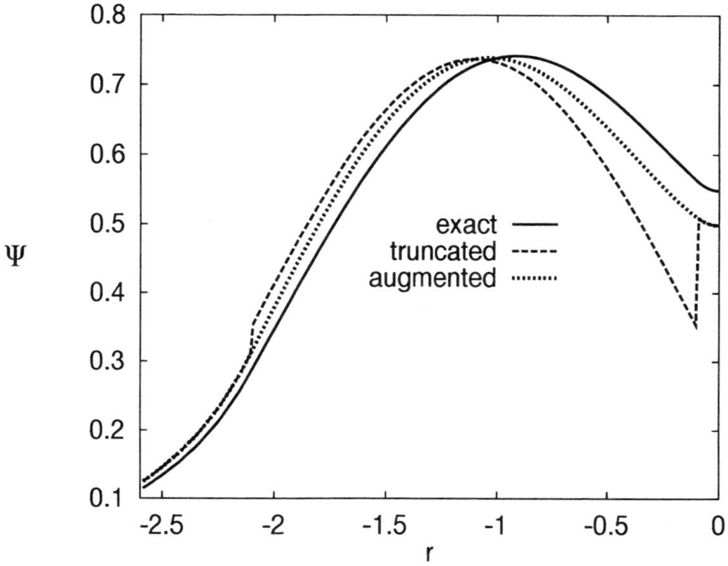

FIGURE 7.2. Comparison of truncated and augmented wave functions for two MT spheres.

expression for the density of states given by the Lloyd formula. Furthermore, the wave functions and their derivatives may be discontinuous at the cell boundaries, and physical properties requiring matrix elements with dipole symmetry will be incorrect or at best less accurate than they could be for comparable computational effort. These effects can be alleviated essentially completely through the augmentation procedures presented in this chapter.

It is also important to state again the general rule that has emerged form the discussion in the last three chapters for obtaining converged expressions for the band structure and the wave function (or the Green function) and hence the charge density in an electronic structure calculation. Such a calculation can be carried out in terms of the cell t-matrices, and *modified* structure constants of the form

$$G_{LL'}(\mathbf{R}_{nm}) = \sum_{L''} g_{LL''}(\mathbf{b}_{nm}) G_{L''L'}(\mathbf{R}_{nm} + \mathbf{b}_{nm}), \tag{7.87}$$

where the sum over L'' must follow that over L or L' when these structure constants are used. Numerical calculations in terms of the Poisson equation, see Chapter 10, suggest that the magnitude of the vector $\mathbf{b}_{nm}$ should be of the order of a near- neighbor vector along $\mathbf{R}_{nm}$. Like the canonical structure constants of the KKR method, these quantities depend only on the structural aspects (geometry) of the material and not on the potential. Like their KKR counterparts, they can be calculated once for all for a given lattice and used as required in applications of full-cell MST. Finally, the modified-structure constant approach can be combined directly with the augmented formulation of MST discussed above to yield properly continuous values and derivatives of the wave function (and Green function) across cell boundaries.

The full-potential version of MST has not been implemented fully within a self-consistent study of the electronic structure of a solid. Initial results [11] indicate that an apparently fully converged band structure of cubic systems (fcc, bcc) such as Cu can be obtained with $L = 4$ through a secular equation of the MT form, involving the cell t-matrices and *unmodified* KKR structure constants. These results are indeed encouraging, but should be kept in mind in view of the convergence properties discussed above. As is illustrated more explicitly in Chapter 10, in the case of cubic systems the MT form (with unmodified structure constants) shows a strong tendency to converge for values of L up to 4. Then it can begin to diverge exhibiting asymptotic behavior. This behavior can always be prevented through the use of modified structure constants when solving the secular equation or constucting the charge density. In any case, it would be prudent when carrying out calculations with the canonical MT form to check convergence directly by increasing the value of L.

References

[1] P. Lloyd, Proc. Phys. Soc. London **90**, 207 (1967); **90**, 217, (1967).

[2] A. R. Williams, J. F. Janak, and V. L. Moruzzi, Phys. Rev. B**6**, 4509 (1972).

[3] V. L. Moruzzi, J. F. Janak, and A. R. Williams, *Calculated Electronic Properties of Metals* (Pergamon Press, New York, 1978).

[4] J. S. Faulkner, Prog. in Mat. Science **27**, 1 (1982).

[5] W. H. Butler and X.-G. Zhang, Phys. Rev. B**44**, 969, (1991).

[6] A. R. Williams and J. van W. Morgan, J. Phys. C**7**, 37 (1974).

[7] J. Friedel, Phil. Mag. **43**, 153 (1952); Adv. Phys. **3**, 446 (1953).

[8] M. G. Krein, Matem. Sborn **33**, 597 (1953); Soviet Math. Doclady **3**, 707 (1962).

[9] J. S. Faulkner, J. Phys. C., Solid State Phys. **10**, 4661 (1977).

[10] S. Kaprzyk and A. Bansil, Phys. Rev. **42**, 7358 (1990).

[11] A. Gonis, *Green Functions for Ordered and Disordered Systems* (Elsevier, Amsterdam, New York, 1992).

8
Relativistic Formalism

8.1 General Comments

Our discussion of multiple scattering theory up to this point is applicable to a broad range of problems, but is nonetheless limited in two important respects. It is restricted to those scattering events that do not involve the internal structure of the colliding bodies, and in which the velocities of the particles are small compared to the velocity of light. Specifically, the effects of the electron spin, as well as other relativistic effects, have been neglected. Such effects can play an important role in determining the electronic properties of materials when the masses and charges of the nuclei in a solid are large enough to cause the electrons, at least those close to the nuclei, to move at speeds that are comparable to that of light. And even in the case of lighter elements the study of many physical properties, for example, magnetic moments, requires consideration of the spin of the electrons, which is a *bona fide* relativistic concept.

In this chapter, we generalize our discussion of MST for cell (non-MT) potentials to take into account relativistic effects. Thus, we develop the formalism of single-cell and multiple scattering theory with respect to the Dirac equation. It is important to note at the outset that the relativistic formulation of full-cell MST contains no new features as far as the "geometric" aspects of the theory are concerned. One still needs to construct convergent expansions in an angular-momentum basis, and the basic features characterizing such a construction are no different than those present in the nonrelativistic case.

On the other hand, as mentioned above, there are very important reasons for developing MST based on a treatment of the Dirac equation in a solid. This entails finding the eigenvectors and eigenvalues of the Dirac Hamiltonian[1]

$$H_D = -\mathrm{i}\mathrm{c}\underline{\alpha} \cdot \nabla + \mathrm{c}^2\underline{\beta} + V(\mathbf{r}) \tag{8.1}$$

where c is the velocity of light. For the sake of ease in developing the formalism, we shall take the potential $V(\mathbf{r})$ to be a scalar. More general forms of the potential, requiring a matrix description can be incorporated into the formalism [1]. The four-dimensional matrices $\underline{\alpha}$ and $\underline{\beta}$ are given in Dirac's notation as

$$\underline{\alpha}_i = \begin{bmatrix} 0 & \underline{\sigma}_i \\ \underline{\sigma}_i & 0 \end{bmatrix}, \quad i = x, y, z, \quad \underline{\beta} = \begin{bmatrix} \underline{I}_2 & 0 \\ 0 & -\underline{I}_2 \end{bmatrix} \tag{8.2}$$

where

$$\underline{\sigma}_x = \begin{bmatrix} 0 & 1 \\ 1 & 0 \end{bmatrix}, \quad \underline{\sigma}_y = \begin{bmatrix} 0 & -\mathrm{i} \\ \mathrm{i} & 0 \end{bmatrix}, \quad \text{and } \underline{\sigma}_z = \begin{bmatrix} 1 & 0 \\ 0 & -1 \end{bmatrix} \tag{8.3}$$

are the Pauli spin matrices, and I_2 is the two-dimensional unit matrix. As is well known, the Pauli spin matrices along with I_2 form a complete set in two-dimensional matrix space, so that for any 2×2 matrix A we have

$$\underline{A} = A_0 I_2 + A_x \sigma_x + A_y \sigma_y + A_z \sigma_z. \tag{8.4}$$

Whether one is concerned with the Schrödinger or with the Dirac equation, the basic elements of scattering theory in general and of multiple scattering theory in particular remain the same. One must calculate the effects on the evolution of a state governed by the Dirac Hamiltonian, Eq.(8.1), starting from a free-particle state which is a solution of the free-particle Dirac Hamiltonian,

$$H_D^0 = -\mathrm{i}\mathrm{c}\underline{\alpha} \cdot \nabla + \mathrm{c}^2\underline{\beta}. \tag{8.5}$$

In multiple scattering theory, we may built the solutions of the Dirac equation,

$$H_D|\Psi\rangle = E|\Psi\rangle, \tag{8.6}$$

for the solid from partial solutions associated with individual cells. These solutions contain all possible scattering events experienced by an electron as it propagates through the multicell assembly. We are interested in providing a description of these solutions in terms of the computationally convenient angular-momentum representation.

---

[1]We are using atomic units in which $\mathrm{e}^2 = \mathrm{m} = \hbar = 1$, and $\mathrm{c} = 137.036$. In addition, we shall also set $\mathrm{c} = 1$ in much of the following discussion.

It follows from the matrix nature of the Dirac Hamiltonian that the solutions of Eq.(8.6) are four-dimensional vectors, or *bi-spinors*. The description of these solutions in terms of an angular-momentum basis requires an extension of the spherical harmonics to *spinors* in order to incorporate the presence of spin. This leads to a generalized angular-momentum or partial-wave basis, which is discussed in Section 8.2. In Section 8.3, we show how the free-particle solutions associated with the free-space Hamiltonian of Eq.(8.5) can be expressed in an angular-momentum basis in terms of bi-spinors. Expressions for the structure constants in the presence of arbitrary spin, as well as for the case of electron spin, are also derived there. The solution of Eq.(8.6) for an arbitrarily shaped cell potential, as well as the form that that solution takes in the case of spherical potentials, are presented Section 8.4. Finally, relativistic MST is briefly described in the last section of this chapter.

8.2 Generalized Partial Waves

First, we discuss the generalization of the angular-momentum basis required for the treatment of spin. The spin-angular functions derived below will be used in the construction of the *bi-spinor* wave functions in the fully relativistic version of MST.

When two particles of spins, S_1 and S_2, collide, the spin of the system, S, can vary from $|S_1 - S_2|$ to $|S_1 + S_2|$, in units of $\hbar$. This spin in turn will combine with the total orbital angular momentum, l, to produce the total angular momentum, J, of the system, whose values vary from $|l - S|$ to $|l + S|$, in units of $\hbar$. In general, neither S nor l will be conserved separately in a scattering event. For rotationally invariant systems, however, the total angular momentum, J, is a constant of the motion and allows the construction of a basis set for the generalized partial wave description of scattering. We show how this basis can be set up for the case of arbitrary spin. The reduction to $S = 1/2$, appropriate to electrons, of interest to us here, will be effected subsequently.

The basis to be used in a generalized partial-wave description of scattering in the presence of spin is the direct (tensor) product of the bases corresponding to orbital angular momentum, $L = (l, m)$, and to spin S. Denoting by μ the projection of (the operator) $\hat{J}$ along an axis of quantization, say the z-axis, we use the laws of addition of angular momentum to introduce the functions [2]

$$\mathrm{Y}_{lS}^{J\mu}(\hat{r}) = \sum_{\substack{m,\nu \\ m+\nu=\mu}} \langle lmS\nu|J\mu\rangle Y_{lm}(\hat{r})\chi_{S\nu}, \tag{8.7}$$

where the symbol $\langle lmS\nu|J\mu\rangle$ denotes the so-called *vector addition* or Clebsch–Gordan coefficients. We also have the inverse relation,

$$Y_{lm}(\hat{r})\chi_{S\nu} = \sum_{J=|l-S|}^{l+S} \sum_{\mu=-J}^{J} \langle lmS\nu|J\mu\rangle \mathrm{Y}_{lS}^{J\mu}(\hat{r}), \tag{8.8}$$

which follows from the orthogonality of the Clebsch–Gordan coefficients. In particular, we have

$$\sum_{\mu\nu} \langle lmS\nu|J\mu\rangle\langle lmS\nu|J'\mu'\rangle = \delta_{JJ'}\delta_{\mu\mu'}. \tag{8.9}$$

These coefficients also satisfy the relation,

$$\sum_{J\mu} \langle l_1m_1l_2m_2|J\mu\rangle\langle l_1m_1'l_2m_2'|J\mu\rangle = \delta_{m_1m_1'}\delta_{m_2m_2'}. \tag{8.10}$$

Here, $\chi_{S\nu}$ denotes the eigenvectors of the spin operator. For $S = 1/2$, we can write $\chi_{\frac{1}{2}\frac{1}{2}} = \begin{pmatrix} 1 \\ 0 \end{pmatrix}$ and $\chi_{\frac{-1}{2}\frac{-1}{2}} = \begin{pmatrix} 0 \\ 1 \end{pmatrix}$, or appropriately chosen linear combinations of these vectors. The generalized spherical harmonic functions, $\mathrm{Y}_{lS}^{J\mu}(\hat{r})$, defined in Eq.(8.7), are eigenfunctions of the operators, $\hat{J}^2$,

$$\hat{J}^2\mathrm{Y}_{lS}^{J\mu}(\hat{r}) = J(J+1)\mathrm{Y}_{lS}^{J\mu}(\hat{r}), \tag{8.11}$$

of $\hat{J}_z$,

$$\hat{J}_z\mathrm{Y}_{lS}^{J\mu}(\hat{r}) = \mu\mathrm{Y}_{lS}^{J\mu}(\hat{r}), \tag{8.12}$$

of $\hat{l}^2$,

$$\hat{l}^2\mathrm{Y}_{lS}^{J\mu}(\hat{r}) = l(l+1)\mathrm{Y}_{lS}^{J\mu}(\hat{r}), \tag{8.13}$$

and of $\hat{S}^2$,

$$\hat{S}^2\mathrm{Y}_{lS}^{J\mu}(\hat{r}) = S(S+1)\mathrm{Y}_{lS}^{J\mu}(\hat{r}). \tag{8.14}$$

Furthermore, if we let $\hat{P}$ denote the parity operator (a linear, unitary operator whose action changes the sign of polar vectors, for example, coordinates, linear momentum, but leaves that of axial vectors invariant, for example, angular momentum, spin), we have

$$\hat{P}\mathrm{Y}_{lS}^{J\mu}(\hat{r}) = (-1)^l\mathrm{Y}_{lS}^{J\mu}(\hat{r}). \tag{8.15}$$

In order to carry out an analysis of scattering processes in the presence of spin, we must also augment the free-wave basis set to include spin states. To this end, we define the (initial) free-plane-wave states as the tensor product

$$|\phi\rangle \equiv |\mathbf{k}, S, \nu\rangle = |\mathbf{k}\rangle\chi_{S\nu}. \tag{8.16}$$

In the position (coordinate) representation, we have

$$\langle\mathbf{r}|\phi\rangle \equiv \langle\mathbf{r}|\mathbf{k}_i, S, \nu\rangle = \phi_{\mathbf{k}_i,S,\nu}(\mathbf{r}) = (2\pi)^{-3/2}\exp[i\mathbf{k}_i\cdot\mathbf{r}]\chi_{S\nu}. \tag{8.17}$$

These states satisfy the conditions of normalization,

$$\langle \mathbf{k}_f, S', \nu' | \mathbf{k}_i, S, \nu \rangle = \delta(\mathbf{k}_f - \mathbf{k}_i)\delta_{SS'}\delta_{\nu\nu'}, \tag{8.18}$$

and completeness,

$$\sum_{S,\nu} \int \mathrm{d}^3k |\mathbf{k}, S, \nu\rangle\langle \mathbf{k}, S, \nu| = 1. \tag{8.19}$$

The left-hand side of Eq.(8.18) involves a space integral, while the position representation of Eq.(8.19) takes the form,

$$\sum_{S,\nu} \int \mathrm{d}^3k \phi_{\mathbf{k},S,\nu}(\mathbf{r})\phi^\dagger_{\mathbf{k},S,\nu}(\mathbf{r}') = \delta(\mathbf{r} - \mathbf{r}'). \tag{8.20}$$

We also note the completeness relation for the spin states alone

$$\sum_{S,\nu} \chi_{S,\nu}\chi^\dagger_{S,\nu} = 1. \tag{8.21}$$

Now, a free wave can be decomposed into spherical waves in the form

$$\phi_{\mathbf{k},S,\nu}(\mathbf{r}) = (2/\pi)^{1/2} \sum_{l=0}^{\infty} \sum_{m=-l}^{l} \mathrm{i}^l j_l(kr)\bar{Y}_{lm}(\hat{k})Y_{lm}(\hat{r})\chi_{S\nu}, \tag{8.22}$$

where a bar denotes the complex conjugation. (A bar rather than an asterisk is used to denote a complex conjugate because of convenience of notation in the following discussion.) The use of the inverse relation, Eq.(8.8), and the definition of the auxiliary functions,

$$Q^{J\mu}_{lS\nu}(\hat{k}) = \sum_{-l}^{l} \langle lmS\nu|J\mu\rangle Y_{lm}(\hat{k}), \tag{8.23}$$

allows us to write

$$\phi_{\mathbf{k}_i,S,\nu}(\mathbf{r}) = (2/\pi)^{1/2} \sum_{lJ\mu} \mathrm{i}^l j_l(kr) Y^{J\mu}_{lS}(\hat{r})\bar{Q}^{J\mu}_{lS\nu}(\hat{k}). \tag{8.24}$$

We can also rewrite this expression as

$$\phi_{\mathbf{k}_i,S,\nu}(\mathbf{r}) = (2/\pi)^{1/2} \sum_{l,J} \mathrm{i}^l j_l(kr) \langle \hat{r}lS\nu|\mathrm{Y}^J|\hat{k}lS\nu\rangle\chi_{S\nu} \tag{8.25}$$

in terms of the quantity

$$\langle \hat{r}l'S'\nu'|\mathrm{Y}^J|\hat{k}lS\nu\rangle = \sum_{m,m',\mu} Y_{l'm'}(\hat{r})\bar{Y}_{lm}(\hat{k})\langle l'm'S'\nu'|J\mu\rangle\langle lmS\nu|J\mu\rangle. \tag{8.26}$$

8.2.1 The Free-Particle Propagator in the Presence of Spin

In addition to the free-particle wave functions, a partial-wave analysis requires the use of the free-particle propagator, or Green function, along with its representation in (partial) spherical waves. Using the completeness relation, Eq.(8.19), we can write the outgoing free-particle propagator in the form,

$$\mathrm{G}_0^{(+)}(\mathbf{k}_i,\mathbf{r},\mathbf{r}') = \sum_{S\nu}\int \mathrm{d}^3k \frac{\langle \mathbf{r}|\phi_{\mathbf{k},S,\nu}\rangle\langle\phi_{\mathbf{k},S,\nu}|\mathbf{r}'\rangle}{E(\mathbf{k}_i)-E(\mathbf{k})+\mathrm{i}\epsilon}, \tag{8.27}$$

which, through the use of Eq.(8.21), can be written as

$$\begin{aligned}
\mathrm{G}_0^{(+)}(\mathbf{k}_i,\mathbf{r},\mathbf{r}') &= (2\pi)^{-3}\sum_{S\nu}\int \mathrm{d}^3k \frac{\mathrm{e}^{\mathrm{i}\mathbf{k}\cdot(\mathbf{r}-\mathbf{r}')}}{E(\mathbf{k}_i)-E(\mathbf{k})+\mathrm{i}\epsilon}\chi_{S\nu}\chi_{S\nu}^{\dagger} \\
&= (2\pi)^{-3}\int \mathrm{d}^3k \frac{\mathrm{e}^{\mathrm{i}\mathbf{k}\cdot(\mathbf{r}-\mathbf{r}')}}{E(\mathbf{k}_i)-E(\mathbf{k})+\mathrm{i}\epsilon} \\
&= -\frac{\mathrm{e}^{\mathrm{i}\mathbf{k}\cdot(\mathbf{r}-\mathbf{r}')}}{4\pi|\mathbf{r}-\mathbf{r}'|}.
\end{aligned} \tag{8.28}$$

As expected, this is the well-known expression for the free-particle Green function. We can now expand $\mathrm{G}_0^{(+)}(\mathbf{k},\mathbf{r},\mathbf{r}')$ in terms of spherical waves,

$$\begin{aligned}
\mathrm{G}_0^{(+)}(\mathbf{k},\mathbf{r},\mathbf{r}') &= -ik\sum_{lm} j_l(kr_<)j_l(kr_>)\bar{Y}_{lm}(\hat{r}')Y_{lm}(\hat{r}) \\
&= \sum_{JmlS} g_l^{(+)}(k,r,r')\bar{\mathrm{Y}}_{lS}^{J\mu}(\hat{r}')\mathrm{Y}_{lS}^{J\mu}(\hat{r}),
\end{aligned} \tag{8.29}$$

where

$$g_l^{(+)}(k,r,r') = -ikj_l(kr_<)h_l(kr_>). \tag{8.30}$$

Here, the symbols $r_<$ ($r_>$) represent the smaller (larger) of the lengths r and r', and the second line in Eq.(8.29) follows from the first upon use of the orthonormality relations of the Clebsch–Gordan coefficients. Finally, defining the generalized spherical functions,

$$\mathrm{A}_{lS}^{J\mu} = \alpha_l(kr)\mathrm{Y}_{lS}^{J\mu}(\hat{r}), \tag{8.31}$$

where α_l denotes either the spherical Bessel (j_l), Neumann (n_l), or Hankel ($h_l^{(+)}$ or $h_l^{(-)}$) functions, we can write

$$\mathrm{G}_0^{(+)}(k,\mathbf{r}-\mathbf{r}') = -ik\sum_{J,\mu,l,S}\bar{\mathrm{J}}_{lS}^{J\mu}(\mathbf{r})\mathrm{H}_{lS}^{J\mu}(\mathbf{r}'), \quad r'>r, \tag{8.32}$$

which is formally identical to the expansion of the free-particle propagator in the absence of spin.

It may be worthwhile to point out explicitly the meaning of the sum in Eq.(8.32). The variable μ takes on all values from $-J$ to J (in units of $\hbar$), while J assumes all values from $|l-S|$ to $|l+S|$ for each l and S. The sum over l extends over all positive integers including zero, while S varies from the minimum to the maximum value of the spin. Finally, it follows from the orthonormality of the generalized plane waves, Eq.(8.18), that the generalized spherical waves are also orthonormal,

$$\langle \mathrm{J}_{lS}^{J\mu} | \mathrm{J}_{l'S'}^{J'\mu'} \rangle = \delta_{JJ'}\delta_{\mu\mu'}\delta_{ll'}\delta_{SS'}, \tag{8.33}$$

or, in the position representation,

$$\int r^2 \mathrm{d}r \int \mathrm{d}\Omega_r j_l(kr) j_{l'}(k'r) \bar{\mathrm{Y}}_{lS}^{J\mu}(\hat{r}) \mathrm{Y}_{l'S'}^{J'\mu'}(\hat{r})$$
$$= s\frac{\pi}{2k^2}\delta(k-k')\delta_{JJ'}\delta_{\mu\mu'}\delta_{ll'}\delta_{SS'}. \tag{8.34}$$

It can also be shown in a straightforward way that the generalized spherical harmonics satisfy the completeness relation,

$$\sum_{J,\mu,l,S} \mathrm{Y}_{lS}^{J\mu}(\hat{r}) \mathrm{Y}_{lS}^{J\mu}(\hat{r}') = \delta(\Omega_r - \Omega_{r'}), \tag{8.35}$$

which is a straightforward generalization of the corresponding relation for ordinary spherical harmonics. Thus, with the minor complication of expanded notation, the generalized states (plane or spherical waves) to be used in the description of scattering in the presence of spin behave in a manner formally similar to that of spinless states.

8.2.2 The (κ, μ) or Λ Representation

It is convenient to introduce a representation that facilitates the discussion of the relativistic application of MST to calculate the electronic structure of solids. This so-called (κ, μ) or Λ representation is tailored to the case of electrons scattered by a spinless target, so that now $S = 1/2$. We introduce the number κ through the relation,

$$\kappa = \begin{cases} l, & \text{for } J = l - \frac{1}{2}, \\ -\kappa + 1, & \text{for } J = l + \frac{1}{2}, \end{cases} . \tag{8.36}$$

The number κ takes on all values, except zero, and gives the values of both l and J,

$$l = \begin{cases} \kappa, & \text{for } \kappa > 0, \\ -\kappa + 1, & \text{for } \kappa < 0, \end{cases} \tag{8.37}$$

and

$$J = |\kappa| - \frac{1}{2}. \tag{8.38}$$

For future reference, we also introduce the signature of κ,

$$S_\kappa = \kappa / |\kappa|. \tag{8.39}$$

It follows from this that l is a function of κ. Explicitly, for $\kappa = -1, 1, -2, 2, \ldots$, the associated values of l and J give rise to the following series of spectroscopic terms,

$$s^{1/2}(l=0, J=1/2),\ p^{1/2}(l=1, J=1/2),$$
$$p^{3/2}(l=1, J=3/2),\ d^{3/2}\ldots, \tag{8.40}$$

as can be easily verified. It is also convenient to introduce the quantity $\hat{l}$

$$\hat{l} = \begin{cases} \kappa - 1, & \text{for } \kappa > 0, \\ -\kappa, & \text{for } \kappa < 0. \end{cases} \tag{8.41}$$

We see that for a given value of J the possible values of κ are $\pm(J \pm \frac{1}{2})$. We also have,

$$l - \hat{l} = S_\kappa \tag{8.42}$$

and

$$J = l - \frac{1}{2} S_\kappa. \tag{8.43}$$

Now, the generalized spherical harmonics for $S = \frac{1}{2}$ can be conveniently expressed in the (κ, μ) or Λ representation in the form

$$\begin{aligned} \mathrm{Y}^\mu_\kappa(\hat{r}) &= \sum_\nu \left\langle \mu - \nu, \frac{1}{2}\nu \middle| J\mu \right\rangle Y_{l,\mu-\nu}(\hat{r}) \chi_{S\nu} \\ &= \chi^\mu_\kappa, \end{aligned} \tag{8.44}$$

where the Clebsch–Gordan coefficients for spin 1/2 are shown in Table 8.1. The functions χ^μ_κ have the following properties of orthogonality:

$$\langle \chi^\mu_\kappa | \chi^{\mu'}_{\kappa'} \rangle = \delta_{\kappa\kappa'} \delta_{\mu\mu'}, \tag{8.45}$$

and completeness:

$$\sum_{\kappa,\mu} \langle \mathbf{r} | \chi^\mu_\kappa \rangle \langle \chi^{\mu'}_{\kappa'} | \mathbf{r}' \rangle = \delta(\mathbf{r} - \mathbf{r}') I_2. \tag{8.46}$$

J	$\nu = \frac{1}{2}$	$\nu = -\frac{1}{2}$
$\ell + \frac{1}{2}$	$\left(\frac{\ell+\mu+\frac{1}{2}}{2\ell+1}\right)^{1/2}$	$\left(\frac{\ell-\mu+\frac{1}{2}}{2\ell+1}\right)^{1/2}$
$\ell - \frac{1}{2}$	$-\left(\frac{\ell-\mu+\frac{1}{2}}{2\ell+1}\right)^{1/2}$	$\left(\frac{\ell+\mu+\frac{1}{2}}{2\ell+1}\right)^{1/2}$

TABLE 8.1. The Clebsch–Gordan coefficients for $S = \frac{1}{2}$.

8.3 Generalized Structure Constants

We saw above, Eq.(8.32), that in the generalized spherical harmonic representation the free-particle propagator is given by the expression,

$$\begin{aligned}\mathtt{G}_0(\mathbf{r}-\mathbf{r}') \equiv G_0(\mathbf{r}-\mathbf{r}') &= -ik\sum_L J_L(\mathbf{r})H_L(\mathbf{r}') \\ &= -ik\sum_{L,\mu}\sum_{l,S} \bar{\mathtt{J}}^{J\mu}_{lS}(\mathbf{r})\mathtt{H}^{J\mu}_{lS}(\mathbf{r}') \quad \text{for} \quad r' > r. \end{aligned} \tag{8.47}$$

We use this expression to derive the form of the structure constants that enter the relativistic formulation of MST. We begin by obtaining the expansion of the generalized Hankel function about a shifted origin (with $|\mathbf{R}| > |\mathbf{r}|$),

$$\begin{aligned}\mathtt{H}^{J\mu}_{lS}(\mathbf{r}-\mathbf{R}) &= \sum_{\substack{m,\nu \\ m+\nu=\mu}} \langle lmS\nu|J\mu\rangle h_l(k|\mathbf{r}-\mathbf{R}|)Y_{lm}(r \hat{-} R)\chi_{S\nu} \\ &= \sum_{m,\nu}\langle lmS\nu|J\mu\rangle\chi_{S\nu}\left\{\sum_{l'm'} G_{lm,l'm'}(\mathbf{R})J_{l'm'}(\mathbf{r})\right\}. \end{aligned} \tag{8.48}$$

Now, for three vectors, $\boldsymbol{\rho}_1, \boldsymbol{\rho}_2$, and $\mathbf{R}$, satisfying the MT conditions, and through the use of Eqs.(8.7) and (8.8), we obtain,

$$\begin{aligned}\mathtt{G}_0(-\boldsymbol{\rho}_1+\mathbf{R}+\boldsymbol{\rho}_2) &= -ik\sum_{J\mu}\sum_{lS} \bar{\mathtt{J}}^{J\mu}_{lS}(\boldsymbol{\rho}_1)\mathtt{H}^{J\mu}_{lS}(\mathbf{R}+(\boldsymbol{\rho}_2) \\ = -ik\sum_{J\mu}\sum_{lS}\bar{\mathtt{J}}^{J\mu}_{lS}(\boldsymbol{\rho}_1)&\left[\sum_{m\nu}\sum_{l_1m_1}\langle lmS\nu|J\mu\rangle\chi_{S\nu}\right. \\ \times \sum_{l_1m_1} &G_{lm,l_1m_1}(\mathbf{R})J_{l_1m_1}(-\boldsymbol{\rho}_2)] \\ = -ik\sum_{J\mu}\sum_{lS}\bar{\mathtt{J}}^{J\mu}_{lS}(\boldsymbol{\rho}_1)&\left[\sum_{m\nu}\sum_{l_1m_1}\langle lmS\nu|J\mu\rangle\right. \\ \times\, G_{lm,l_1m_1}(\mathbf{R})&\left.\left\{\sum_{J'\mu'}\langle l_1m_1S\nu|J'\mu'\mathtt{J}^{J'\mu'}_{l_1S}(\boldsymbol{\rho}_2)\rangle\right\}\right] \\ = -ik\sum_{J\mu}\sum_{J'\mu'}\sum_{S}\bar{\mathtt{J}}^{J\mu}_{lS}(\boldsymbol{\rho}_1)&\left[\sum_{ll_1}\sum_{m,m_1,\nu}\langle lmS\nu|J\mu\rangle\right. \\ \times\, G_{lm,l_1m_1}(\mathbf{R})\langle l_1m_1S\nu|J'\mu'\rangle&\Big]\mathtt{J}^{J'\mu'}_{l_1S}(\boldsymbol{\rho}_2). \end{aligned} \tag{8.49}$$

Defining the structure constants in analogy with the spinless case, i.e., as the quantity in the expansion of the free-particle propagator that connects the two regular solutions, we have

$$\mathsf{G}^{J\mu,J'\mu'}_{lS,l_1S_1}(\mathbf{R}) = \delta_{SS_1} \sum_{m,m_1,\nu} \langle lmS\nu|J\mu\rangle G_{lm,l_1m_1}(\mathbf{R})\langle l_1m_1S\nu|J'\mu'\rangle. \quad (8.50)$$

This is the desired expression for the structure constants in the case of arbitrary spin. Note that they are given as linear combinations of the ordinary (spinless) structure constants multiplied by Clebsch–Gordan coefficients. For the case $S = \frac{1}{2}$, use of Table 8.1 gives

$$\left[G^{ij}_{KK'}\right] = \sum_{\nu=\pm 1/2} \left\langle l\mu - \nu, \frac{1}{2}\nu \middle| J\mu \right\rangle G^{ij}_{LL'} \left\langle l'\mu' - \nu, \frac{1}{2}\nu \middle| J'\mu' \right\rangle, \quad (8.51)$$

where K is a combined index, $K = (\kappa, \mu)$.

8.4 Free-Particle Solutions

In developing MST within a relativistic formalism, we follow essentially the same lines as in the case of nonrelativistic scattering. In this section, we discuss the free-particle solutions of the Dirac equation, Eq.(8.6), and the corresponding free-particle Green function. The solution of the Dirac equation for a single potential by the use of phase functions is presented in the following section. We summarize the cases of both spherically symmetric and generally shaped potential cells. Multiple scattering theory is then presented in Section 9.6. In this discussion we concentrate on the case of electron scattering, so that $S = \frac{1}{2}$.

8.4.1 Free-Particle Solution of the Dirac Equation

The Dirac equation is a four-dimensional matrix equation whose solutions, correspondingly, are four-dimensional vectors, or *bi-spinors* [3, 4]. In the free-particle case, it can be easily verified that the solution regular at the origin takes the form

$$|J^{\mu}_{\kappa}\rangle = \begin{pmatrix} j_l(pr)\chi^{\mu}_{\kappa}(\hat{r}) \\ \dfrac{ipS_{\kappa}}{E+1} j_{\bar{l}}(pr)\chi^{\mu}_{-\kappa}(\hat{r}) \end{pmatrix}, \quad (8.52)$$

while the irregular solution becomes,

$$|H^{\mu}_{\kappa}\rangle = \begin{pmatrix} h_l(pr)\chi^{\mu}_{\kappa}(\hat{r}) \\ \dfrac{ipS_{\kappa}}{E+1} h_{\bar{l}}(pr)\chi^{\mu}_{-\kappa}(\hat{r}) \end{pmatrix}. \quad (8.53)$$

It is to be noted that the square root, p, of the kinetic energy, rather than the total energy of the electron appears in the argument of the spherical functions.

8.4.2 The Free-Particle Propagator

In order to proceed with a discussion of scattering theory within a relativistic formalism, it is convenient to obtain the Green function operator for the Dirac Hamiltonian, both in the absence and the presence of a potential. When the potential vanishes, the corresponding free-particle propagator satisfies the equations,

$$(\underline{\alpha} \cdot \mathbf{p} + \underline{\beta} - E)\mathtt{G}_0(\mathbf{r}, \mathbf{r}') = \delta(\mathbf{r} - \mathbf{r}')I_4, \tag{8.54}$$

or

$$\mathtt{G}_0(\mathbf{r}, \mathbf{r}') = (\underline{\alpha} \cdot \mathbf{p} + \underline{\beta} - E)^{-1}\delta(\mathbf{r} - \mathbf{r}')I_4. \tag{8.55}$$

We now multiply the right-hand side of the last equation by unity in the form $(\underline{\alpha} \cdot \mathbf{p} + \underline{\beta} + E)(\underline{\alpha} \cdot \mathbf{p} + \underline{\beta} + E)^{-1}$, and use the fact that

$$H_D^2 = (\underline{\alpha} \cdot \mathbf{p} + \underline{\beta})^2 = (\nabla^2 + 1), \tag{8.56}$$

to obtain the result

$$\begin{aligned} \mathtt{G}_0(\mathbf{r}, \mathbf{r}') &= (\underline{\alpha} \cdot \mathbf{p} + \underline{\beta} + E)(\nabla^2 - p^2)^{-1}\delta(\mathbf{r} - \mathbf{r}')I_4 \\ &= (\underline{\alpha} \cdot \mathbf{p} + \underline{\beta} + E)G_0^n(\mathbf{r} - \mathbf{r}')I_4, \end{aligned} \tag{8.57}$$

where the definition of the nonrelativistic free-particle propagator, $G_0^n(\mathbf{r} - \mathbf{r}')$, has been used on the right-hand side. The last expression can be written in the form [4]

$$\begin{aligned} \mathtt{G}_0(\mathbf{r}, \mathbf{r}') &= \begin{pmatrix} E+1 & \underline{\sigma} \cdot \mathbf{p} \\ \underline{\sigma} \cdot \mathbf{p} & E-1 \end{pmatrix} G_0^n(\mathbf{r} - \mathbf{r}') \\ &= \begin{pmatrix} G_0^{11} & G_0^{12} \\ G_0^{21} & G_0^{22} \end{pmatrix}. \end{aligned} \tag{8.58}$$

Now, the free-particle propagator $G_0^n(\mathbf{r} - \mathbf{r}')$ can be expanded in terms of spherical harmonics, *or* in terms of spinor functions,

$$\begin{aligned} G_0^n(\mathbf{r} - \mathbf{r}') &= -\frac{e^{\mathrm{i}p|\mathbf{r} - \mathbf{r}'|}}{4\pi|\mathbf{r} - \mathbf{r}'|} \\ &= \mathrm{i}p \sum_{J\mu} \sum_{lS} h_l(pr_>) j_l(pr_<) \mathtt{Y}_{lS}^{J\mu}(\hat{r}) \bar{\mathtt{Y}}_{lS}^{J\mu}(\hat{r}') \\ &= \mathrm{i}p \sum_{J\mu} \sum_{lS} h_l(pr_>) j_l(pr_<) \mathtt{Y}_{\kappa}^{\mu}(\hat{r}) \bar{\mathtt{Y}}_{\kappa}^{\mu}(\hat{r}'). \end{aligned} \tag{8.59}$$

The last line follows from the definition of the Λ representation, use of the identity

$$\underline{\sigma} \cdot \mathbf{p}|\mathtt{Y}_{\kappa}^{\mu}\rangle = \mathrm{i}pS_{\kappa}\mathtt{Y}_{-\kappa}^{\mu}, \tag{8.60}$$

and the fact that we are only considering a single spin. Now, the free-particle propagator can be constructed as the tensor product of the regular and irregular solutions of the free-particle Dirac equation, Eqs.(8.52) and (8.53). Setting $|\mathbf{r}| > \mathbf{r}'$, we have

$$\mathrm{G}_0(\mathbf{r}-\mathbf{r}') = \sum_{\kappa,\mu} |\tilde{J}_\kappa^\mu((\mathbf{r}))\rangle\langle H_\kappa^\mu((\mathbf{r}'))| = \begin{pmatrix} G_0^{11}(\mathbf{r}-\mathbf{r}') & G_0^{12}(\mathbf{r}-\mathbf{r}') \\ G_0^{21}(\mathbf{r}-\mathbf{r}') & G_0^{22}(\mathbf{r}-\mathbf{r}') \end{pmatrix}, \tag{8.61}$$

where each of the elements is a two-dimensional matrix. Explicitly, we have,

$$G_0^{11}(\mathbf{r}-\mathbf{r}') = \left[\frac{p(E+1)}{\pi}\right] \sum_{\kappa,\mu} h_l(pr) j_l(pr') \chi_\kappa^\mu(\hat{r}) \bar{\chi}_\kappa^\mu(\hat{r}'), \tag{8.62}$$

$$G_0^{22}(\mathbf{r}-\mathbf{r}') = \frac{E-1}{E+1} G_0^{11}(\mathbf{r}-\mathbf{r}'), \tag{8.63}$$

and

$$G_0^{12}(\mathbf{r}-\mathbf{r}') = -G_0^{21}(\mathbf{r}-\mathbf{r}') = -p^2 \sum_{\kappa\mu} S_\kappa \sum_{\kappa,\mu} h_l(pr) j_l(pr') \chi_{-\kappa}^\mu(\hat{r}) \bar{\chi}_\kappa^\mu(\hat{r}'). \tag{8.64}$$

Because of the symmetry of $G_0(\mathbf{r}-\mathbf{r}')$ with respect to its arguments, the expressions for $G_0^{\alpha\beta}$ when $|\mathbf{r}'| > |\mathbf{r}|$ can be obtained from those above through an interchange of the arguments. We note also that the expression for the stationary, rather than the outgoing, propagator can be obtained through the replacement of h_l by n_l in the expressions quoted above.

8.5 Relativistic Single-Site Scattering Theory

In this section, we begin our exposition of relativistic scattering theory. First, we examine the case of a spherically symmetric, spatially bounded potential (MT potential), and subsequently generalize to cell potentials of arbitrary shape. As in the nonrelativistic case, we base our discussion on the Lippmann–Schwinger equation, and the method of phase functions. Thus, we seek to obtain the solution of the equation,

$$\psi^+(\mathbf{r}) = \phi^+(\mathbf{r}) + \int \mathrm{d}r^3 G_0(\mathbf{r}-\mathbf{r}') V(\mathbf{r}') \psi^+(\mathbf{r}'), \tag{8.65}$$

where the various quantities are to be understood within a relativistic, bi-spinor, context.

8.5.1 Spherically Symmetric Potentials

When the potential is spherically symmetric, each l (or κ) channel behaves independently of the others, and Eq.(8.65) can be written down for each channel (partial wave) separately,

$$\psi_\kappa^+(\mathbf{r}) = J_\kappa(\mathbf{r}) + \int \mathrm{d}r^3 G_0(\mathbf{r}-\mathbf{r}')V(\mathbf{r}')\psi_\kappa^+(\mathbf{r}'). \tag{8.66}$$

Here, the index μ is suppressed as it does not influence the solutions of this equation. We have also defined the outgoing free-particle propagator in terms of partial propagators corresponding to the κ channel,

$$G_0(\mathbf{r}-\mathbf{r}') = \sum_{\kappa\kappa'} G_0^{\kappa\kappa'}(r,r')Y_\kappa(\hat{r})Y_\kappa(\hat{r}'), \tag{8.67}$$

where

$$G_0^{\kappa\kappa}(r,r') = \left[\frac{(E+1)p}{\pi}\right] \times \begin{pmatrix} j_l(pr)h_l(pr') & \dfrac{\mathrm{i}pS_\kappa}{(E+1)} j_l(pr)h_{\bar{l}}(pr') \\ \dfrac{\mathrm{i}pS_\kappa}{(E+1)} j_{\bar{l}}(pr)h_l(pr') & \dfrac{(\mathrm{i}p)^2}{(E+1)^2} j_{\bar{l}}(pr)h_{\bar{l}}(pr') \end{pmatrix} \tag{8.68}$$

As in the case of nonrelativistic scattering theory, the solution of Eq.(8.65) for the case of spherically symmetric potentials assumes the form,

$$\phi_\alpha = \bar{C}_{\alpha\alpha}(r)J_\alpha(\mathbf{r}), \tag{8.69}$$

and

$$\psi_\alpha^{(+)} = \left[c_{\alpha\alpha}(r)J_\alpha(\mathbf{r}) - s_{\alpha\alpha}(r)H_\alpha(\mathbf{r})\right]. \tag{8.70}$$

Specific expressions for the radial wave functions are given in the literature [4].

Quantities such as the t-matrix and S-matrix can be defined in exact analogy with nonrelativistic scattering theory, in terms of the phase functions. For example, the S-matrix takes the form,

$$S_\kappa(p) = \mathrm{e}^{-2\mathrm{i}\delta_\kappa}. \tag{8.71}$$

As a simple illustration, Fig.(8.1) shows the nonrelativistic and the relativistic phase shifts for Ni and Pt. We note that relativistic effects are much stronger in Pt than in Ni, resulting in greater splitting of the parts associated with opposite spins (spin–orbit interaction).

8.5.2 Spin-Orbit Coupling

It is useful to continue our discussion of central fields to derive a detailed form of the Dirac radial equation. This will also reveal explicitly the role played by the spin–orbit interaction.

For central fields, the square of the total angular momentum, $\mathbf{J} = \mathbf{L}+\mathbf{S}$, its projection, J_z, along the z-axis, the square of the spin operator, S^2,

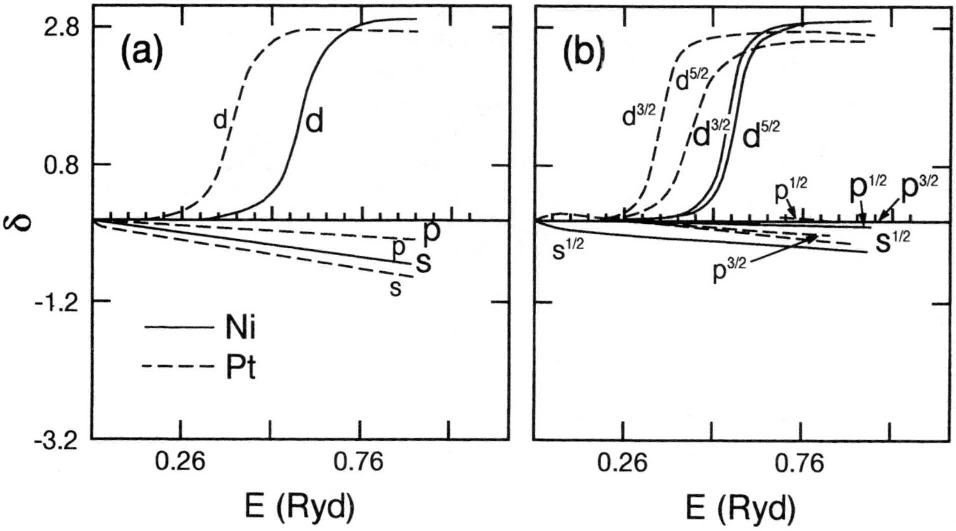

FIGURE 8.1. Nonrelativistic (a) and relativistic (b) phase shifts for Ni and Pt relevant to spherically symmetric potentials for a $Ni_{50}Pt_{50}$ alloy. (J. Staunton, private communication).

where $\mathbf{S} = (1/2)\boldsymbol{\sigma}$, and $K = \beta(\boldsymbol{\sigma} \cdot \mathbf{L} + 1)$ are constants of the motion, so that

$$\left[H, J^2\right] = \left[H, J_z\right] = \left[H, S^2\right] = \left[H, K\right] = 0. \tag{8.72}$$

With the wave function being a bi-spinor of the form,

$$|\psi\rangle = |\phi, \chi\rangle, \tag{8.73}$$

we have the eigenvalue equations,

$$\begin{aligned} J^2|\phi\rangle &= j(j+1)|\phi\rangle, \quad J^2|\chi\rangle = j(j+1)|\chi\rangle \\ J_z|\phi\rangle &= \mu|\phi\rangle, \quad J_z|\chi\rangle = \mu|\chi\rangle \\ (\boldsymbol{\sigma} \cdot \mathbf{L} + 1)|\phi\rangle &= -\kappa|\phi\rangle, \quad (\boldsymbol{\sigma} \cdot \mathbf{L} + 1)|\chi\rangle = -\kappa|\chi\rangle, \end{aligned} \tag{8.74}$$

where μ is the azimuthal quantum number and $j = 1/2, 3/2, \ldots$, with $-j \leq \mu \leq j$.

The wave function, $|\psi_Q\rangle$, with $Q = (\kappa\mu)$, is given by an expression analogous to that in Eq.(8.52),

$$\langle \mathbf{r}|\psi_Q\rangle = \begin{pmatrix} \phi_Q(\mathbf{r}) \\ \chi_q(\mathbf{r}) \end{pmatrix} = \begin{pmatrix} g_\kappa(r)\chi_\kappa^\mu(\hat{r}) \\ \mathrm{i}f_\kappa(r)\chi_{-\kappa}^\mu(\hat{r}) \end{pmatrix}, \tag{8.75}$$

where $g_\kappa(r)$ and $f_\kappa(r)$ are radial amplitudes. Again, the phase i makes these amplitudes explicitly real. In order to derive equations for the radial amplitudes, we make use of the radial momentum,

$$p_r = \frac{\hbar}{\mathrm{i}} \frac{1}{r} \frac{\partial}{\partial r} = \frac{\hbar}{\mathrm{i}} \left(\frac{\partial}{\partial r} + \frac{1}{r} \right), \tag{8.76}$$

and the so-called "radial velocity," α_r

$$\alpha_r = \boldsymbol{\alpha} \cdot \hat{r}. \tag{8.77}$$

In terms of these quantities, we can write

$$\begin{aligned} \boldsymbol{\alpha} \cdot \mathbf{p} &= \alpha_r \left(p_r + \mathrm{i}\beta K/r \right) = -\mathrm{i}\alpha_r \left(\frac{\partial}{\partial r} + \frac{1}{r} - \frac{\beta K}{r} \right) \\ &= \mathrm{i}\gamma_5 \sigma_r \left(\frac{\partial}{\partial r} + \frac{1}{r} - \frac{\beta K}{r} \right), \end{aligned} \tag{8.78}$$

where

$$\gamma_5 = \begin{pmatrix} 0 & -I_2 \\ -I_2 & 0 \end{pmatrix}, \quad \sigma_r \boldsymbol{\sigma} \cdot \mathbf{L} = \mathrm{i}\boldsymbol{\sigma} \cdot \mathbf{r} \times \mathbf{L}. \tag{8.79}$$

We can now write the Dirac equation for central fields in the form of two coupled equations for the radial amplitudes,

$$\begin{aligned} \frac{\mathrm{d}f_\kappa(r)}{\mathrm{d}r} &= \frac{\kappa - 1}{r} f_\kappa(r) - \left[E - U(r) - 1 \right] g_\kappa(r) \\ \frac{\mathrm{d}g_\kappa(r)}{\mathrm{d}r} &= \left[E - U(r) - 1 \right] f_\kappa(r) - \frac{\kappa - 1}{r} g_\kappa(r). \end{aligned} \tag{8.80}$$

These two equations are often referred to as "the radial Dirac equation." We note that in contrast to the single Schrödinger equation, which is second order, the radial Dirac equation consists of two-coupled equations each of first order. In these equations, the term involving $\boldsymbol{\sigma} \cdot \mathbf{L}$ in Eqs.(8.78) and (8.79) couples the spin and orbital angular momenta of a particle giving rise to the so-called *spin–orbit interaction.*

8.5.3 Scalar Relativistic Expressions

The terms *scalar relativistic* or *pseudorelativistic* are often used to denote simplified forms of the Dirac equation. We now elucidate the meaning of these terms.

Through the use of the bi-spinor wave function, the Dirac Hamiltonian, Eq.(8.1) gives rise to two equations (in atomic units),

$$\begin{aligned} \mathrm{c}\boldsymbol{\sigma} \cdot \mathbf{p}|\chi\rangle - V|\phi\rangle &= \epsilon|\phi\rangle \\ \mathrm{c}\boldsymbol{\sigma} \cdot \mathbf{p}|\phi\rangle - (U - 2\mathrm{c}^2)|\chi\rangle &= \epsilon|\chi\rangle. \end{aligned} \tag{8.81}$$

In these equations, the spinor $|\chi\rangle$ can be eliminated in terms of $|\phi\rangle$

$$|\chi\rangle = \left(\frac{1}{2\mathrm{c}} \right) B^{-1} \boldsymbol{\sigma} \cdot \mathbf{p}|\phi\rangle, \tag{8.82}$$

where

$$B = 1 + \left(1/2c^2\right)(\epsilon - V), \tag{8.83}$$

leading to the single equation for $|\phi\rangle$,

$$D|\phi\rangle = \epsilon|\phi\rangle, \tag{8.84}$$

$$D = \frac{1}{2}\boldsymbol{\sigma} \cdot \mathbf{p}B^{-2}\boldsymbol{\sigma} \cdot \mathbf{p} + V. \tag{8.85}$$

We can also express the normalization condition of the wave function, $|\psi\rangle$, in terms of the single spinor, $|\phi\rangle$,

$$\begin{aligned}\langle\psi|\psi\rangle &= \langle\phi|\phi\rangle + \langle\chi|\chi\rangle R_\kappa(r) \\ &= \langle\phi|\left\{1 + (1/4c^2)\right\}\boldsymbol{\alpha} \cdot \mathbf{p}B^{-2}\boldsymbol{\alpha} \cdot \mathbf{p}|\phi\rangle.\end{aligned} \tag{8.86}$$

For central fields, Eq.(8.84) is separable in the radial and angular coordinates. Writing the radial amplitude $|\phi\rangle$ in the form $R_\kappa(r)/r$, we obtain [5],

$$\begin{aligned}&\left[\frac{1}{2}\left(-\frac{\mathrm{d}^2}{\mathrm{d}r^2} + \frac{l(l+1)}{r^2}\right) + V(r) - \epsilon\right] \\ &= \left\{\frac{1}{4c^2}B^{-2}\frac{\mathrm{d}V(r)}{\mathrm{d}r}\frac{\kappa}{r}\frac{1}{4c^2}\left[\left[\epsilon - V(r)\right]B^{-1}\left(-\frac{\mathrm{d}^2}{\mathrm{d}r^2} + \frac{l(l+1)}{r^2}\right)\right]\right. \\ &\left. + \frac{1}{4c^2}B^{-2}\frac{\mathrm{d}V(r)}{\mathrm{d}r}\frac{\mathrm{d}}{\mathrm{d}r}\right\}R_\kappa(r).\end{aligned} \tag{8.87}$$

This equation has a number of outstanding features.

1. In the limit $c \to \infty$ it reduces to the Schrödinger equation.

2. Setting $B = 1$, we obtain the "Pauli–Schrödinger" equation, in which we can identify the three terms on the right-hand side as the "spin–orbit coupling" term, depending explicitly on κ, the "mass-velocity" term, and the "Darwin shift".

3. For $B \neq 1$, relativistic corrections of order c^{-4} enter via the normalization condition, Eq.(8.86).

4. Omitting the term in κ leads to "modified Schrödinger" or "scalar relativistic" equation. This equation includes some relativistic effects, namely, the mass velocity term and the Darwin shift while being of simple enough form to be amenable to study by means used with the ordinary Schrödinger equation.

8.5.4 Generally Shaped, Scalar Potentials

In the case of nonspherical, generally shaped potentials, the solutions of Eq.(8.65) are of the general forms,

$$\phi_\alpha = \sum_{\alpha'} C_{\alpha'\alpha}(r)J_{\alpha'}(\mathbf{r}) \tag{8.88}$$

and

$$\psi_{\alpha}^{(+)} = \sum_{\alpha'} \left[c_{\alpha'\alpha}(r) J_{\alpha'}(\mathbf{r}) - s_{\alpha'\alpha}(r) H_{\alpha'}(\mathbf{r}) \right]. \tag{8.89}$$

Keeping in mind the four-dimensional nature of the various quantities entering the last two equations, and in complete analogy with the nonrelativistic case, we obtain the expressions

$$\mathtt{c}_{\kappa\kappa'}(r) = 1 - \int_{r'>r} \mathrm{d}^3r' |\bar{H}_{\kappa'}(\mathbf{r}')\rangle V(\mathbf{r}') \langle \psi^{(+)}{}_{\kappa}(\mathbf{r}')| \tag{8.90}$$

and

$$\mathtt{s}_{\kappa\kappa'}(r) = - \int_{r'<r} \mathrm{d}^3r' |\bar{J}_{\kappa'}(\mathbf{r}')\rangle V(\mathbf{r}') \langle \psi^{(+)}{}_{\kappa}(\mathbf{r}')|, \tag{8.91}$$

for the phase functions. One can proceed from these expressions and obtain the integral or differential equations determining the phase functions, $\mathtt{C}_{\kappa\kappa'}(r)$ and $\mathtt{S}_{\kappa\kappa'}(r)$ in a manner completely analogous to the nonrelativistic case.

At points outside a sphere bounding the potential, $V(\mathbf{r})$, the phase functions assume their asymptotic, constant values, $\mathtt{C}(\infty)$ and $\mathtt{S}(\infty)$, in terms of which we can define the four-dimensional single-site scattering matrix [6],

$$\mathtt{t} = - \left[\mathtt{C}(\infty) S^{-1}(\infty) - \mathrm{i} \right] / (E+1)p. \tag{8.92}$$

This relation can also be expressed in terms of surface integrals of Wronskian relations, again in complete analogy with the nonrelativistic case. Finally, the single-particle Green function is given by the expression ($r' > r$),

$$G(\mathbf{r},\mathbf{r}') = \sum_{KK'} |Z_K(\mathbf{r})\rangle \mathtt{t}_{KK'} \langle Z_{K'}| - \sum_{K} |Z_K(\mathbf{r})\rangle \langle S_K(\mathbf{r}')|, \tag{8.93}$$

where $K = (\kappa, \mu)$ is a combined index, and Z_K and S_K are those regular and irregular solutions of the Dirac equation that on the surface of the bounding sphere join smoothly to the functions,

$$|Z_K(\mathbf{r})\rangle \to \sum_{K} \mathtt{t}^{-1}_{KK'} |\bar{J}_{K'}(\mathbf{r}) - \mathrm{i}p(E+1) |\bar{H}_{K'}(\mathbf{r})\rangle \tag{8.94}$$

and

$$|S_K(\mathbf{r})\rangle \to |\bar{J}_K(\mathbf{r})\rangle, \tag{8.95}$$

respectively.

8.6 Relativistic Multiple Scattering Theory

The reader may have realized by now that relativistic scattering theory bears great formal similarity to its nonrelativistic counterpart. To obtain the former from the latter, at least in the case of scalar fields[2] it is only necessary to interpret the quantities in the nonrelativistic formalism in terms of four-dimensional vectors and matrices. Nonscalar fields, notably involving the presence of a vector potential have been discussed in the literature [8-11]. This similarity extends to MST. We trust that the reader can write down all appropriate generalizations without undue effort. For the sake of illustration, we quote the form of the relativistic, multiple scattering Green function

$$G(\mathbf{r}, \mathbf{r}') = \sum_{KK'} |Z_K(\mathbf{r})\rangle \tau^{ij}_{KK'} \langle Z_{K'}| - \delta_{ij} \sum_{K} |Z_K(\mathbf{r})\rangle \langle S_K(\mathbf{r}')|, \tag{8.96}$$

where we chose $r' > r$. For a more complete discussion we refer the reader to the literature [12].

References

[1] E. Tamura, Phys. Rev. B**45**, 3271 (1992).
[2] C. J. Joachain, *Quantum Collision Theory* (North-Holland, Amsterdam, 1975).
[3] P. M. Dirac, *Quantum Mechanics* (Oxford, 1961).
[4] M. E. Rose, *Relativistic Electron Theory* (Wiley, New York, 1961).
[5] F. Rosicky. P. Weinberger, and F. Mark, J. Phys. B, Molecular Phys. **9**, 2871 (1976).
[6] A. P. Shen, Phys. Rev. B**9**, 1328 (1974).
[7] A. K. Rajagopal and J. Callaway, Phys. Rev. B**7** 1912 (1973).
[8] R. Feder, F. Rosicky, and B. Ackerman, Z. Phys. B**52** 31 (1983); **53**, 144 (1983).
[9] P. Strange. J. B. Staunton, and B. L. Györffy, J. Phys. C**17**, 3355 (1984).
[10] G. Shadler, P. Weinberger, A. M. Boring, and R. C. Albers, Phys. Rev. B**34**, 713 (1986).
[11] P. Strange, H. Ebert, J. B. Staunton, and B. L. Gyorffy, J. Phys. Condens. Matt. **1**, 2959 (1989).
[12] P. Weinberger, *Electron Scattering Theory for Ordered and Disordered Matter* (Clarendon Press, Oxford, 1990).

---

[2]The presence of vector fields, such as arise in the presence of a magnetic field, requires a more detailed consideration.

9
The Poisson Equation

9.1 General Comments

Up to this point in our discussion we have considered the application of multiple scattering theory in obtaining the solution of the one-particle Schrödinger equation in solid materials. We have seen that given a potential, MST allows the exact treatment of the associated Schrödinger equation even in the general case of cell potentials that fill all space.

In this chapter, we turn our attention to the determination of the electronic potential that enters the Schrödinger equation by considering the solution of Poisson's equation in solid materials. In the language of classification of partial differential equations, the Poisson equation is of the same, elliptic type as the Schrödinger equation, and one might reasonably expect that these two equations can be treated by similar means. We will see that in fact the methods of multiple scattering theory can be used essentially intact in the solution of the Poisson equation and the calculation of the electrostatic interaction in solids described by arbitrarily shaped, space-filling charges.

9.2 Multipole Moments

Realistic calculations of the electronic structure of a material are commonly based on the self-consistent calculation of the potential within the framework of the local-density approximation (LDA) [1–3] to density functional theory [3, 4]. In this approach, one solves iteratively an interrelated system of Poisson and Schrödinger equations, where an input potential is used to generate a wave function and hence a charge density which in turn yields a new value of the potential for the next iteration. In addition to the potential, the determination of the total energy requires an accurate evaluation of the electrostatic interaction associated with the calculated charge. Clearly, in proceeding to self-consistency, the Schrödinger and the Poisson equations are of equal importance.

Our aim in this chapter is to develop methods for the exact treatment of the Poisson equation associated with a charge density, $\rho(\mathbf{r})$,

$$\nabla^2 V = -4\pi\rho, \tag{9.1}$$

or, equivalently, the evaluation of the integral

$$V(\mathbf{r}) = \int \frac{\rho(\mathbf{r}')}{|\mathbf{r} - \mathbf{r}'|} \mathrm{d}^3 r', \tag{9.2}$$

which is the solution of Eq.(9.1) in free space [5]. Also, we wish to consider the related problem of calculating the electrostatic energy of the charge $\rho(\mathbf{r})$,

$$\begin{aligned} E &= \int V(\mathbf{r})\rho(\mathbf{r})\mathrm{d}^3 r \\ &= \int\int \frac{\rho(\mathbf{r})\rho(\mathbf{r}')}{|\mathbf{r} - \mathbf{r}'|} \mathrm{d}^3 r \, \mathrm{d}^3 r'. \end{aligned} \tag{9.3}$$

Over the years, a number of methods have been proposed for the evaluation of the integral in Eq.(9.2), most of them geared towards the special features of the system under consideration. Perhaps the simplest form of electrostatic interaction is that associated with point charges of opposite sign arranged on the sites of a lattice. The self-potential of such an ionic crystal was originally considered by Madelung [6]. A general procedure for obtaining the self-interaction of a translationally invariant material consisting of cell charges, $\rho^i(\mathbf{r})$, was subsequently given by Ewald [7], whose method gave rise to a number of specific applications [8–10], modifications and generalizations [11–17]. Works based on Ewald's method have encompassed charge distributions of the muffin-tin (MT) form, that is, spherically symmetric cell charges confined inside nonoverlapping spheres situated in a constant interstitial potential (often chosen to be zero), or more generally shaped, nonspherical arrangements of charge. The general approach in these works has been to consider the short-range Coulomb interactions arising from neutral cell charges, and to take account of the long-range interaction by lattice sums using Ewald's method. In addition, approaches based on the use of finite differences [18] and on multipole expansions [19–25] of the charge $\rho(\mathbf{r})$ have been used.

Of the approaches based on multipole expansions, that of Morgan [25] must be singled out as the first attempt to solve the Poisson equation for a solid entirely in terms of the multipole moments of the charge inside the WS cells. In this approach, one expresses the potential due to a spatially bounded charge distribution in the form

$$V(\mathbf{r}_0) = \sum_{l,m} \frac{4\pi}{2l+1} \frac{Y_{lm}(\widehat{\mathbf{r}_0})}{r_0^{l+1}} Q_{lm}, \tag{9.4}$$

where the multipole moments for the charge are given [5] by the integral

$$Q_{lm} = \int r^l Y^*_{lm}(\hat{\mathbf{r}})\rho(\mathbf{r})d^3r. \tag{9.5}$$

Here, $Y_{lm}(\hat{\mathbf{r}})$ is a spherical harmonic of order $L(= l, m)$, and an asterisk denotes complex conjugation. It is to be noted that Eq.(9.4) is rigorously valid only outside a sphere bounding the charge, for example, for charges that satisfy the muffin-tin conditions. Thus, for a system of MT charges the electrostatic potential at point $\mathbf{r}_0$ inside cell j is given by the expression

$$V(\mathbf{r}_0) = \sum_{n\neq j}\sum_{l,m} \frac{4\pi}{2l+1} \frac{Y_{lm}(\widehat{\mathbf{r}_0})}{r_0^{l+1}} Q^n_{lm} + V^j_0(\mathbf{r}_0) \tag{9.6}$$

where Q^n_{lm} are the multipole moments of the charge in cell n centered at point $\mathbf{R}_n$, and $\mathbf{r}_n = \mathbf{r} - \mathbf{R}_n$. The quantity $V^j_0(\mathbf{r}_0)$ denotes the potential inside cell j, which can be obtained by a number of means [23]. Morgan's method consists in using Eq.(9.6) even in cases in which the potential did not conform to the MT description, such as the space-filling, Wigner–Seitz cells in a solid. However, for points near the face of such a cell, Eq.(9.6) diverges, an effect noted by Morgan. In the discussion which follows, we will formulate an expansion for the potential in terms of the multipole moments which is free of divergences and leads to well-defined values of the potential and the electrostatic interaction everywhere in space.

9.3 Comparison with the Schrödinger Equation

In spite of the difficulties noted by Morgan, the idea of expressing the potential in terms of individual multipole moments is very appealing for a number of reasons. First, it would be in nearly perfect accord with the multiple-scattering theory approach to electronic structure, which also relies on the evaluation of cell multipole expansions, in this case the cell t-matrix. This similarity can be seen clearly upon comparison of Eq.(9.5) with Eq.(4.77). With the formal identification of $\rho(\mathbf{r})$ in Eq.(9.5) with $V(\mathbf{r})\psi_L(\mathbf{r})$ in Eq.(4.77), and the realization that the quantity $r^\ell Y_L(\hat{\mathbf{r}})$ is the zero-energy limit of $J_L(\mathbf{r})$ [26], we see that $Q_{\ell m}$ is the formal equivalent of the t-matrix. In fact, as $E \to 0$ $(k \to 0)$, the free-particle propagator that appears in Eq.(4.75) reduces to the form $\frac{1}{|\mathbf{r}-\mathbf{r}'|}$ and the Poisson equation becomes in appearance equivalent to the Schrödinger equation. Of course, the source term in the Schrödinger equation is proportional to the wave function that results in a homogeneous equation still to be solved, whereas the integral form of the Poisson equation represents a formal solution of Eq.(9.1). Nevertheless, the similarity is strong enough that the two equations can be attacked by formally similar means.

Second, it would be ideally suited for the treatment of point defects, either substitutional or interstitial, as is MST, obviating the need for using artificial constructions such as supercells in the treatment of the embedding problem. Third, it would conceivably apply to systems without full translational invariance, because of the possibility of summing Eq.(9.6) directly in real space in many cases of physical interest, for example, cubic systems. This last feature would be of great relevance to those applications of MST to the calculation of the electronic structure [26–31] that allow the treatment of systems with severely reduced or no translational invariance. It would also fit in well with applications of MST to potentials partitioned into space-filling cells. Provided that these features were made part of a computationally practical procedure, the use of multipole expansions in solving both the Schrödinger and Poisson equations would provide a unique and transparent solution of the self-consistent calculation of the electronic structure of solids, with essentially no restrictions in the shape of the cells or the underlying lattice structure.

Now, a serious problem in implementing such a procedure is the divergence of the multipole expansion of the MT form when applied to space-filling cell charges. The MT expression, Eq.(9.6), and the corresponding form for the electrostatic interaction diverge when applied to points that lie in the charge-free region inside a sphere bounding the charge, that is, in the moon region of the cell.

This behavior is analogous to that obtained when the asymptotic form of the wave function is used inside the moon region of a cell, as was discussed in Section 3.5. Conceivably, one could attempt a Padé-approximant–like continuation of the multipole expansion into the moon region. Besides being conceptually unattractive, such a continuation would be computationally rather cumbersome, would have to be examined for convergence on a case-by-case basis, and would not provide a reliable measure of the possible errors introduced by its use. Obtaining the potential in the moon region through direct integration over the domains of adjacent cells might improve the convergence aspect of the calculation. However, that also would detract from the overall unifying features of the method, and might also be computationally sluggish in treating impurities, as one would have to treat the moon regions differently from the rest of the material.

In attempting an evaluation of the potential in a crystal, one may work either with the Poisson Eq.(9.1) or with the integral representation of its solution, Eq.(9.2). We will follow the latter approach. As mentioned above, the integral equation provides a proper solution of the differential equation and leads directly to expressions for the potential and the electrostatic interaction in terms of the multipole moments. The resulting expressions, which are generalizations of those corresponding to MT charges, are analytic and provide a transparent explanation both for the reasons of the divergences noted by Morgan [21], as well as of the resolution of the difficulty. Moreover, these expressions are fairly straightforward to code, involving in addition to the multipole moments the well-known structure constants of the underlying lattice. These can be evaluated once and for all for a given, translationally invariant structure, and used repeatedly as necessary. Even in cases with reduced or no translational symmetry, these expressions provide a means of calculating the electrostatics of space-filling cells through direct real-space summations. Finally, they provide a point of comparison against which numerical procedures, for example, Padé approximants, can be judged.

For convex-shaped cells bounded by planar surfaces, for example, Wigner–Seitz cells, the method presented below becomes particularly straightforward. It allows the calculation of the potential as well as of the electrostatic interaction of a material in terms of easily evaluated quadratures. The method can be extended to the treatment of concave cells. Thus, even in its simplest form, the method is characterized by all the desirable features quoted above and can be used readily in the performance of charge-self-consistent, total-energy calculations in solids. In its most general form, and one that would be admittedly rather difficult to implement computationally at present, it allows the calculation of the potential not only in charge-free regions but also inside the charge distribution itself in terms of properly defined multipole moments of the charge. Furthermore, it allows the calculation of the electrostatic interaction between interpenetrating or overlapping charge densities, and even the self-interaction of a charge in terms of these same multipole moments.

9.4 Convex Polyhedral Cells

9.4.1 Mathematical Preliminaries

The reason that Eq.(9.6) leads to divergent results is that the points $\mathbf{r}_0$ in the moon region may fail to satisfy the $r_</r_>$ criterion with respect to points $\mathbf{r}$ inside the cell in the expansion of the Green function $G_0(\mathbf{r}_0 - \mathbf{r}) = 1/|\mathbf{r}_0 - \mathbf{r}|$. This criterion is rigorously satisfied in the formalism presented below. In order to see how this is accomplished, we begin with a review of those expansions of the Green function that are relevant to our discussion.

We begin with the expansion of the potential at $\mathbf{r}_0$ due to a unit point charge at point $\mathbf{r}$, that is, the Green function $G_0(\mathbf{r}_0 - \mathbf{r})$ of electrostatics,

$$\frac{1}{|\mathbf{r} - \mathbf{r}_0|} = \sum_{l=0}^{\infty} \frac{r_<^l}{r_>^{l+1}} P_l(\cos\gamma), \tag{9.7}$$

where $r_<(r_>)$ is the smaller (larger) of $|\mathbf{r}_0|$ and $|\mathbf{r}|$, and γ is the angle between $\mathbf{r}_0$ and $\mathbf{r}$. We note that $G_0(\mathbf{r}-\mathbf{r}')$ satisfies the Laplace equation

$$\nabla^2 G_0(\mathbf{r}-\mathbf{r}') = -\delta(\mathbf{r}-\mathbf{r}'). \tag{9.8}$$

Upon using the expansion of Legendre polynomials in terms of spherical harmonics [5], we can also write

$$\frac{1}{|\mathbf{r}-\mathbf{r}_0|} = \sum_{l=0}^{\infty}\sum_{m=-l}^{l} \frac{r_<^l}{r_>^{l+1}} Y_{lm}(\hat{\mathbf{r}}_0) Y_{lm}^*(\hat{\mathbf{r}}), \tag{9.9}$$

where $\hat{\mathbf{r}}$ denotes a unit vector in the direction of $\mathbf{r}$. It can be easily established [26] that this expression constitutes the zero-energy limit of Eq.(5.12). Thus, Eq.(9.9) represents the potential in completely factorized form in the coordinates $\mathbf{r}_0$ and $\mathbf{r}$. Furthermore, for two vectors $\mathbf{r}$ and $\mathbf{R}$ such that $R > r$, we have the expansion [26]

$$\begin{aligned}\frac{\mathrm{i}^l}{|\mathbf{r}-\mathbf{R}|^{l+1}} Y_{lm}(\widehat{\mathbf{r}-\mathbf{R}}) = \; & 4\pi \sum_{l'}\sum_{m=-l'}^{l'} r^{l'} \mathrm{i}^{l'} Y_{l'm'}(\hat{\mathbf{r}}) \\ & \times \frac{(2l''-1)!!}{(2l-1)!!(2l'+1)!!} C(LL'L'') \\ & \times \frac{1}{R^{l''}} \left[\mathrm{i}^{l''} Y_{l''m''}(\hat{\mathbf{R}})\right]^*, \end{aligned} \tag{9.10}$$

where $l'' = l + l'$, $m'' = m' - m$, $C(L, L', L'')$ is a Gaunt number (the integral of the product of three spherical harmonics), and for n odd the symbol $n!!$ denotes the product $1 \times 3 \times 5 \times \cdots \times n$. Equations (9.7)–(9.10) are the only ones needed to obtain the solution of Poisson's equation for any distribution of cell charges. As is well known, and as was quoted in the previous section, the use of these equations allows one to write the potential outside a sphere bounding a charge distribution in a form that involves the multipole moments of the charge density, Eq.(9.5). Crucial in the derivation of Eq.(9.4) and its multicell counterpart, Eq.(9.6), is that the "observation" point $\mathbf{r}_n$ lies outside a sphere circumscribing the cell at $\mathbf{R}_n$, so that $r_n > r$ for any intracell vector $\mathbf{r}$. Thus, at any point $\mathbf{r}$ *outside* all bounding spheres in a collection of cells centered at sites $\mathbf{R}_n$, the potential can be written as the sum

$$V(\mathbf{r}) = \sum_n \sum_{l,m} \frac{4\pi}{2l+1} \frac{Y_{lm}(\widehat{\mathbf{r}_n})}{r_n^{l+1}} Q_{lm}^n. \tag{9.11}$$

This is essentially Eq.(9.6) but with the potential $V_0^j(\mathbf{r})$ also expressed as a sum over multipole moments. Thus, Eq.(9.11) allows one to express a collective property of an assembly of cells, the potential $V(\mathbf{r})$ at points outside all bounding spheres, as a linear combination of individual and independently determined cell quantities, the multipoles Q_{lm}^n. For MT potentials, Eq.(9.11) can be modified to the form of Eq.(9.6) in which form it holds for all points $\mathbf{r}$ in the material. These MT forms of the multipole expansion are particularly convenient for computational purposes. For example, in order to find the change in the potential at $\mathbf{r}$ due to the change in the charge distribution in the cell at $\mathbf{R}_n$, one simply replaces the original Q^n_{lm} in Eqs.(9.6) or (9.11) by the new ones. The solution of Poisson's equation for a solid under the MT approximation has been described in the literature [32], and has been used extensively [33] in the performance of electronic-structure calculations.

9.4.2 Non-MT, Space-Filling Cells of Convex Shape

Unfortunately, a great number of problems of physical importance, for example, nonspherical charge distributions, potentials near surfaces and interfaces and around impurities, may not conform to the MT model. In such cases, Eq.(9.6) is no longer valid in the "moon" regions, that is, between the charge distribution inside a cell and the circumscribing sphere. It is now no longer evident *a priori* that a knowledge of the multipole moments suffices to determine the potential in this region. To address these problems, one requires a method for treating electrostatic interactions for complex systems consisting of space-filling cells, and a number of attempts [24, 25] have been made in this direction. However, these attempts have not led to a satisfactory solution to the problem. Because Weinert's [24] approach utilizes only the multipole moments of the charge inside a sphere inscribed in the cell, the straightforward replacement of the cell multipole moments Q^n_{lm} to treat impurities or other localized imperfections is no longer possible. Morgan's approach [25], based on the use of the MT expansions to treat space-filling cell charges, does utilize the full-cell moments, but runs into severe convergence difficulties that effectively make it unreliable.

In this section, we provide the solution to this problem. We show rigorously that it is indeed formally and computationally possible to obtain the intercell contributions to the potential and the electrostatic interaction of an assembly of space-filling cells in terms of the individual multipole moments of these cells. For the sake of clarity, we confine our discussion to cells of convex shape bounded by planar surfaces, such as Wigner–Seitz cells. The extension to more general, even interpenetrating cell charges will be given in a later section. For the case of convex cells, we derive explicit expressions for both the solution of Poisson's equation as well as the electrostatic interaction in a solid in terms of the cell multipole moments. We illustrate our formal expressions through the results of numerical calculations that are reported in a following section. These calculations also clarify the origin of the divergences noted by Morgan, and demonstrate the analytic process that can be used to circumvent such behavior. As will be seen, the resulting formulae are simple enough to be incorporated into most existing computer codes for the calculation of electronic structure.

Let us consider the potential outside a convex cell, but inside the sphere circumscribing the cell. At a point outside a sphere bounding the cell, the potential is given by Eq.(9.4). We inquire as to whether the potential at a point inside the moon region can similarly be expressed in terms of the multipole moments of the charge density $\rho(\mathbf{r})$ inside the cell.

Let us denote by $\mathbf{r}_0$ the radius vector from the center of the cell to point P_2. We note that a straightforward expansion of the denominator in Eq.(9.2), aimed at recovering the form of Eq.(9.4), is not possible because $\mathbf{r}_0$ is not larger than all intracell vectors $\mathbf{r}$. Thus, Eq.(9.4) is not valid for $\mathbf{r}_0$ inside the moon region, and in fact it leads to divergent results there as is illustrated by the results of numerical calculations. However, an expansion in terms of the cell's multipole moments Q^n_{lm} can be obtained even in this case by means of the following procedure: We add and subtract a vector $\mathbf{b}$ in the denominator of the integrand in Eq.(9.2) such that for all $\mathbf{r}$ in the cell we have $b < |\mathbf{r} - (\mathbf{r}_0 + \mathbf{b})|$. Such a vector can always be constructed[1] for *each planar face of a convex cell,* and for all $\mathbf{r}$ and $\mathbf{r}_0$. In fact, it suffices to choose any vector $\mathbf{b}$ normal to the face intersected by a line along $\mathbf{r}_0$, and pointing away from the cell. In view of Eq.(9.9), we can now write

$$V(\mathbf{r}_0) = \sum_{l,m} \frac{4\pi}{2l+1} b^l Y_{lm}(\widehat{-\mathbf{b}}) \int_\Omega \rho(\mathbf{r}) \frac{Y^*_{lm}(\mathbf{r} - \widehat{(\mathbf{r}_0 + \mathbf{b})})}{|\mathbf{r} - (\mathbf{r}_0 + \mathbf{b})|^{l+1}} \mathrm{d}^3 r, \tag{9.12}$$

where the sum and the integral can now be safely interchanged because the sum converges for all values of the integrand. At this point the magnitude of $\mathbf{b}$ is still arbitrary. We now impose the further condition that $\mathbf{b}$ have a magnitude larger than the maximum distance between the bounding sphere and the face of the cell intersected by $\mathbf{r}_0$. In other words, $\mathbf{b}$ is such that for all $\mathbf{r}_0$ in the moon region adjacent to a face, $(\mathbf{r}_0 + \mathbf{b})$ lies outside the bounding sphere. Then, $\mathbf{r}$ in Eq.(9.12) satisfies the condition $r < |\mathbf{r}_0 + \mathbf{b}|$, so that we can use Eq.(9.10) to expand once again and obtain the result

$$V(\mathbf{r}_0) = \sum_{l,m} \frac{4\pi}{2l+1} b^l Y_{lm}(\widehat{-\mathbf{b}}) \left[\sum_{l',m'} (-)^l S_{lm;l'm'}(\mathbf{r}_0 + \mathbf{b}) Q_{l',m'} \right], \tag{9.13}$$

where the structure constants $\mathbf{S}$ are defined by the expression

$$\begin{aligned} S_{lm;l'm'}(\mathbf{R}) &= 4\pi \frac{(2l+2l'-1)!!}{(2l-1)!!(2l'+1)!!} C(lm;l'm';l+l',m'-m) \\ &\quad \times \frac{Y_{l+l',m'-m}(\hat{\mathbf{R}})}{R^{l+l'+1}} \\ &= \frac{\alpha(lm;l'm') Y_{l+l',m'-m}(\hat{\mathbf{R}})}{R^{l+l'+1}}. \end{aligned} \tag{9.14}$$

[1]See corresponding discussion and figures in Chapter 3.

These structure constants are closely related to those used in the Korringa–Kohn–Rostoker [34, 35] (KKR) and linear band-structure methods [26], and can be readily calculated for arbitrary values of l and m.

It is to be emphasized that Eq.(9.12) is valid for all convex-shaped cells, because it is always possible to choose $\mathbf{b}$ such that $b < |\mathbf{r} - (\mathbf{r}_0 + \mathbf{b})|$ and $r < |\mathbf{r}_0 + \mathbf{b}|$. Thus, Eq.(9.13) leads to the potential in the moon region, as well as everywhere else outside the cell, *entirely in terms of the multipole moments of the cell charge*. This is analogous to the result established in Chapter 3 with respect to the Schrödinger equation, that the wave function associated with a spatially bounded potential can be expressed in terms of the t-matrix (or the phase functions) everywhere outside the potential. The double summation indicated in Eq.(9.13) has been derived in such a way that, if the internal sum is carried out first, the double sum converges. This double summation is the foundation of the formalism in this section. Therefore, it may be instructive to gain a "pictorial" understanding of the derivation of Eq.(9.13), along with its convergence characteristics.

First, note that the single sum in Eq.(9.4), which diverges in the moon region, has been replaced by a double sum in Eq.(9.13). The brackets in Eq.(9.13) indicate that the internal sum must be performed first. This is because that sum was obtained after the outer sum, involving $\mathbf{b}$, and the order of summations must reflect that fact. Proceeding backward from Eq.(9.13), with the internal sum carried out first, one readily obtains the integral formula, Eq.(9.2). Pictorially, it is as though point P_2 is moved outside the bounding sphere through a translation by the vector $\mathbf{b}$, in a region where the potential can be represented in terms of the cell multipole moments. This expression corresponds to the sum inside the brackets in Eq.(9.13), and by itself yields the (l, m)-component of the potential at $(\mathbf{r}_0 + \mathbf{b})$. Next, the potential at $\mathbf{r}_0 + \mathbf{b}$ is analytically continued (moved back) to the original point $\mathbf{r}_0$ by means of the outer, or second, sum. Alternatively, one may consider the cube as being displaced by $-\mathbf{b}$, causing all points in the moon region to fall outside the bounding sphere, and leading to the multipole expansion of the internal sum in Eq.(9.13). Then, the outer sum returns the cube back to its original position. Whether one invokes the passive or active interpretation of the double sum in Eq.(9.13), that is, moving the points or moving the cell, the order of summations must be maintained as indicated to guarantee convergence. Also, it is to be noted that the Q_{lm}'s in Eq.(9.13) are the multipole moments of the cell about its original, undisplaced center. In spite of the points or the cell "moving around," no moments about displaced centers enter Eq.(9.13).

Equation (9.13) is manifestly different from Eq.(9.4), but reduces to that equation as $\mathbf{r}_0$ moves outside the bounding sphere. For such values of $\mathbf{r}_0$ the order of the sums in Eq.(9.13) can be reversed, causing that equation to collapse to the simpler form of Eq.(9.4). At the same time, in applying this formalism to realistic systems it may be convenient to use Eq.(9.13) for all points outside a cell, avoiding the cumbersome process of switching formulae on a point-by-point basis. Thus, the potential inside a given cell in a material is the sum of three terms: a term $V_0(\mathbf{r})$ arising from the charge in the cell itself, one term contributed by those cells whose bounding spheres do not intersect the cell in question, and a term contributed by a set, $\{\delta\}$, of nearby cells whose bounding spheres do intersect the cell. The first term, V_0, can be found by direct integration or other means, the second is given by Eq.(9.4) for each cell not in the set $\{\delta\}$, and the third by an equation with a double sum, as in Eq.(9.13), for each of the cells in $\{\delta\}$. Explicitly, we have the expression

$$
\begin{aligned}
V(\mathbf{r}) = V_0(\mathbf{r}) &+ \sum_{n\notin\{\delta\}}\sum_{l,m}\frac{4\pi}{2l+1}\frac{Y_{lm}(\widehat{\mathbf{r}_n})}{r_n^{l+1}}Q_{lm}^n \\
&+ \sum_{n\in\{\delta\}}\sum_{l,m}\frac{4\pi}{2l+1}R_n^l Y_{lm}(\widehat{\mathbf{R}_n}) \\
&\times\left[\sum_{l',m'}(-)^l S_{lm;l'm'}(\mathbf{r}_n-\mathbf{R}_n)Q_{l',m'}^n\right], \qquad (9.15)
\end{aligned}
$$

where the symbol $\in$ ($\notin$) denotes the relation of belonging (not belonging) to a set, and we have set $-\mathbf{b}$ equal to the intercell vector $\mathbf{R}_n$. Now, we see that the divergences noted by Morgan arise from the use of the MT expressions, consisting of only the first two terms in Eq.(9.15) being used for all cells, including those in $\{\delta\}$. This, of course, violates the conditions for expanding the Green function in the moon region and indeed leads to divergent results (see next section). We end the present section with the derivation of an expression for the Coulomb interaction of any cell in a material, say the one at the origin, with all other cells. Upon expanding the structure constants $\mathbf{S}$ as prescribed in Eq.(9.10), we readily obtain the result

$$
\begin{aligned}
U = U_0 &+ \sum_{n\notin\{\delta\}}\sum_{l,m}\frac{4\pi}{2l+1}Q_{lm}^{0*}\sum_{l',m'}(-)^l S_{lm;l'm'}(\mathbf{R}_n)Q_{l',m'}^n \\
&+ \sum_{n\in\{\delta\}}\sum_{l,m}\frac{4\pi}{2l+1}R_n^l Y_{lm}(\widehat{\mathbf{R}_n}) \\
&\times\left[\sum_{l'm'l''m''}(-)^{l+l'}Q_{l'm'}^0\alpha(lm;l'm')S_{l+l',m'-m;l''m''}^*(2\mathbf{R}_n)Q_{l'',m''}^{n*}\right]. \qquad (9.16)
\end{aligned}
$$

Even though they look rather involved, Eqs.(9.15) and (9.16) are rather easy to use. More importantly, they lead to converged results, provided that the various sums contained in them are performed in the proper order as is explicitly indicated by the brackets.

9.5 Numerical Results for Convex Cells

In this section, we report the results of calculations based on the formalism of the previous section. We treat explicitly the case of constant charge densities, set equal to one in arbitrary units, confined inside cells of convex shape. As a prototype of such cells, we consider both unit cubes ($1 \times 1 \times 1$) and prisms ($1 \times 1 \times 2$). In each case, the exact potential and electrostatic interaction were obtained analytically using an algebraic-manipulation program [36], and were compared to the results of the MT expansions and the modified expressions Eqs.(9.13) and (9.16).

In the following figures, we plot the ratios $\Delta V/V$ and $\Delta E/E$, where ΔV or (ΔE) denotes the difference in potential (energy) between the exact values, obtained analytically, of the potential (energy) and those computed through various multipole expansions. In all figures, a dotted line depicts the results of the MT-like expansions, while a solid line represents the results of the newly derived expressions Eqs.(9.13) and (9.16). The ratio $\Delta V/V$ versus the value of the outer variable l in Eq.(9.13) for the case of the unit cube of unit charge density is shown in Fig. 9.1 at various distances z along the z-axis away from the center of the cube. In each case, the value of the internal sum l' in Eq.(9.13) was truncated at 30. Outside the bounding sphere ($z > \sqrt{3/2}$), both Eqs.(9.4) and (9.13) converge rapidly to the exact value of the potential, as shown in panel (a) of the figure. In applying Eq.(9.13), we used a value of $b = 0.5$, as indicated in the inset. (All vectors $\mathbf{b}$ are chosen perpendicular to the cell face, that is, along the z-axis.) On the other hand, inside the moon region, where it is invalid, Eq.(9.4) leads to an oscillatory series with an ever-increasing amplitude as l increases, or as the observation point gets closer and closer to the face of the cube, as is indicated by the dotted curves in panels (b) and (c). In contrast to this behavior, Eq.(9.13) converges smoothly (solid curves) to the exact results for an appropriate choice of b (see below) and of internal summation cutoff ($l' = 30$). The results in panels (b) and (c) correspond to points just inside the bounding sphere and just outside the cube face, respectively. The latter position of the observation point clearly violates the MT condition rather severely. (We choose b such that $z + b = 3/2$ lies outside the bounding sphere, which has radius $r^2 = 3/2$). For the case in panel (b), point P_2 lies $1/10$ inside the bounding sphere and the error produced by the use of Eq.(9.13) amounts to 0.002% of the exact result. Even when P_2 is moved to a distance of $1/200$ outside the cube face (panel(c)), violating the MT condition even more severely, the error only increases to 0.07%! If greater accuracy is desired, both sums should be increased, with the internal sum being carried out to larger values of the angular momentum than the outer sum. The computational labor involved in performing these sums, even to values of l and l' that give the impression of being "large," is not excessive, especially after the various expansions have been coded for a particular cell shape. After this is done, these expressions can be used repeatedly in

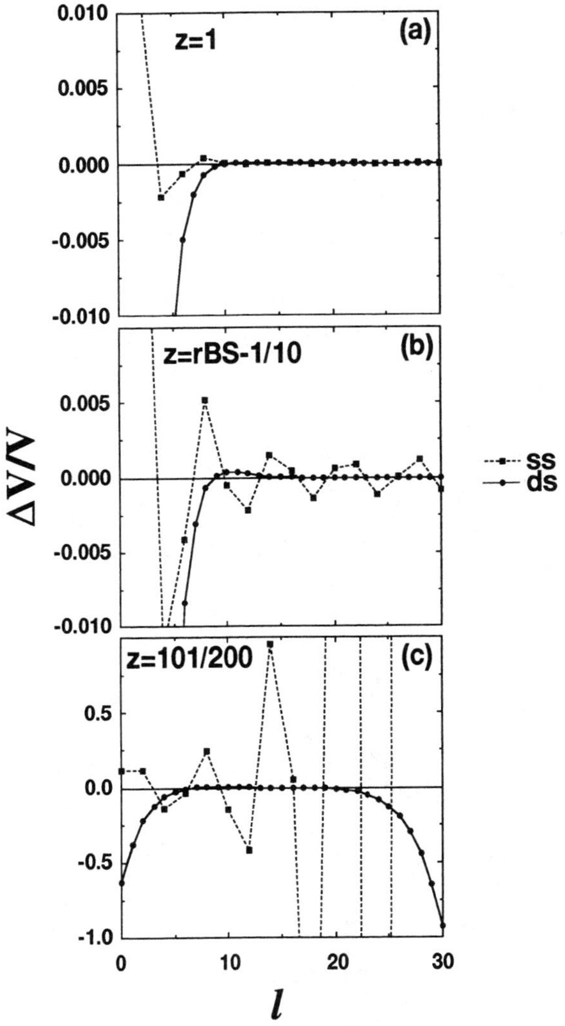

FIGURE 9.1. Convergence of multipole expansions for a uniform charge confined inside a cubic cell

different applications.

Results analogous to those for the $(1 \times 1 \times 1)$ cube are shown in Fig. 9.2 for the $(1 \times 1 \times 2)$ prism. Here, the observation point is taken outside a long face of the prism, the value of z is indicated explicitly in the various panels, b was chosen equal to 2, and the internal sum in Eq.(9.13) was again truncated at 30. Only the results for points P_2 inside the moon region are shown, as both Eqs.(9.4) and (9.13) converge rapidly to the exact result outside the bounding sphere. We note that in this case the failure of the

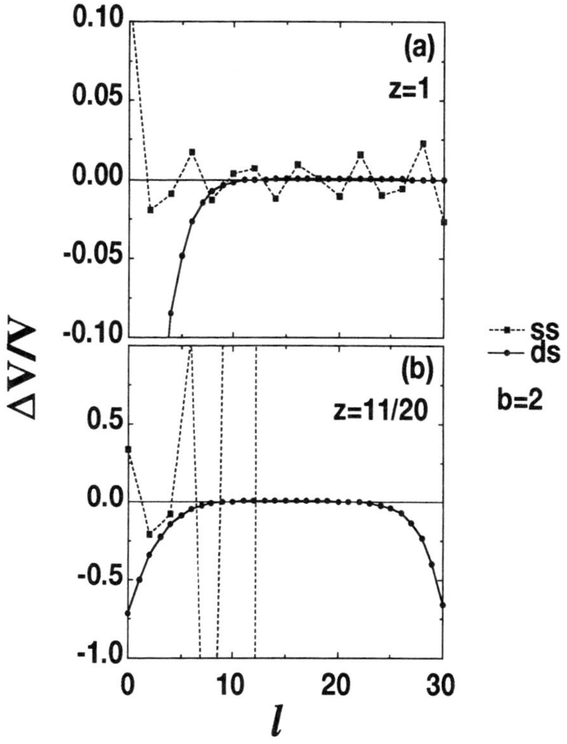

FIGURE 9.2. Convergence of multipole expansions for a uniform charge confined inside a prism.

MT expression, Eq.(9.4), becomes much more pronounced than in the case of the cube as the observation point moves closer to the prism. As can be seen in panel (b), the results obtained through Eq.(9.4) (dotted lines) oscillate wildly even at relatively small values of $l \cong 10$ when $z = 11/20$, that is, when the observation point lies 1/20 above the long face of the prism. On the other hand, Eq.(9.13) converges rapidly to the exact result over a substantial range of the outer l-sum (solid curves). At the same time, as the maximum value of l used in the outer sum increases, even Eq.(9.13) begins to yield results that deviate from the exact ones, as is indicated in Figs. 9.2 and 9.3. This behavior can be readily understood on the basis of the convergence properties of the double sum in Eq.(9.13).

As already discussed following Eq.(9.13), the sums over l and l' are infinite in extent, and the internal sum over l', which corresponds to the second expansion of the denominator in Eq.(9.2), should be carried out first, while the outer sum over l, corresponding to the first expansion, should be carried out last. Under these conditions, Eq.(9.13) converges to the integral expression, Eq.(9.2). However, in practical applications both sums must be

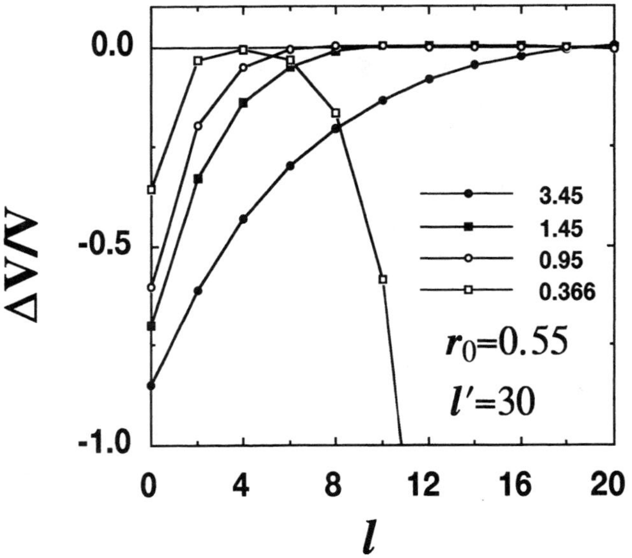

FIGURE 9.3. Convergence of multipole expansions in the double sum as a function of b.

truncated. Under these circumstances, the internal sum must be carried out to values larger than those of the outer sum in order to maintain the correct priority between them. In this way, the convergence of the outer sum is subjugated to that of the inner one. Thus, for a fixed value of b and the angular momentum cutoff of the inner sum (l'_{max}), Eq.(9.13) will converge over a range of values of the outer sum before beginning to diverge. This effect arises because, as l approaches l', the distinction between inner and outer sum is obscured, and the expansion behaves more and more like Eq.(9.4), which diverges in the moon region. This behavior is clearly exhibited in Figs. 9.2 and 9.3 for high values of $l(\geq 20)$. On the other hand, provided that the inner sum is carried out to sufficiently large values of l' so that the outer sum converges up to some l, the two sums can be interchanged, as they are now both finite.

The convergence of the double sum in Eq.(9.13) also depends on the choice of **b**. As indicated in Fig. 9.3, for small values of b the sum rapidly approaches the exact result, but then quickly diverges as l gets larger. As b increases, the sum approaches the exact result more slowly, but matches it over a wider range of l. Practical applications then depend on a suitable choice of **b** and the maximum values of l and l'. As Fig. 9.4 indicates, an intermediate b, approximately the size of the radius of the bounding sphere of a cell, leads to an acceptable range and rate of convergence.

At this point, it should be emphasized that the relatively large values of l at which the internal summation must be truncated are due to the

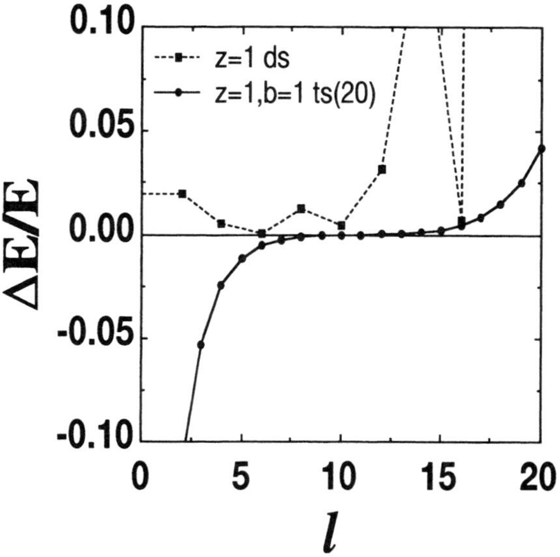

FIGURE 9.4. Convergence of multipole expansions of the electrostatic interaction between two cubic cells.

expansion of a nonspherical shape into angular-momentum eigenfunctions that are orthogonal on the surface of a sphere. This feature is particularly relevant to the performance of electronic structure calculations. In using the formalism presented here to calculate the electrostatic potential and energy in a solid, one need not necessarily calculate the charge density to correspondingly high values of l. It suffices to obtain only the first few elements in an angular-momentum expansion, say $l = 6$ or 8 (corresponding to wave function expansions to $l = 3$ or 4) and use the spherical harmonic expansion of the cell to generate the higher multipole moments. The expansion of the cell shape can be calculated once and for all for each cell shape and used repeatedly in the calculation of the potential (and the energy) of a material. The relative ratios $\Delta E/E$ of the electrostatic interaction between two cubes in contact along a face, and between two $(1 \times 1 \times 2)$ prisms touching along a long face, are shown in Figs. 9.4 and 9.5, respectively. In these calculations, l'_{max} was set equal to 30, and the values of b were chosen equal to 1 and 2 for the cube and prism, respectively. We see once again that Eq.(9.16) leads to converged results (solid curves), while the MT expression (last term in Eq.(9.16) omitted) diverges (dotted curves). As might be expected, the divergence is more pronounced for the prism, Fig. 9.5 , while the convergent expression, Eq.(9.16), is not affected greatly by the geometry (provided that a suitable choice of b is made).

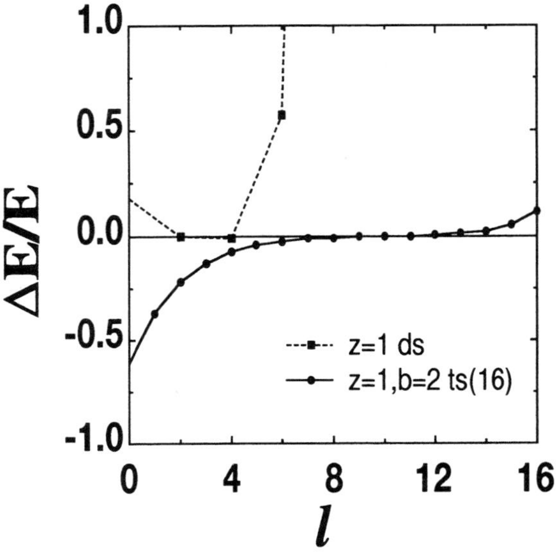

FIGURE 9.5. Convergence of multipole expansions of the electrostatic interaction between two prisms.

9.6 Concave Cells

9.6.1 Analytic Continuation

The method of Section 9.3 is rigorously applicable only to cells of convex shape. For concave cells, no single vector $\mathbf{b}$ can be found that satisfies all the conditions leading to the expansions in Eqs.(9.11) and (9.12) with respect to points $\mathbf{r}_0$ in the concave region. For example, the inequality $b < |\mathbf{r}-(\mathbf{r}_0+\mathbf{b})|$ is violated in the case of the concave cell shown in Figs. 3.5 and 3.6. As was the case with the Schrödinger equation, discussed in Chapter 3, a sequence of vectors $\mathbf{b}_n$ can be found that does allow one to obtain the potential in terms of the multipole moments, even inside the cavity region of a concave cell. The idea is to move the observation point $\mathbf{r}_0$ outside the bounding sphere by a series of steps involving the vectors $\mathbf{b}_n$, rather than in a single step. This process represents an analytic continuation of the potential and amounts to an expansions around the pole of the Green function $1/|\mathbf{r}_0-\mathbf{r}|$. Provided that the $\mathbf{b}_n$, $n = 1, 2, 3, ..., N$, satisfy the conditions

$$b_m < \left|\mathbf{r} - \left(\mathbf{r}_0 + \sum_{i=1}^{m} \mathbf{b}_i\right)\right| \tag{9.17}$$

for all $m = 1, 2, 3, ..., N$, the following sequence of steps becomes possible:

$$V(\mathbf{r}_0) = \int_\Omega \frac{\rho(\mathbf{r})}{|\mathbf{r}-\mathbf{r}_0|}$$

$$
\begin{aligned}
&= \sum_{L_1} \frac{4\pi}{2l+1} b_1^{l_1} Y_{L_1}(-\widehat{\mathbf{b}_1}) \int_\Omega \frac{\rho(\mathbf{r}) Y^*_{l_1 m_1}(\mathbf{r} - \widehat{(\mathbf{r}_0 + \mathbf{b}_1)})}{|\mathbf{r} - (\mathbf{r}_0 + \mathbf{b}_1)|^{l_1+1}} \mathrm{d}^3 r \\
&= \sum_{L_1} \frac{4\pi}{2l+1} b_1^{l_1} Y_{L_1}(-\widehat{\mathbf{b}_1}) \left[\sum_{L_2} g_{L_1 L_2}(\mathbf{b}_2) \right. \\
&\quad \left. \times \int_\Omega \frac{\rho(\mathbf{r}) Y^*_{l_1+l_2, m_1+m2}(\mathbf{r} - \widehat{(\mathbf{r}_0 + \mathbf{b}_1 + \mathbf{b}_2)})}{|\mathbf{r} - (\mathbf{r}_0 + \mathbf{b}_1 + \mathbf{b}_2)|^{l_1+l_2+1}} \mathrm{d}^3 r \right] \\
&= \sum_{L_1} \frac{4\pi}{2l+1} b_1^{l_1} Y_{L_1}(-\widehat{\mathbf{b}_1}) \\
&\quad \times \left[\sum_{L_2} g_{L_1, L_2}(\mathbf{b}_2) \left[\cdots \sum_{L_n} g_{L_{n-1}, L_n}(\mathbf{b}_n) \right. \right. \\
&\quad \left. \left. \times \int_\Omega \frac{\rho(\mathbf{r}) Y^*_{\sum_i l_i + \sum_i m_i}(\mathbf{r} - \widehat{(\mathbf{r}_0 + \mathbf{b})})}{|\mathbf{r} - (\mathbf{r}_0 + \mathbf{b})|^{\sum_i l_i + 1}} \mathrm{d}^3 r \right] \right],
\end{aligned}
\tag{9.18}
$$

where

$$
\mathbf{b} = \sum_{n=1}^{N} \mathbf{b}_n, \tag{9.19}
$$

$$
g_{LL'}(\mathbf{a}) = 4\pi a^{l'} Y_{l'm'}(\hat{\mathbf{a}})(-1)^l \frac{[2(l+l') - 1]!!}{(2l-1)!!(2l''+1)!!} C(L, L', L''), \tag{9.20}
$$

$l'' = l + l'$, $m'' = m' - m$, and the order of operations (summations) must be kept as indicated in the brackets in Eq.(9.18) to guarantee convergence.

The expansion process in Eq.(9.18) has a straightforward pictorial representation, as is illustrated in Fig. 3.6. The algebraic process is the same as that used in the case of the Schrödinger equation, and leads to the expression,

$$
\begin{aligned}
V(\mathbf{r}_0) = \sum_{L_1} \frac{4\pi}{2l+1} b_1^{l_1} Y_{L_1}(-\widehat{\mathbf{b}_1}) &\left[\sum_{L_2} g_{L_1, L_2}(\mathbf{b}_2) \right. \\
&\cdots \left. \left[\sum_{L'} (-1)^{l'} S_{L_n L'}(\mathbf{r}_0 + \mathbf{b}) Q_{L'} \right] \right],
\end{aligned}
\tag{9.21}
$$

for the potential at a point $\mathbf{r}_0$ in the concave region. An expression for the electrostatic interaction between two interpenetrating, but not overlapping, cells can be readily obtained through the replacement of the summation over L' in Eq.(9.21) by a double sum, as was done to derive Eq.(9.15). The derivation is straightforward, involving an expansion of r^0/b, for a proper choice of $\mathbf{b}$, and an integration over $\mathbf{r}_0$.

9.7 Direct Analogy with MST

In this section we provide an alternative derivation of the solution of the Poisson equation that is directly related to the formalism of multiple scattering theory. In this approach, we combine the solutions (potentials) associated with individual cell charges so that the solution for a collection of cell charges is continuous in value and derivative across cell boundaries. In order to emphasize the formal relation to MST, we will continue to use the notation that was established in previous chapters. However, it should be kept in mind that we are now dealing with the $E \longrightarrow 0$ $(k \longrightarrow 0)$ limit of the various functions used in connection with MST. For example, in this limit the free-particle propagator, Eq.(3.110), is given by Eq.(9.9), whereas the expansion formula for the Hankel function, Eq.(D.10), takes the form of Eq.(9.10) [26]. Thus, in the following discussion we will use the symbols $J_L(\mathbf{r})$ and $H_L(\mathbf{r})$ to indicate the functions,

$$J_L(\mathbf{r}) = r^l Y_L(\hat{\mathbf{r}}) \quad \text{and} \quad H_L(\mathbf{r}) = r^{-(l+1)} Y_L(\hat{\mathbf{r}}). \tag{9.22}$$

As the formalism of this section is based on the knowledge of potential functions associated with individual, spatially bounded cell charge, we begin with a brief review [5] of the general form of these potentials.

9.7.1 *Single Cell Charges*

We consider a charge distribution, $\rho(\mathbf{r})$, which vanishes outside a bounding sphere of radius r_s. The potential, $v(\mathbf{r})$, associated with this charge is given by the integral

$$\begin{aligned} v(\mathbf{r}) &= \int \frac{\rho(\mathbf{r}')\mathrm{d}^3 r}{|\mathbf{r} - \mathbf{r}'|} \\ &= \sum_L \left\{ J_L(\mathbf{r}) \int_{r'>r} H_L^*(\mathbf{r}')\rho(\mathbf{r}')\mathrm{d}^3 r' \right. \\ &\quad \left. + H_L(\mathbf{r}) \int_{r'<r} J_L^*(\mathbf{r}')\rho(\mathbf{r}')\mathrm{d}^3 r' \right\}, \end{aligned} \tag{9.23}$$

where we have used the expansion of the Green function given in Eq.(9.9), and the definition in Eq.(9.22). With obvious definitions, we can write

$$v(\mathbf{r}) = \sum_L \{ J_L(\mathbf{r}) C_L(r) + H_L(\mathbf{r}) S_L(r) \} \tag{9.24}$$

or

$$v(\mathbf{r}) = \sum_L v_L(\mathbf{r}), \tag{9.25}$$

where

$$v_L(\mathbf{r}) = J_L(\mathbf{r}) C_L(r) + H_L(\mathbf{r}) S_L(r). \tag{9.26}$$

We note that outside the bounding sphere, $r > r_s$, we have

$$v_L(\mathbf{r}) = H_L(\mathbf{r})S, \tag{9.27}$$

where

$$S_L \equiv Q_L = \int J_L^*(\mathbf{r}')\rho(\mathbf{r}')\mathrm{d}^3r' \tag{9.28}$$

is the asymptotic value of the function defined implicitly in Eq.(9.23), and is precisely the multipole moment of order L, Eq.(9.5), associated with the charge $\rho(\mathbf{r})$. It follows that for $r > r_s$ the potential is given by the multipole expansion, Eq.(9.11).

9.7.2 *Multiple Scattering Solutions*

We seek to obtain the potential, $V(\mathbf{r})$, associated with a collection of spatially bounded but nonoverlapping cell charges, $\rho^n(\mathbf{r})$, centered at positions specified by vectors $\mathbf{R}_n$. For the sake of clarity, we begin by assuming that the cells are far enough apart from one another as to satisfy the MT conditions, that is, the sphere circumscribing any one cell does not intersect similar spheres around other cells. Space-filling charges are considered in the following subsection.

We proceed in a manner typical of MST: We obtain solutions within each cell which contain a set of coefficients that are to be chosen so that these solutions are matched smoothly to solutions obtained for other cells $\mathbf{r}_n = \mathbf{r} - \mathbf{R}_n$. The solution inside cell n may be written in the form (where the vector/matrix notation of Appendix D is used to denote sums over angular momentum states),

$$\begin{aligned} V_{In}(\mathbf{r}) &= v_n(\mathbf{r}_n) + \sum_L c_L^n J_L(\mathbf{r}_n) \\ &= v_n(\mathbf{r}_n) + \langle J|c^n\rangle. \end{aligned} \tag{9.29}$$

Here, the function $v_n(\mathbf{r}_n)$ is a local solution of the Poisson equation associated with the charge $\rho^n(\mathbf{r}_n)$ and satisfies Eq.(9.1). The general form of these local solutions is given by Eqs.(9.24)–(9.29). We assume that $v_n(\mathbf{r}_n)$ in specific cases can be obtained through the use of standard techniques so that the remaining problem is the determination of the coefficients, c_L^n, in Eq.(9.29).

The determination of these coefficients requires the use of two additional forms of the global solution, $V(\mathbf{r})$. Outside a sphere bounding the charge in cell n, the potential can be written either as a single-center or a multicenter expansion. In analogy with Eq.(3.66) we obtain the single-center expansion

$$\begin{aligned} V_{IIn}(\mathbf{r}) &= \sum_L \left[a_L^n J_L(\mathbf{r}_n) + b_L^n H_L(\mathbf{r}_n)\right] \\ &= \langle a^n|J(\mathbf{r}_n)\rangle + \langle b^n|H(\mathbf{r}_n)\rangle, \end{aligned} \tag{9.30}$$

while the multicenter expansion takes the form,

$$\begin{aligned} V_{II}(\mathbf{r}) &= \sum_n \sum_L b_L^n H_L(\mathbf{r}_n) \\ &= \langle b^n | H(\mathbf{r}_n) \rangle, \end{aligned} \tag{9.31}$$

which corresponds to Eq.(5.91).

Equations (9.29), (9.30) and (9.31) contain three sets of coefficients, a_L^n, b_L^n, and c_L^n which are still to be determined. The number of unknowns can be reduced to two through the observation that b_L^n is related to the multipole moments of the charge in cell n

$$b_L^n = \frac{4\pi Q_L^n}{2l+1}. \tag{9.32}$$

Furthermore, use of the fundamental condition of MST that the single-center and multicenter expansion be consistent leads to the relation

$$\begin{aligned} \langle a^n | J(\mathbf{r}_n) \rangle &= \sum_{n' \neq n} \langle b^{n'} | H(\mathbf{r}_n) \rangle \\ &= \sum_{n' \neq n} \langle b^{n'} | S^{nn'} | J(\mathbf{r}_n) \rangle, \end{aligned} \tag{9.33}$$

where the structure-constant matrix, $\mathbf{S}$, is given by Eq.(9.14) and its use here is justified because the cell charges conform to the MT geometry. Finally, the requirement that V_{In} and V_{IIn} be consistent leads to the following expression for the coefficients c_L^n,

$$c_L^n = \sum_{n' \neq n} \sum_{L'} b_{L'}^{n'} S_{L'L}^{nn'}. \tag{9.34}$$

The use of this expression in Eq.(9.29) and the use of Eq.(9.10) leads to the expression for the potential given in Eq.(9.6).

9.7.3 Space-Filling Charges of Arbitrary Shape

The generalization of the formalism give above to space-filling cell charges can be effected in a number of ways. For example, one could replace the single sum over L in Eq.(9.34) by a double, or a multiple, conditionally convergent sum as was demonstrated in Section 9.4. Alternatively, one may follow the procedures of Chapter 3, which allows the proper expansion of the Green function even in the case of interpenetrating cell shapes. In an effort to amplify the connection with MST we give a summary of the latter approach.

¿From the defining equation for the Green function, Eq.(9.8), we obtain the expression

$$V(\mathbf{r}) = -\int \nabla'^2 G_0(\mathbf{r} - \mathbf{r}') V(\mathbf{r}') \mathrm{d}^3 r'. \tag{9.35}$$

When this expression for the potential is subtracted from Eq.(9.2), it yields the relation

$$\int d^3r' \left[G_0(\mathbf{r}-\mathbf{r}')\nabla'^2 V(\mathbf{r}') - \left(\nabla'^2 G_0(\mathbf{r}-\mathbf{r}')\right) V(\mathbf{r}')\right] = 0, \tag{9.36}$$

where we have used Poisson's equation to replace $\rho(\mathbf{r})$ with $-\nabla^2 V(\mathbf{r})$. Now, the integral over all space in Eq.(9.36) can be converted to integrals over the interiors of individual cells and a subsequent sum over cells,

$$\sum_n \int_{\Omega_n} d^3r'_n \left[G_0(\mathbf{r}-\mathbf{r}')\nabla'^2 V(\mathbf{r}') - \left(\nabla'^2 G_0(\mathbf{r}-\mathbf{r}')\right) V(\mathbf{r}')\right] = 0. \tag{9.37}$$

Furthermore, through the use of Green's theorem each cell integral can be converted to an integral over the surface of the cell, or any surface that contains the cell

$$\sum_n \int_{S_n} dS'\hat{n} \cdot \left[G_0(\mathbf{r}-\mathbf{r}')\nabla'^2 V(\mathbf{r}') - \left(\nabla'^2 G_0(\mathbf{r}-\mathbf{r}')\right) V(\mathbf{r}')\right] = 0. \tag{9.38}$$

There are at least two different procedures that can be followed in obtaining a valid expansion for the Green function in the last expression. We can confine the cell vector $\mathbf{r}$ to a spherical neighborhood of the origin of the cell that lies outside the bounding spheres of adjacent cells; or, we can allow $\mathbf{r}$ to vary over the entire volume of the cell while choosing the surface of integration, S_n, to be that of a sphere that contains the entire assembly of charges which in principle can have an infinite radius. The latter construction has the added advantage that it applies in *all* cases of nonoverlapping cell charges, even when a sphere bounding a given cell overlaps the entire region of an adjacent cell. Both constructions can be applied to an infinite number of cell charges. In either case, we can write Eq.(9.38) in the form

$$\sum_{n'} \sum_{L'} J_{L'}(\mathbf{r}_n) \int_{S_{n'}} dS'\hat{n} \cdot [H_{L'}(\mathbf{r}'_n), V(\mathbf{r}'_n)] = 0, \tag{9.39}$$

where the brackets denote the Wronskian of the functions contained in them. As the integral in the last expression is independent of $\mathbf{r}_n$, we have

$$\sum_{n'} \int_{S_{n'}} dS'\hat{n} \cdot [H_{L'}(\mathbf{r}'_n), V(\mathbf{r}'_n)] = 0. \tag{9.40}$$

Now, we replace $V(\mathbf{r}_n)$ by the expression in Eq.(9.29) and use the Wronskian relation,

$$\int_{S_n} dS'\hat{n} \cdot [H_{L'}(\mathbf{r}'_n), J_L(\mathbf{r}'_{n'})] = \delta_{nn'}\delta_{LL'}, \tag{9.41}$$

to obtain the formal expression,

$$c^n_L = \int_{S_n} dS'\hat{n} \cdot [H_L(\mathbf{r}'_n), v_n(\mathbf{r}'_n)]$$

$$+\sum_{n'\neq n}\int_{S_{n'}} \mathrm{d}S'\hat{n}\cdot[H_L(\mathbf{r}'_n), v_{n'}(\mathbf{r}'_{n'})]\,. \tag{9.42}$$

The first integral on the right-hand side of this expression vanishes because on the surface of any sphere bounding a cell charge the local potential is given in terms of Hankel functions, and the Wronskian between two regular or two irregular solutions of a second order differential equation vanishes. Thus, we obtain

$$c_L^n = \sum_{n'\neq n} C_L^{nn'}, \tag{9.43}$$

where

$$C_L^{nn'} = \int_{S_{n'}} \mathrm{d}S'\hat{n}\cdot[H_L(\mathbf{r}'_n), v_{n'}(\mathbf{r}'_{n'})]\,. \tag{9.44}$$

Note that in this integral, the argument of the Hankel function varies over the surface of cell n' but is measured from the center of cell n. If the intercell vectors $\mathbf{R}_{nn'}$ are larger than all intracell vectors, $\mathbf{r}_n$, then we can expand the Hankel function, Eq.(9.10), and obtain the expression,

$$\begin{aligned} C_L^{nn'} &= \sum_{L'} G_{LL'}^{nn'} \int_{S_{n'}} \mathrm{d}S'\hat{n}\cdot[J_{L'}(\mathbf{r}'_{n'}), v_{n'}(\mathbf{r}'_{n'})] \\ &= \sum_{L'} G_{LL'}^{nn'} Q_{L'}^{n'}. \end{aligned} \tag{9.45}$$

By use of Green's theorem one can readily verify that the surface integral of the Wronskian in the last equation is equal to the multipole moment of the charge in cell n', which yields the second line of the equation. It remains only to obtain the final expression for the potential through a substitution of an expression for c_L^n into Eq.(9.29). We note that if Eq.(9.43) is used it yields a valid, convergent expression for $V(\mathbf{r})$ everywhere in space. This is because in this expressions internal summations are carried to convergence through the use of surface integrals over the surfaces of adjacent cells. More recognizable forms for the potential can be obtained by using Eq.(9.45) between cells that are far enough part from one another that the sum $\sum_{L'}\sum_L J_L(\mathbf{r}n)G_{LL'}^{nn'}Q_{L'}^n$ converges. Summing this last expression over L we obtain the expression

$$V(\mathbf{r}) = v_n(\mathbf{r}^n) + \sum_{n'\notin\{\delta\}}{}' \sum_L H_L(\mathbf{R}_{nn'}+\mathbf{r}_n)Q_L^{n'} + \sum_{n'\in\{\delta\}}\sum_L C_L^{nn'} J_L(\mathbf{r}_n), \tag{9.46}$$

where $\{\delta\}$ denotes the set of neighbors of the cell n intersected by the sphere circumscribing that cell. We note that this last expression for the potential corresponds closely to Eq.(9.15). The only difference is that the double sums in that expression are replaced by integrals over the surfaces of adjacent cells leading in principle to fully converged results.

References

[1] P. Hohenberg and W. Kohn, Phys. Rev. **136B**, 864 (1964).
[2] W. Kohn and L. J. Sham, Phys. Rev. **140A**, 1133 (1965).
[3] Robert C. Parr and Weitao Yang, *Density-Functional Theory of Atoms and Molecules* (Oxford University Press, New York, 1989).
[4] R. M. Dreizler and E. K. U. Gross, *Density Functional Theory* (Springer-Verlag, New York, 1990).
[5] J. D. Jackson, *Classical Electrodynamics* (John Wiley and Sons, New York, 1975).
[6] E. Madelung, Physik. Z. **19**, 524 (1918).
[7] P. P. Ewald, Ann. Phys. **64**, 253 (1921).
[8] P. D. De Cicco, Phys. Rev. **153**, 931 (1967).
[9] G. S. Painter, Phys. Rev. B**7**, 3520 (1973).
[10] N. Elyasher and D. D. Koelling, Phys. Rev. B**13**, 5362 (1976).
[11] B. R. A. Nijboer and F. W. de Wette, Physica **23**, 309 (1957).
[12] B. R. A. Nijboer and F. W. de Wette, Physica **24**, 422 (1958).
[13] F. W. de Wette and B. R. A. Nijboer, Physica **24**, 1105 (1958).
[14] F. G. Fumi and M. P. Tosi, Phys. Rev. **117**, 1466 (1960).
[15] W. E. Rudge, Phys. Rev. **181**, 1020 (1969).
[16] J. L. Birman, J. Phys. Chem. Solids **6**, 65 (1958).
[17] M. P. Tosi, Solid State Phys. **16**, 1 (1964).
[18] T. L. Loucks, *Augmented Plane Wave Method* (Benjamin, New York, 1967).
[19] E. J. Baerends, D. E. Ellis, and P. Ross, Chem. Phys. **2**, 41 (1973).
[20] B. I. Dunlap, J. W. D. Connolly, and J. R. Sabin, J. Chem. Phys. **71**, 4993 (1979).
[21] J. W. Mintmire, Int. J. Quantum Chem. Symp. **13**, 163 (1979).
[22] J. Harris and G. S. Painter, Phys. Rev. B **22**, 2614 (1980).
[23] G. S. Painter, Phys. Rev. B**23**, 1624 (1981).
[24] M. Weinert, J. Math. Phys. **22**, 2433 (1981).
[25] J. van W. Morgan, J. Phys. C**10**, 1181 (1977).
[26] H. L. Skriver, *The LMTO Method* (Springer-Verlag, Berlin, 1984).
[27] X. -G. Zhang and A. Gonis, Phys. Rev. Lett. **61**, 1161 (1989).
[28] X. -G. Zhang, A. Gonis, and James M. MacLaren, Phys. Rev. B**40**, 3694 (1989).
[29] X. -G. Zhang, A. Gonis, and D. M. Nicholson, Phys. Rev. B**40**, 947 (1989).
[30] R. K. Nesbet, Phys. Rev. B**41**, 4948 (1990).
[31] W. H. Butler and R. K. Nesbet, Phys. Rev. B**42**, 1518 (1990).
[32] J. F. Janak, Phys. Rev. B**9**, 3985 (1974).
[33] V. L. Moruzzi, J. F. Janak, and A. R. Williams, *Calculated Electronic Properties of Metals* (Pergamon Press, New York, 1978).
[34] J. Korringa, Physica **13**, 392 (1947).
[35] W. Kohn and N. Rostoker, Phys. Rev. **94**, 1111 (1954).
[36] *Mathematica* 1.2 (Wolfram Research, Inc., Champaign, IL, 1989).

Appendix A
Time-Dependent Green Functions

The time-dependent Green function or propagator is defined as the solution of the equation,

$$\left[i\frac{\partial}{\partial t} - H_0 \right] G_0(\mathbf{r}, t; \mathbf{r}', t') = \delta(\mathbf{r} - \mathbf{r}')\delta(t - t'). \tag{A.1}$$

Through the definitions, $\mathbf{R} = \mathbf{r} - \mathbf{r}'$ and $\tau = t - t'$, we can write this equation in the form,

$$\left[i\frac{\partial}{\partial \tau} + \nabla^2_{\mathbf{R}} \right] G_0(\mathbf{R}, \tau) = \delta(\mathbf{R})\delta(\tau). \tag{A.2}$$

It is convenient to introduce the quantity $\tilde{G}_0(\mathbf{k}, \omega)$, the Fourier transform of $G_0(\mathbf{r}, t)$, through the relation,

$$G_0(\mathbf{R}, t) = (2\pi)^{-4} \int e^{i\mathbf{k}\cdot\mathbf{R}} e^{-i\omega\tau} \tilde{G}_0(\mathbf{k}, \omega) d^3k \, d\omega. \tag{A.3}$$

Upon substituting Eq.(A.3) and the relation,

$$\delta(\mathbf{R})\delta(t) = (2\pi)^{-4} \int e^{i\mathbf{k}\cdot\mathbf{R}} e^{-i\omega\tau} d^3k \, d\omega \tag{A.4}$$

into Eq.(A.2) we obtain the expression,

$$(2\pi)^{-4} \int \left[(\omega - k^2)\tilde{G}_0(\mathbf{k}, \omega) - 1 \right] e^{i(\mathbf{k}\cdot\mathbf{R} - \omega\tau)} d^3k \, d\omega = 0, \tag{A.5}$$

so that

$$\tilde{G}_0(\mathbf{k}, \omega) = \frac{1}{\omega - k^2}. \tag{A.6}$$

It now follows that $G_0(\mathbf{R}, \tau)$ has the integral representation,

$$G_0(\mathbf{R}, \tau) = (2\pi)^{-4} \int \frac{\exp\{\mathrm{i}(\mathbf{k} \cdot \mathbf{R} - \omega\tau)\}}{\omega - k^2} \, \mathrm{d}^3k \, \mathrm{d}\omega. \tag{A.7}$$

In order to evaluate this integral, we first consider the integral over ω,

$$I = \int_{\infty}^{-\infty} \frac{\mathrm{e}^{-\mathrm{i}\omega\tau}}{\omega - \omega_0} \mathrm{d}\omega, \tag{A.8}$$

where $\omega_0 = k^2$. The integrand has a pole at $\omega = \omega_0$ and we need a prescription for avoiding this singularity by choosing an appropriate path of integration in the complex plane. The criterion for selecting this path is the physical requirement of causality, that is, the Green function must lead to physically meaningful behavior. As is well known, proper paths for the integration of Eq.(A.8) are obtained by considering the limiting process,

$$I^{\pm} = \lim_{\epsilon \to 0^+} \int_{-\infty}^{\infty} \frac{\mathrm{e}^{-\mathrm{i}\omega\tau}}{\omega - \omega_0 \pm \mathrm{i}\epsilon} \mathrm{d}\omega. \tag{A.9}$$

Clearly, the introduction of the term $\pm\mathrm{i}\epsilon$ has the effect of shifting the poles of the integrand below (+) or above (−) the real ω-axis by a small positive quantity. Because

$$\exp(-\mathrm{i}\omega\tau) = \exp(-\mathrm{i}\tau\Re\omega) \exp(\tau\Im\omega), \tag{A.10}$$

for negative values of τ ($\tau < 0$ or $t < t'$) we should close the contour in the upper half-plane. Choosing the semicircle C_1 indicated in Fig. A.1, and letting the radius of C_1 tend to infinity, we obtain,

$$I^+(\tau) = 0, \qquad \tau < 0, \tag{A.11}$$

and

$$I^-(\tau) = 2\pi\mathrm{i} \exp(-\mathrm{i}\omega_0\tau) \exp(\epsilon\tau), \quad \tau < 0, \tag{A.12}$$

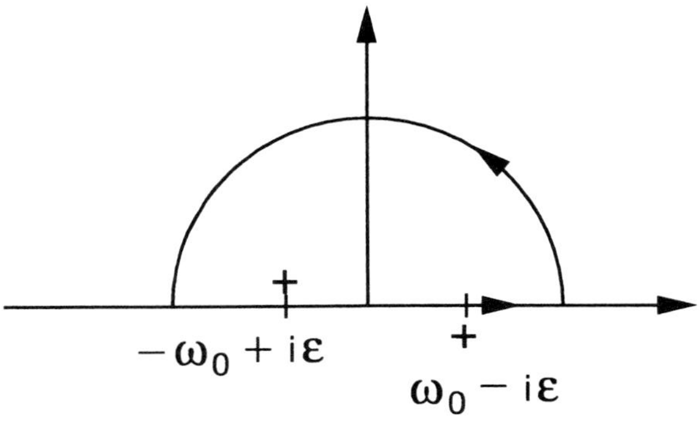

FIGURE A.1. Contours for Green function integrations.

where the limiting process $\epsilon \to 0$ is to be understood. Similarly, we have

$$I^+(\tau) = -2\pi \mathrm{i} \exp(-\mathrm{i}\omega_0\tau) \exp(-\epsilon\tau), \quad \tau > 0, \tag{A.13}$$

while

$$I^-(\tau) = 0, \qquad \tau > 0. \tag{A.14}$$

We can rewrite Eqs.(A.11) through (A.14) in more condensed form by introducing the step function $\Theta(\tau)$ such that

$$\Theta(\tau) = \begin{cases} 1, & \text{if } \tau > 0, \\ 0, & \text{if } \tau < 0. \end{cases} \tag{A.15}$$

We now obtain

$$I^+(\tau) = -2\pi \mathrm{i} \exp(-\mathrm{i}\omega_0\tau) \exp(-\epsilon\tau)\Theta(\tau), \tag{A.16}$$

and

$$I^-(\tau) = 2\pi \mathrm{i} \exp(-\mathrm{i}\omega_0\tau) \exp(-\epsilon\tau)\Theta(-\tau), \tag{A.17}$$

These equations can be inserted into the integral representation for G_0, Eq.(A.7), to yield the expression,

$$G_0^+(\mathbf{R},\tau) = -(2\pi)^{-3}\mathrm{i} \exp(-\epsilon\tau)\Theta(\tau) \int \exp(\mathrm{i}\mathbf{k}\cdot\mathbf{R}) \exp(-\mathrm{i}k^2\tau)\mathrm{d}^3k, \tag{A.18}$$

corresponding to I^+, and

$$G_0^-(\mathbf{R},\tau) = (2\pi)^{-3}\mathrm{i} \exp(-\epsilon\tau)\Theta(-\tau) \int \exp(\mathrm{i}\mathbf{k}\cdot\mathbf{R}) \exp(-\mathrm{i}k^2\tau)\mathrm{d}^3k \tag{A.19}$$

corresponding to I^-. Note that the Green function G_0^+ vanishes for past times ($t' < t$), while G_0^- vanishes into the future ($t' > t$). The function G_0^+ is called the *causal* or *retarded* Green function or propagator, and G_0^- the *advanced* Green function.

It now remains to evaluate the k-space integrals in Eqs.(A.7). Using Cartesian coordinates such that $\mathbf{R} = (X, Y, Z)$ and $\mathbf{k} = (k_x, k_y, k_z)$, we note that we must evaluate integrals of the kind,

$$\begin{aligned} &\int_{-\infty}^{\infty} \exp(\mathrm{i}k_x X) \exp(\mathrm{i}k^2\tau)\mathrm{d}k_x \\ &= \exp\left(\mathrm{i}\frac{X^2}{\tau}\right) \int_{-\infty}^{\infty} \exp\left\{-\mathrm{i}\tau\left(k_x - \frac{X}{2\tau}\right)^2\right\} \mathrm{d}k_x \\ &= \exp\left(\mathrm{i}\frac{X^2}{\tau}\right) \left(\frac{1}{\tau}\right)^{1/2} \int_{-\infty}^{\infty} \exp(-\mathrm{i}q_x^2)\mathrm{d}q_x, \end{aligned} \tag{A.20}$$

where $q_x = \sqrt{\tau}(k_x - X/2\tau)$. Although not strictly convergent, the last integral can be evaluated by including an integration factor, $\exp(-\alpha q_x^2)$, for α real and positive, and taking the limit $\alpha \to 0$ at the end of the calculation. Because

$$\int_{-\infty}^{\infty} \exp\{-(\alpha + \mathrm{i})q_x^2\}\mathrm{d}q_x = \left(\frac{\pi}{\alpha + \mathrm{i}}\right)^{1/2}, \tag{A.21}$$

we find

$$\int_{-\infty}^{\infty} \exp\{\mathrm{i}k_x X\} \exp\{-\mathrm{i}k_x^2 \tau\}\mathrm{d}k_x = \left(\frac{\pi}{\mathrm{i}\tau}\right)^{1/2} \exp\left\{\mathrm{i}\frac{X^2}{4\tau}\right\}. \tag{A.22}$$

With similar results for the integrals over k_y and k_z we obtain,

$$G_0^+(\mathbf{R}, \tau) = -\mathrm{i}(2\pi)^{-3} \exp(-\epsilon\tau)\Theta(\tau) \left(\frac{\pi}{\mathrm{i}\tau}\right)^{1/2} \exp\left\{\mathrm{i}\frac{R^2}{4\tau}\right\}. \tag{A.23}$$

Returning to the original variables, $(\mathbf{r}, t; \mathbf{r}', t')$, and taking the limit $\epsilon \to 0$, we can write the retarded Green function in the form,

$$G_0^+(\mathbf{r}, t; \mathbf{r}', t') = -\mathrm{i}\left[\frac{1}{4\pi\mathrm{i}(t - t')}\right]^{3/2} \Theta(t - t') \exp\left\{\mathrm{i}\frac{|\mathbf{r} - \mathbf{r}'|^2}{4(t - t')}\right\}, \tag{A.24}$$

with a similar expression for the advanced propagator,

$$G_0^-(\mathbf{r}, t; \mathbf{r}', t') = \mathrm{i}\left[\frac{1}{4\pi\mathrm{i}(t - t')}\right]^{3/2} \Theta(t' - t) \exp\left\{\mathrm{i}\frac{|\mathbf{r} - \mathbf{r}'|^2}{4(t - t')}\right\}. \tag{A.25}$$

Appendix B
Time-Independent Green Functions

We seek the determination of the free-particle propagator, $G_0^+(\mathbf{r}-\mathbf{r}')$, defined by the integral in Eq.(2.38). First, through a change of variables, $\mathbf{R}=\mathbf{r}-\mathbf{r}'$, we can write Eq.(2.38) in the form,

$$G_0^+(R) = -(2\pi)^{-3}\int_0^\infty \mathrm{d}^2k'k' \int_0^\pi \mathrm{d}\theta' \sin\theta' \int_0^{2\pi} \mathrm{d}\phi' \frac{\exp\{\mathrm{i}k'R\cos\theta'\}}{k'^2-k^2-\mathrm{i}\epsilon}, \tag{B.1}$$

where we have chosen polar coordinates so that the z-axis is along the vector $\mathbf{R}$. Upon performing the angular integration in Eq.(B.1), we obtain

$$G_0^+(R) = -(4\pi^2R)^{-1}\int_{-\infty}^\infty \frac{k'\sin k'R}{k'^2-k^2-\mathrm{i}\epsilon}\mathrm{d}k'. \tag{B.2}$$

Because the poles of the integrand are displaced from the real axis into the upper half-plane, the contour of integration can be closed in the upper plane, yielding the result

$$\begin{aligned} G_0^+(|\mathbf{r}-\mathbf{r}'|) &= -\frac{1}{4\pi}\frac{\mathrm{e}^{\mathrm{i}kR}}{R} \\ &= -\frac{1}{4\pi}\frac{\mathrm{e}^{\mathrm{i}k|\mathbf{r}-\mathbf{r}'|}}{|\mathbf{r}-\mathbf{r}'|}. \end{aligned} \tag{B.3}$$

The reader is urged to derive the corresponding expression for G_0^-. The derivation just given provides a clear justification for the retention of the terms retarded and advanced when referring to the propagator even in the energy rather than the time representation.

Appendix C
Spherical Functions

C.1 The Spherical Harmonics

The solutions of the generalized Legendre equation

$$\frac{\mathrm{d}}{\mathrm{dx}}\left[(1-x^2)\frac{\mathrm{d}P}{\mathrm{d}x}\right]+\left[\ell(\ell+1)-\frac{m^2}{1-x^2}\right]P=0, \tag{C.1}$$

are the associated Legendre polynomials, P_ℓ^m, which for m positive are defined by the formula

$$P_\ell^m(x)=(-1)^m(1-x^2)^{m/2}\frac{\mathrm{d}^m}{\mathrm{d}x^m}P_\ell(x), \tag{C.2}$$

where the $P_\ell(x) = P_\ell^0(x)$ are the Legendre polynomials. These are given by *Rodrigues's formula*

$$P_\ell(x)=\frac{1}{2^\ell \ell!}\frac{\mathrm{d}^\ell}{\mathrm{d}x}(x^2-1)^\ell, \tag{C.3}$$

which in combination with Eq.(C.3) yields a formula for the associated Legendre polynomials that is valid for both positive and negative m, $m = -\ell, -\ell+1, \ldots, \ell-1, \ell$,

$$P_\ell^m(x)=\frac{(-1)^m}{2^\ell \ell!}(1-x^2)^{m/2}\frac{\mathrm{d}^{\ell+m}}{\mathrm{d}x^{\ell+m}}(x^2-1)^\ell. \tag{C.4}$$

It can be shown that

$$P_\ell^{-m}(x)=(-1)^m\frac{(\ell-m)!}{(\ell+m)!}P_\ell^m(x). \tag{C.5}$$

The associated Legendre polynomials satisfy the orthogonality relation

$$\int_{-1}^{1} P_\ell^{-m}(x)P_{\ell'}^{-m}(x)\mathrm{d}x = \frac{2}{2\ell+1}\frac{(\ell+m)!}{(\ell-m)!}\delta_{\ell\ell'}. \tag{C.6}$$

Now, as is well known, the functions $Q_m(\phi) = \mathrm{e}^{\mathrm{i}m\phi}$ form a complete set of orthogonal functions in the index m on the interval $0 \leq \phi \leq 2\pi$. From the orthogonality of the $P_\ell^m(\cos\theta)$ in the interval $-1 \leq \cos\theta \leq -1$ it follows that the product $P_\ell^m Q_m$ forms a complete orthogonal set of functions in the indices ℓ, m on the surface of the unit sphere. The *spherical harmonics* $Y_{\ell m}(\theta,\phi)$ are suitably normalized functions on the surface of the unit sphere, and are given explicitly by the expression

$$Y_{\ell m}(\theta,\phi) = \sqrt{\frac{2\ell+1}{4\pi}\frac{(\ell-m)!}{(\ell+m)!}}P_\ell^m(\cos\theta)\mathrm{e}^{\mathrm{i}m\phi}. \tag{C.7}$$

¿From Eq.(C.5) we see that

$$Y_{\ell,-m}(\theta,\phi) = (-1)^m Y_{\ell m}^*(\theta,\phi), \tag{C.8}$$

with A^* denoting the complex conjugate of A. Using L as a combined index, $L = (\ell, m)$, we can also write

$$Y_{\ell m}(\theta,\phi) \equiv Y_L(\theta,\phi) \equiv Y_{\ell m}(\hat{r}), \tag{C.9}$$

where $\hat{r}$ is a unit vector in the direction (θ,ϕ).

The spherical harmonics are eigenfunctions of the square of the angular momentum operator, $\hat{L}$,

$$\hat{L}^2 Y_{\ell m}(\theta,\phi) = \ell(\ell+1)Y_{\ell m}(\theta,\phi) \tag{C.10}$$

and of the component of $\hat{L}$ along the z-axis,

$$L_z Y_{\ell m}(\theta,\phi) = mY_{\ell m}(\theta,\phi). \tag{C.11}$$

In the interval $0 \leq \theta \leq \pi,\ 0 \leq \phi \leq 2\pi$ the $Y_{\ell m}(\theta,\phi)$ satisfy, respectively, the orthogonality and completeness relations,

$$\int_0^{2\pi}\mathrm{d}\phi\int_0^{\pi}\sin\theta\mathrm{d}\theta Y_{\ell' m'}^*(\theta,\phi)Y_{\ell m}(\theta,\phi) = \delta_{\ell\ell'}\delta_{mm'} \tag{C.12}$$

and

$$\sum_{\ell=0}^{\infty}\sum_{-\ell}^{\ell} Y_{\ell m}^*(\theta',\phi')Y_{\ell m}(\theta,\phi) = \delta(\phi-\phi')\delta(\cos\theta-\cos\theta'). \tag{C.13}$$

The first few spherical harmonics are listed below.

$\ell = 0$:

$$Y_{00} = \frac{1}{\sqrt{4\pi}} \tag{C.14}$$

$\ell = 1$:

$$Y_{11} = -\sqrt{\frac{3}{8\pi}} \sin\theta e^{i\phi}$$

$$Y_{10} = \sqrt{\frac{3}{4\pi}} \cos\theta \tag{C.15}$$

$\ell = 2$:

$$Y_{22} = \frac{1}{4}\sqrt{\frac{15}{4\pi}} \sin^2\theta e^{2i\phi}$$

$$Y_{21} = -\sqrt{\frac{15}{8\pi}} \sin\theta\cos\theta e^{i\phi} \tag{C.16}$$

$$Y_{20} = \sqrt{\frac{5}{4\pi}}\left(\frac{3}{2}\cos^2\theta - \frac{1}{2}\right)$$

$\ell = 3$:

$$Y_{33} = -\frac{1}{4}\sqrt{\frac{35}{4\pi}} \sin^3\theta e^{3i\phi}$$

$$Y_{32} = \frac{1}{4}\sqrt{\frac{105}{2\pi}} \sin^2\theta\cos\theta e^{2i\phi}$$

$$Y_{31} = -\frac{1}{4}\sqrt{\frac{21}{4\pi}} \sin\theta(5\cos^2\theta - 1)e^{i\phi} \tag{C.17}$$

$$Y_{30} = \sqrt{\frac{7}{4\pi}}\left(\frac{5}{2}\cos^3\theta - \frac{3}{2}\cos\theta\right)$$

We see that for $m = 0$

$$Y_{\ell 0} = \sqrt{\frac{2\ell+1}{4\pi}} P_\ell(\cos\theta). \tag{C.18}$$

A detailed discussion of the properties of the functions $Y_{\ell m}(\theta, \phi)$ is given in Condon and Shortley [1]. Of particular interest is the relation known as *the addition theorem for spherical harmonics.* If two unit vectors, $\hat{r}$ and $\hat{r}'$, are defined by the angles (θ, ϕ) and (θ', ϕ'), and make between them an angle Θ, then

$$P_\ell(\Theta) = \frac{4\pi}{2\ell+1} \sum_{m=-\ell}^{\ell} Y^*_{\ell m}(\theta', \phi') Y_{\ell m}(\theta, \phi). \tag{C.19}$$

This relation can also be expressed in terms of the associated Legendre polynomials

$$\begin{aligned} P_\ell(\Theta) = {} & P_\ell(\theta)P_\ell(\theta') \\ & +2\sum_{m=-\ell}^{\ell} \frac{(\ell-1)!}{(\ell+m)!} P_\ell^m(\cos\theta)P_\ell^m(\cos\theta')\cos\left[m(\phi-\phi')\right]. \end{aligned} \tag{C.20}$$

The addition theorem is important in obtaining many useful expressions, such as the expansions of the free-particle propagator discussed in Appendix D.

C.2 The Bessel, Neumann, and Hankel Functions

The Bessel functions of order ν are the solutions of the Bessel equation

$$\frac{\mathrm{d}^2 R}{\mathrm{d}x^2} + \frac{1}{x}\frac{\mathrm{d}R}{\mathrm{d}x} + \left(1 - \frac{\nu^2}{x^2}\right) R = 0. \tag{C.21}$$

There are two solutions for each value of ν, given in terms of the gamma function,

$$J_\nu(x) = \left(\frac{x}{2}\right)^\nu \sum_{j=0}^{\infty} \frac{(-1)^j}{j!\Gamma(j+\nu+1)} \left(\frac{x}{2}\right)^{2j} \tag{C.22}$$

and

$$J_{-\nu}(x) = \left(\frac{x}{2}\right)^{-\nu} \sum_{j=0}^{\infty} \frac{(-1)^j}{j!\Gamma(j-\nu+1)} \left(\frac{x}{2}\right)^{2j} \tag{C.23}$$

called Bessel functions of the first kind of order $\pm\nu$. If ν is an integer, these two solutions are linearly dependent

$$J_{-\nu}(x) = (-1)^\nu J_\nu(x), \tag{C.24}$$

in which case a second, linearly independent solution is given by the Bessel function of the second kind, or *Neumann function*,

$$N_\nu = \frac{J_\nu(x)\cos\nu\pi - J_{-\nu}}{\sin\nu\pi}. \tag{C.25}$$

The Bessel functions of the third kind, or *Hankel functions*, are defined as linear combinations of J_ν and N_ν,

$$\begin{aligned} H_\nu^{(+)}(x) &= J_\nu(x) + \mathrm{i}N_\nu(x) \\ H_\nu^{(-)}(x) &= J_\nu(x) - \mathrm{i}N_\nu(x). \end{aligned} \tag{C.26}$$

Each of the two combinations, $\{J_\nu, N_\nu\}$ and $\{H_\nu^{(+)}, H_\nu^{(-)}\}$ forms a linearly independent set of solutions to Bessel's equation, Eq.(C.21).

When ν is not an integer, the functions $\{J_\nu, J_{-\nu}\}$ form a linearly independent set of solutions to Bessel's equation. For $\nu = \ell + \frac{1}{2}$, with ℓ being an integer, it is customary to define spherical Bessel, Neumann, and Hankel functions through the relations,

$$j_\ell(x) = \sqrt{\frac{\pi}{2x}} J_{\ell+\frac{1}{2}}(x)$$

$$n_\ell(x) = \sqrt{\frac{\pi}{2x}} N_{\ell+\frac{1}{2}}(x)$$

$$h_\ell^{(\pm)}(x) = \sqrt{\frac{\pi}{2x}} H_{\ell+\frac{1}{2}}^{(\pm)}(x). \tag{C.27}$$

For x real, we have

$$h_\ell^{(\pm)}(x) = h_\ell^{(\mp)^*}(x). \tag{C.28}$$

It follows from the series representations, Eqs.(C.22) and (C.23), that

$$j_\ell(x) = (-x)^\ell \left(\frac{1}{x}\frac{\mathrm{d}}{\mathrm{dx}}\right)^\ell \left(\frac{\sin x}{x}\right)$$

$$n_\ell(x) = -(-x)^\ell \left(\frac{1}{x}\frac{\mathrm{d}}{\mathrm{dx}}\right)^\ell \left(\frac{\cos x}{x}\right). \tag{C.29}$$

For the first few values of ℓ, the functions j_ℓ, n_ℓ, and $h_\ell^{(+)}$ are given by the expressions

$$j_0(x) = \frac{\sin x}{x}; \quad j_1(x) = \frac{\sin x}{x^2} - \frac{\cos x}{x};$$

$$j_2(x) = \left(\frac{3}{x^3} - \frac{1}{x}\right) \sin x - 3\frac{\cos x}{x}$$

$$n_0(x) = -\frac{\cos x}{x}; \quad n_1(x) = -\frac{\cos x}{x^2} - \frac{\sin x}{x};$$

$$n_2(x) = -\left(\frac{3}{x^3} - \frac{1}{x}\right) \cos x - 3\frac{\sin x}{x}$$

$$h_0^{(+)}(x) = \frac{\mathrm{e}^{\mathrm{i}x}}{\mathrm{i}x}; \quad h_1^{(+)}(x) = -\frac{\mathrm{e}^{\mathrm{i}x}}{x}\left(1 + \frac{\mathrm{i}}{x}\right);$$

$$h_2^{(+)}(x) = \frac{\mathrm{i}\mathrm{e}^{\mathrm{i}x}}{x}\left(1 + \frac{3\mathrm{i}}{x} - \frac{3}{x^2}\right). \tag{C.30}$$

The spherical Bessel, Neumann, and Hankel functions, denoted collectively by z_ℓ, satisfy the following recursion formulae,

$$\frac{2\ell+1}{x} z_\ell(x) = z_{\ell-1}(x) + z_{\ell+1}(x) \tag{C.31}$$

and

$$z'_\ell(x) = \frac{1}{2\ell+1}\left[\ell z_{\ell-1}(x) - (\ell+1) z_{\ell+1}(x)\right], \tag{C.32}$$

where $z'(x) = \mathrm{d}z(x)/\mathrm{d}x$. Finally, in the limit of small argument, $x << \ell$, we have the asymptotic forms,

$$j_\ell(x) \to \frac{x^\ell}{(2\ell+1)!!}$$

$$n_\ell(x) \to -\frac{(2\ell-1)!!}{x^{\ell+1}}, \tag{C.33}$$

where $(2\ell+1)!! = (2\ell+1)(2\ell-1)(2\ell-3)\cdots(5)(3)(1)$. In the limit of large argument $x >> \ell$, we have

$$j_\ell(x) \to \frac{1}{x}\sin\left(x - \frac{\ell\pi}{2}\right)$$
$$n_\ell(x) \to -\frac{1}{x}\cos\left(x - \frac{\ell\pi}{2}\right)$$
$$h_\ell^{(+)}(x) \to (-\mathrm{i})^{\ell+1}\frac{\mathrm{e}^{\mathrm{i}x}}{x}. \tag{C.34}$$

Finally, the spherical functions satisfy a set of useful Wronskian relations. With

$$W(f(x), g(x)) = fg' - gf', \qquad f' = \frac{\mathrm{d}f}{\mathrm{d}x}, \tag{C.35}$$

we have

$$W(j_\ell(x), n_\ell(x)) = \frac{1}{i}W(j_\ell(x), h_\ell(x)) = -W(n_\ell(x), h_\ell(x)) = \frac{1}{x^2}. \tag{C.36}$$

As a simple illustration of the use of the spherical harmonics, we present the solutions of the force-free Schrödinger equation, the Helmholtz equation, in one, two, and three dimensions.

C.3 Solutions of the Helmholtz Equation

In this appendix we give multipolar solutions to the Helmholtz equation in one, two, and three dimensions as well as some useful identities. We have attempted to choose a notation and normalization for the regular and irregular solutions such that most of the formulae in the text are valid in one, two, or three dimensions. The expression $J_L(\mathbf{r})$ is used to denote the regular solution that is proportional to r^ℓ near the origin. It is useful to define two irregular solutions. Similarly, $N_L(\mathbf{r})$ represents the irregular solution appropriate to standing-wave boundary conditions, and $H_L(\mathbf{r})$ represents the irregular solution compatible with outgoing waves at large distances. The spherical function $H_L(\mathbf{r})$ is also the irregular solution that vanishes at infinity when analytically continued to negative energies.

The Helmholtz equation, $[\nabla^2 + E]\psi(x) = 0$, has solutions of the form $\psi(x) = \exp(\mathbf{ik}\cdot\mathbf{r})$ where $k^2 = E$. We desire, however, solutions in the form of radial functions multiplying angular functions, that is, $J_L(\mathbf{r}) = f_L(r)Y_L(\hat{r})$. To achieve this end we introduce angular harmonics, $Y_L(\mathbf{r})$, which are orthonormal and complete over the angular variables

$$\int \mathrm{d}\hat{r}\, Y_L(\hat{r})Y_{L'}^{(*)}(\hat{r}) = \delta_{LL'} \tag{C.37}$$

$$\sum_L Y_L(\hat{r})Y_L^{(*)}(\hat{r}') = \delta(\hat{r}, \hat{r}'). \tag{C.38}$$

In one dimension there are only two angular harmonics corresponding to functions that are either symmetric or antisymmetric about the origin,

$$Y_0(\hat{r}) = \frac{1}{\sqrt{2}}, \quad Y_1(\hat{r}) = \frac{x}{r\sqrt{2}}, \qquad (d=1), \tag{C.39}$$

with $r = |x|$. In two dimensions the angular harmonics may be defined as,

$$Y_L(\hat{r}) = \frac{\exp(iL\theta)}{\sqrt{2\pi}} \qquad (L = 0, \pm 1, \pm 2 \pm 3, \ldots), \qquad (d=2) \tag{C.40}$$

but we can equally well take them to be real,

$$Y_L(\hat{r}) = \begin{cases} \frac{1}{\sqrt{\pi}} \cos L\theta, & L > 0; \\ \frac{1}{\sqrt{2\pi}}, & L = 0; \\ \frac{1}{\sqrt{\pi}} \sin L\theta, & L < 0. \end{cases} \qquad (d=2) \tag{C.41}$$

In three dimensions the angular harmonics are the spherical harmonics,

$$Y_L(\hat{r}) = N_\ell^m P_\ell^{|m|}(\cos\theta)\exp(im\phi), \qquad (d=3), \tag{C.42}$$

$$L = \{\ell, m\}, \qquad \ell = 0, 1, 2, 3, \ldots, \qquad m = (-\ell, -\ell+1, \ldots, \ell-1, \ell) \tag{C.43}$$

where $P_\ell^m(x)$ is an associated Legendre polynomial,

$$P_\ell^m(x) = (1-x^2)^{m/2} \frac{1}{d^{\ell+m}} dx^{\ell+m} (x^2-1)^\ell, \tag{C.44}$$

and N_ℓ^m is a normalization factor

$$N_\ell^m = \left(-\frac{m}{|m|}\right)^m \sqrt{\frac{2\ell+1}{4\pi}} \sqrt{\frac{(\ell-|m|)!}{(\ell+|m|)!}}. \tag{C.45}$$

There is substantial arbitrariness in the phases of the spherical harmonics and it is often convenient to define them so that they are real,

$$Y_L(\hat{r}) = N_\ell^m P_\ell^{|m|}(\cos\theta) \begin{cases} \sqrt{2}\cos m\phi, & (m>0), \\ 1, & (m=0), \\ \sqrt{2}\sin m\phi, & (m<0). \end{cases} \qquad (d=3) \tag{C.46}$$

Functions, $J_L(\mathbf{r})$, of the desired form that are regular (finite, continuous, and smooth) everywhere can be defined as the coefficients in an expansion of $\exp(i\mathbf{k}\cdot\mathbf{r})$,

$$\exp(i\mathbf{k}\cdot\mathbf{r}) = \begin{cases} 2\sum_{L=0,1} i^L J_L(\mathbf{r}) Y_L(\hat{k}), & (d=1), \\ 2\pi \sum_{L=-\infty}^{\infty} i^{|L|} J_{|L|}(\mathbf{r}) Y_L(\hat{k}), & (d=2), \\ 4\pi \sum_{\ell=0}^{\infty} \sum_{m=-\ell}^{\ell} i^\ell J_L(\mathbf{r}) Y_L(\hat{k}), & (d=3), \end{cases} \tag{C.47}$$

or

$$J_L(\mathbf{r}) = \int d\hat{k} \exp(i\mathbf{k}\cdot\mathbf{r}) Y_L(\hat{k}) \begin{cases} \frac{i^{-L}}{2}, & (d=1), \\ \frac{i^{-|L|}}{2\pi}, & (d=2), \\ \frac{i^{-\ell}}{4\pi}, & (d=3). \end{cases} \tag{C.48}$$

These regular, multipolar solutions are easily evaluated in one, two, or three dimensions,

$$J_L(\mathbf{r}) = Y_L(\hat{r}) \begin{cases} \cos(kr - l\pi/2), & (d=1), \\ \mathcal{J}_{|L|}(kr), & (d=2), \\ j_\ell(kr), & (d=3). \end{cases} \tag{C.49}$$

Here we have used $\mathcal{J}_{|L|}(z)$ to represent the cylindrical Bessel function and $j_\ell(z)$ to represent the spherical Bessel function.

It may be helpful to view Eq.(C.47) as the *generating* function for the Bessel functions. The generating function for the cylindrical Bessel coefficients is (usually) given by the expression

$$\exp\left[z\left(t - \frac{1}{t}\right)\right] = \sum_{L=-\infty}^{\infty} t^L \mathcal{J}_L(z). \tag{C.50}$$

This can be converted into Eq.(C.47) by the substitution $t \to \exp i(\theta + \pi/2)$. Similarly an expression equivalent to the $d = 3$ version of Eq.(C.47),

$$\exp(izx) = \sum_{\ell=0}^{\infty}(2\ell+1)P_\ell(x)i^\ell j_\ell(z), \tag{C.51}$$

where $P_\ell(x)$ is a Legendre polynomial may be taken as a definition of the spherical Bessel functions.

The cylindrical and spherical Bessel functions have the following limiting forms for small arguments,

$$\begin{aligned} \mathcal{J}_{|L|}(z) &\to \frac{(z/2)^{|L|}}{|L|!}(1 + O[z^2]) \\ j_\ell(z) &\to \frac{z^\ell}{(2\ell+1)!!}, \end{aligned} \tag{C.52}$$

and vary at large arguments as,

$$\begin{aligned} \mathcal{J}_{|L|}(z) &\to \sqrt{\frac{2}{\pi z}} \cos\left(z - |L|\frac{\pi}{2} - \frac{\pi}{4}\right) \\ j_\ell(z) &\to \sin\left(z - \frac{\ell\pi}{2}\right)/z. \end{aligned} \tag{C.53}$$

In addition to these regular solutions to the Helmholtz equation, irregular solutions are also needed in order to construct the Green function which satisfies the inhomogeneous Helmholtz equation,

$$[\nabla^2 + E]G_0(\mathbf{r}, \mathbf{r}') = \delta(\mathbf{r} - \mathbf{r}'). \tag{C.54}$$

It is easy to verify that this equation is satisfied by the following integral expression for G_0,

$$G_0(\mathbf{r}, \mathbf{r}') = \left(\frac{1}{2\pi}\right)^d \int_0^\infty k^{d-1}\mathrm{d}k \int \mathrm{d}\hat{k} \frac{\exp[i\mathbf{k}\cdot(\mathbf{r} - \mathbf{r}')]}{E - k^2}. \tag{C.55}$$

In addition to satisfying the inhomogeneous Helmholtz equation, Eq.(C.54), the Green function specifies the boundary condition satisfied by the wave functions. We will usually use outgoing wave boundary conditions because they are the easiest to understand physically. They correspond to a causal or retarded Green function; that is, a perturbation at the origin at time t propagates outward as time increases. Sometimes, however, it is convenient to use standing-wave boundary conditions which correspond to the average of the advanced and retarded Green function. With standing-wave boundary conditions a disturbance initiates both incoming and outgoing waves of equal amplitude that set up a standing-wave pattern.

Mathematically, the two types of Green function arise from different ways of handling the singularities at $k = \pm\sqrt{E}$ in Eq.(C.55) when the integral over k is converted into a contour integral. The stationary-wave Green function arises from averaging the residues from both poles while the outgoing-wave Green function arises from assuming the energy to have a small positive imaginary part and closing contours in such a way that the integral remains finite. The outgoing-wave Green function has the advantage that its analytic continuation to negative energies vanishes at large distances so it is appropriate for describing wave functions for bound states.

The outgoing-wave or causal Green function in one, two, or three dimensions can be calculated from Eqs.(C.47) and (C.55) together with the following identities

$$\int_0^\infty \frac{\mathrm{d}k}{E-k^2}\left\{\begin{array}{c}\cos(kr)\\ k\mathcal{J}_0(kr)\\ k^2 j_0(kr)\end{array}\right\} = \left\{\begin{array}{c}-(1/2)\pi i E^{-(1/2)}\exp(ikr)\\ -(1/2)\pi i\mathcal{H}_0(kr)\\ -(1/2)\pi ikrh_\ell(kr)\end{array}\right\}. \tag{C.56}$$

The resultant expression for the Green function is $G_0(\mathbf{r},\mathbf{r}') = H_0(\mathbf{r}-\mathbf{r}')$, where H_0 is an $L = 0$ irregular solution to the Helmholtz equation given by the expressions

$$H_0(\mathbf{r}-\mathbf{r}') = Y_L(\widehat{r}-r')\left\{\begin{array}{cc}-iE^{-(1/2)}\exp(ik|\mathbf{r}-\mathbf{r}'|) & (d=1)\\ -(1/4)i\mathcal{H}_0(k|\mathbf{r}-\mathbf{r}'|) & (d=2)\\ -ikrh_0(k|\mathbf{r}-\mathbf{r}'|) & (d=3)\end{array}\right\}. \tag{C.57}$$

In the text we make extensive use of an expansion for the Green function as a sum of terms each of which is separable in $\mathbf{r}$ and $\mathbf{r}'$. Using Eqs.(C.47) and (C.55) again but substituting an expansion in terms of angular harmonics for both $\exp(i\mathbf{k}\cdot\mathbf{r})$ and $\exp(-i\mathbf{k}\cdot\mathbf{r}')$ and using the orthonormality of the angular harmonics, we have

$$G_0(\mathbf{r},\mathbf{r}') = \frac{\Omega^2}{(2\pi)^3}\int_0^\infty \frac{k^{d-1}\mathrm{d}k}{E-k^2}\sum_L J_L(\mathbf{r};k)J_{L'}^{(*)}(\mathbf{r}';k), \tag{C.58}$$

where Ω is the angular phase space, equal to 2, 2π, and 4π in one, two, and three dimensions, respectively. The dependence of the Helmholtz equation solutions on the energy parameter is shown explicitly for clarity. The integral gives

$$G_0(\mathbf{r},\mathbf{r}') = \sum_L J_L(\mathbf{r};k)H_L^{(*)}(\mathbf{r}';k)\Theta(r'-r) + H_L(\mathbf{r};k)J_L^{(*)}(\mathbf{r}';k)\Theta(r-r'), \tag{C.59}$$

where $\Theta(x)$ is the heaviside step function, which is unity if its argument is greater than zero and vanishes if its argument is less than zero, and where the irregular solution, $H_L(\mathbf{r})$, is given by the expressions

$$H_L(\mathbf{r}) = Y_L(\mathbf{r}) \begin{cases} -iE^{-(1/2)}\exp[i(kr-\ell\pi/2)] & (d=1) \\ -(1/4)i\mathcal{H}_{|L|}(kr) & (d=2) \\ -ikrh_\ell(kr) & (d=3) \end{cases}. \tag{C.60}$$

The notation $H_L^{(*)}$ indicates complex conjugation of the angular harmonic part of the function only. If real angular harmonics are used, all of the $(*)$'s may be omitted. The expression $\mathcal{H}_L = \mathcal{J}_L + i\mathcal{Y}_L$ is the cylindrical Hankel function which is given approximately for small values of its argument by

$$\mathcal{H}_{|L|}(z) \rightarrow \begin{cases} \frac{2}{\pi}(\ln(z/2)+\gamma)(1+O[z^2]) & (L=0) \\ -\frac{(|L|-1)!}{\pi(z/2)^{|L|}}(1+O[z^2]) & (|L|>0) \end{cases}, \tag{C.61}$$

where γ is Euler's constant, $\gamma = 0.5772156649\ldots$. For large values of its argument $\mathcal{H}_L(z)$ varies as

$$\mathcal{H}_{|L|}(z) \rightarrow \sqrt{\frac{2}{\pi z}}\exp i\left(z - \frac{|L|\pi}{2} - \frac{\pi}{4}\right). \tag{C.62}$$

Similarly, the three-dimensional irregular solutions are proportional to spherical Hsankel functions $h_\ell = j_\ell + in_\ell$. At small values of their arguments these functions are approximately

$$h_\ell(z) \rightarrow \frac{(2\ell-1)!!}{z^{\ell+1}}, \tag{C.63}$$

and at large values they vary as

$$h_\ell(z) \rightarrow \exp i\frac{\left(z-\frac{\ell\pi}{2}\right)}{z}. \tag{C.64}$$

It is important that the regular and irregular solutions satisfy the following Wronskian-like relation,

$$\int_S \mathrm{d}^2S\cdot[J_L(\mathbf{r})\nabla H_{L'}(\mathbf{r}) - [\nabla J_L(\mathbf{r})]H_{L'}(\mathbf{r})] = \begin{cases} \delta_{LL'}, & \mathbf{r}=0\in S; \\ 0, & \text{otherwise,} \end{cases} \tag{C.65}$$

where the delta function corresponds to the origin lying inside the volume defined by S. This relation is equivalent to the requirement that the Green function be correctly normalized since if $[\nabla^2+E]G(\mathbf{r},\mathbf{r}') = \delta(\mathbf{r}-\mathbf{r}')$ and $[\nabla^2+E]J_L(x) = 0$ it follows from Green's theorem that

$$\int \mathrm{d}^2S\cdot[J_L(\mathbf{r})\nabla G(\mathbf{r},\mathbf{r}') - [\nabla J_L(\mathbf{r})]G(\mathbf{r},\mathbf{r}')] = J_L(\mathbf{r}') \tag{C.66}$$

provided $\mathbf{r}'$ is within the interval bounded by the surface of integration. Expanding the Green function in the usual way yields

$$\int \mathrm{d}^2 S \cdot \sum_{L'} [J_L(\mathbf{r}) \nabla H_{L'}(\mathbf{r}) - [\nabla J_L(\mathbf{r})] H_{L'}(\mathbf{r}')] J_L(\mathbf{r}') = J_L(\mathbf{r}'), \quad \text{(C.67)}$$

which requires the Wronskian condition (C.65).

Both the regular and irregular functions can be expanded about a shifted origin,

$$J_L(\mathbf{r} + \mathbf{R}) = \sum_{L'} g_{LL'}(\mathbf{R}) J_{L'}(\mathbf{r}) \quad \text{(C.68)}$$

$$H_L(\mathbf{r} + \mathbf{R}) = \sum_{L'} G_{LL'}(\mathbf{R}) J_{L'}(\mathbf{r}). \quad \text{(C.69)}$$

The coefficients, $g^J_{LL'}(\mathbf{R})$, for the expansion of $J_L(\mathbf{r} + \mathbf{R})$ play the role of a translation operator. The coefficients, $g_{LL"}(\mathbf{R})$, for the expansion of the irregular solutions in terms of the regular solutions play a central role in multiple scattering theory. They are given by

$$G_{LL'}(\mathbf{R}) = \begin{cases} 2\sum_{L''} i^{L'-L-L''} C(L, L', L'') H^{(*)}_{L''}(\mathbf{R}) & (d = 1), \\ 2\pi \sum_{L''} i^{|L'|-|L|-|L''|} C(L, L', L'') H^{(*)}_{|L''|}(\mathbf{R}) & (d = 2), \\ 4\pi \sum_{L''} i^{\ell'-\ell-\ell''} C(L, L', L'') H^{(*)}_{L''}(\mathbf{R}) & (d = 3). \end{cases} \quad \text{(C.70)}$$

Here the Gaunt numbers, $C(L, L', L'')$ are defined as integrals over three angular harmonics,

$$C(L, L', L'') = \int \mathrm{d}\hat{r} Y^{(*)}_L(\hat{r}) Y_{L'}(\hat{r}) Y^{(*)}_{L''}(\hat{r}) \quad \text{(C.71)}$$

The expression for $g_{LL'}(\mathbf{R})$ is identical to that for $G_{LL'}(\mathbf{R})$ except that $H^{(*)}_{L''}(\mathbf{R})$ is replaced by $J^{(*)}_{L''}(\mathbf{R})$.

In one dimension, simple analytic expressions for the structure constants both in real and reciprocal space are available,

$$H_L(\mathbf{r} - \mathbf{R}) = \sum_{L'=0,1} G_{LL'}(-\mathbf{R}) J_{L'}(\mathbf{r}) = \sum_{L'=0,1} J_{L'}(\mathbf{r}) G_{L'L}(\mathbf{R}) \quad \text{(C.72)}$$

$$G_{LL'}(\mathbf{R}) = \frac{\exp(i\sqrt{E}r)}{\mathrm{i}\sqrt{E}} \begin{pmatrix} 1 & i^{-1}\mathbf{R}/R \\ \mathrm{i}\mathbf{R}/R & 1 \end{pmatrix} \quad \text{(C.73)}$$

and

$$\tilde{G}_{LL'}(k) = \sum_{n=-\infty}^{\infty} g_{LL'}(na) \mathrm{e}^{\mathrm{i}kna}$$

$$= \frac{1}{\mathrm{i}\sqrt{E}} \begin{pmatrix} -1 + \dfrac{i \sin\phi}{\cos\theta - \cos\phi} & \dfrac{\sin\theta}{\cos\theta - \cos\phi} \\ \dfrac{-\sin\theta}{\cos\theta - \cos\phi} & -1 + \dfrac{i\sin\phi}{\cos\theta - \cos\phi} \end{pmatrix}$$

$$\phi = \sqrt{E}a, \ \theta = ka. \tag{C.74}$$

For $E < 0$ it is convenient to define the regular solutions to be

$$J_L(\mathbf{r}) = Y_L(\hat{r}) \begin{cases} i^{-L}\cos(i\sqrt{-E}r - \ell\pi/2) & (d = 1) \\ i^{-|L|}\mathcal{J}_{|L|}(i\sqrt{-E}r) & (d = 2) \\ i^{-\ell}j_\ell(i\sqrt{-E}r) & (d = 3), \end{cases} \tag{C.75}$$

and the irregular solutions to be

$$H_L(\mathbf{r}) = Y_L(\hat{r}) \begin{cases} -[-E]^{-(1/2)}\exp(-\sqrt{-E}r) & (d = 1) \\ (1/4)i^{|L|-1}\mathcal{H}_L(i\sqrt{-E}r) & (d = 2) \\ i^{\ell}\sqrt{-E}h_\ell(i\sqrt{-E}r) & (d = 3). \end{cases} \tag{C.76}$$

These definitions yield real values for J_L and H_L at real negative energies provided that the angular harmonics are chosen to be real.

References

[1] E. U. Condon and G. H. Shortley *The Theory of Atomic Spectra* (Cambridge University Press, Cambridge 1967).

Appendix D
Displacements of Spherical Functions

In this appendix we give formulae for the expansion of spherical functions about displaced origins. These spherical functions are defined as the products of a spherical Bessel, Neumann or Hankel function with a spherical harmonic. In the following discussion, such products are denoted by $A_L(\mathbf{x}) = a_l(x)Y_L(\hat{\mathbf{x}})$, where $a_l(x)$ denotes either $j_l(x)$, $n_l(x)$, $h_l^+(x)$ or $h_l^-(x)$. The notation $Y_L(\hat{\mathbf{x}})$ denotes a real spherical harmonic (expressions analogous and very similar to those derived below can also be obtained with complex spherical harmonics). Expressions are derived that connect values of A_L about a displaced origin, $A_L(\mathbf{x}+\mathbf{a})$, to the undisplaced values, $A_L(\mathbf{x})$.

We begin with the expansion of a plane wave in terms of spherical functions using the well-known expression (Bauer's identity),

$$\begin{aligned} e^{i\mathbf{k}\cdot\mathbf{r}} &= 4\pi \sum_{l,m} i^l j_l(kr) Y_{l,m}(\hat{\mathbf{r}}) Y_{l,m}(\hat{\mathbf{k}}) \\ &= 4\pi \sum_L i^l J_L(\mathbf{r}) Y_L(\hat{\mathbf{k}}). \end{aligned} \tag{D.1}$$

Thus, for any vectors $\mathbf{R}$ and $\mathbf{r}$ we obtain,

$$\begin{aligned} e^{i\mathbf{k}\cdot(\mathbf{R}+\mathbf{r})} &= (4\pi)^2 \sum_{L_2} \sum_{L_3} i^{l_2+l_3} J_{L_3}(\mathbf{R}) J_{L_2}(\mathbf{r}) Y_{L_2}(\hat{\mathbf{k}}) Y_{L_3}(\hat{\mathbf{k}}) \\ &= 4\pi \sum_{L_1} i^{l_1} J_{L_1}(\mathbf{R}+\mathbf{r}) Y_{L_1}(\hat{\mathbf{k}}). \end{aligned} \tag{D.2}$$

Upon multiplying the equality of the last two expressions by $Y_L(\hat{\mathbf{k}})$ and integrating over the angles of $\mathbf{k}$, we obtain the result

$$\begin{aligned} J_{L_1}(\mathbf{R}+\mathbf{r}) &= 4\pi \sum_{L_2}\sum_{L_3} \mathrm{i}^{l_1-l_2-l_3} C(L_1,L_2,L_3) J_{L_3}(\mathbf{R}) J_{L_2}(\mathbf{r}) \\ &= \sum_{L_2} g_{L_1L_2}(\mathbf{R}) J_{L_2}(\mathbf{r}). \end{aligned} \tag{D.3}$$

Here, we have used the notation

$$C(L_1,L_2,L_3) = \int Y_{L_1}(\Omega) Y_{L_2}(\Omega) Y_{L_3}(\Omega) \mathrm{d}\Omega \tag{D.4}$$

for the Gaunt numbers (the integral over three spherical harmonics). We also used the fact that $\mathrm{i}^{(l_1-l_2-l_3)}$ can replace $\mathrm{i}^{(-l_1+l_2+l_3)}$ because the gaunt numbers vanish if $l_1 + l_2 + l_3$ is not an even integer. The expansion coefficients, $g_{L_1L_2}(\mathbf{R})$, are given by the expression,

$$g_{L_1L_2}(\mathbf{R}) = 4\pi \sum_{L_3} \mathrm{i}^{l_1-l_2-l_3} C(L_1,L_2,L_3) J_{L_3}(\mathbf{R}). \tag{D.5}$$

These coefficients are the matrix elements of the translation operator [1] in the angular-momentum representation and form a unitary matrix as is shown below. Equation (D.5) provides the desired expression for the "shifted" function $J_L(\mathbf{R}+\mathbf{r})$ in terms of the unshifted functions $J_L(\mathbf{r})$, for all vectors $\mathbf{R}$ and $\mathbf{r}$.

In order to derive an analogous result for the Hankel function $H_L(\mathbf{R}+\mathbf{r})$, we recall the usual expression, Eq.(B.3), for the outgoing free-particle propagator,

$$\begin{aligned} G_0(\mathbf{R}-\mathbf{r}) &= -\frac{\mathrm{e}^{\mathrm{i}k|\mathbf{R}-\mathbf{r}|}}{4\pi|\mathbf{R}-\mathbf{r}|} \\ &= -\mathrm{i}k \sum_L j_l(kr) h_l(kR) Y_L(\hat{\mathbf{r}}) Y_L(\hat{\mathbf{R}}) \\ &= -\mathrm{i}k \sum_L J_L(\mathbf{r}) H_L(\mathbf{R}) \quad \text{for } R > r, \end{aligned} \tag{D.6}$$

For vectors $\mathbf{R}$ and $\mathbf{r}$ such that $R > r$, consider vectors $\mathbf{a}$ such that $|\mathbf{R}+\mathbf{r}| > a$ and $R > |\mathbf{a}-\mathbf{r}|$, so that we have

$$\begin{aligned} \frac{\mathrm{e}^{\mathrm{i}k|\mathbf{R}+\mathbf{r}-\mathbf{a}|}}{|\mathbf{R}+\mathbf{r}-\mathbf{a}|} &= 4\pi\mathrm{i}k \sum_{L_1} H_{L_1}(\mathbf{R}+\mathbf{r}) J_{L_1}(\mathbf{a}) \\ &= 4\pi\mathrm{i}k \sum_{L_3} H_{L_3}(\mathbf{R}) J_{L_3}(\mathbf{a}-\mathbf{r}) \end{aligned} \tag{D.7}$$

Thus we have

$$\sum_{L_1} H_{L_1}(\mathbf{R}+\mathbf{r}) J_{L_1}(\mathbf{a})$$

$$= \sum_{L_3} H_{L_3}(\mathbf{R}) 4\pi \sum_{L_1 L_2} \mathrm{i}^{(l_3-l_1-l_2)} C(L_1, L_2, L_3) J_{L_1}(\mathbf{a}) J_{L_2}(-\mathbf{r})$$

$$= 4\pi \sum_{L_1 L_2 L_3} J_{L_1}(\mathbf{a}) \mathrm{i}^{(-l_1+l_2+l_3)} C(L_1, L_2, L_3) H_{L_3}(\mathbf{R}) J_{L_2}(\mathbf{r}). \quad \text{(D.8)}$$

It now follows that

$$H_{L_1}(\mathbf{R}+\mathbf{r}) = \sum_{L_2} G_{L_1 L_2}(\mathbf{R}) J_{L_2}(\mathbf{r}) \quad \text{(D.9)}$$

where we have defined the expansion coefficients,

$$G_{L_1 L_2}(\mathbf{R}) = 4\pi \sum_{L_3} \mathrm{i}^{l_1-l_2-l_3} C(L_1, L_2, L_3) H_{L_3}(\mathbf{R}), \quad R \neq 0. \quad \text{(D.10)}$$

(For values of $\mathbf{R}$ corresponding to the translation vectors of a Bravais lattice, these coefficients give rise to the real-space structure constants of the method of Korringa, Kohn, and Rostoker (KKR)). By using Eqs.(D.3) and (D.5) to expand $J_L(\mathbf{r})$, and Eqs.(D.9) and (D.10) to expand $H_L(\mathbf{r})$ we can obtain the analogous formulae for expanding the Neumann function $N_L(\mathbf{r}) = -i[H_L(\mathbf{r}) - J_L(\mathbf{r})]$,

$$N_{L_1}(\mathbf{R}+\mathbf{r}) = \sum_{L_2} B_{L_1 L_2}(\mathbf{R}) J_{L_2}(\mathbf{r}), \quad R > r, \quad \text{(D.11)}$$

$$B_{L_1 L_2}(\mathbf{R}) = 4\pi \sum_{L_3} \mathrm{i}^{l_1-l_2-l_3} C(L_1, L_2, L_3) N_{L_3}(\mathbf{R}). \quad \text{(D.12)}$$

In summary, we have shown that

$$F_{L_1}(\mathbf{R}+\mathbf{r}) = 4\pi \sum_{L_2 L_3} \mathrm{i}^{(l_1-l_2-l_3)} C(L_1, L_2, L_3) F_{L_3}(\mathbf{R}) J_{L_2}(\mathbf{r}) \quad \text{(D.13)}$$

where F may represent any of the Bessel, Hankel, or Neumann spherical functions J,H, or N. Note that g, G, and B satisfy

$$\begin{aligned} g_{L_1 L_2}(\mathbf{R}) &= g_{L_2 L_1}(-\mathbf{R}) \\ G_{L_1 L_2}(\mathbf{R}) &= G_{L_2 L_1}(-\mathbf{R}) \\ B_{L_1 L_2}(\mathbf{R}) &= B_{L_2 L_1}(-\mathbf{R}). \end{aligned} \quad \text{(D.14)}$$

The expressions derived so far can be conveniently summarized by means of a vector and matrix notation. Denoting by bras and kets row and column vectors indexed by L, we can write Eqs.(D.3), (D.6), and (D.10) in the forms,

$$|J(\mathbf{r}+\mathbf{a})\rangle = g(\mathbf{a})|J(\mathbf{r})\rangle, \quad \text{all } \mathbf{r} \text{ and } \mathbf{a}, \quad \text{(D.15)}$$

or

$$\langle J(\mathbf{r}+\mathbf{a})| = \langle J(\mathbf{r})| g(-\mathbf{a}), \quad \text{(D.16)}$$

$$\begin{aligned} G_0(\mathbf{r}, \mathbf{r}') &= -ik\langle J(\mathbf{r})|H(\mathbf{r}')\rangle, \\ &= -ik\langle H(\mathbf{r}')|J(\mathbf{r})\rangle, \quad r' > r, \end{aligned} \quad \text{(D.17)}$$

$$|H(\mathbf{R}+\mathbf{a})\rangle = G(\mathbf{R})|J(\mathbf{a})\rangle, \quad R > a, \tag{D.18}$$

or

$$\langle H(\mathbf{R}+\mathbf{a})| = \langle J(\mathbf{a})|G(-\mathbf{R}), \quad R > a, \tag{D.19}$$

and

$$\begin{aligned} G(\mathbf{R}+\mathbf{a}) &= G(\mathbf{R})g(\mathbf{a}) \\ &= g(\mathbf{a})G(\mathbf{R}), \quad R > a. \end{aligned} \tag{D.20}$$

Furthermore, one can easily show that the following relations hold,

$$\begin{aligned} g(\mathbf{a}+\mathbf{b}) &= g(\mathbf{a})g(\mathbf{b}) \\ &= g(\mathbf{b})g(\mathbf{a}), \quad \text{all } \mathbf{a} \text{ and } \mathbf{b}, \end{aligned} \tag{D.21}$$

and

$$G(\mathbf{R}+\mathbf{a}+\mathbf{b}) = g(\mathbf{b})G(\mathbf{R}+\mathbf{a}), \quad |\mathbf{R}+\mathbf{a}| > b \tag{D.22}$$
$$= g(\mathbf{b})G(\mathbf{R})g(\mathbf{a}) \tag{D.23}$$
$$= g(\mathbf{a})G(\mathbf{R})g(\mathbf{b}). \tag{D.24}$$

The conditions for convergence of the summations implied by Eqs. (D.23) and (D.24) depend on the order in which the sums are performed, that is, they can be summed if both $R > b$ and $|\mathbf{R}+\mathbf{b}| > a$ or if both $R > a$ and $|\mathbf{R}+\mathbf{a}| > b$.

As a special case of Eq.(D.21), we have the unitarity condition

$$g(\mathbf{a})g(-\mathbf{a}) = 1, \quad \text{or} \quad g(\mathbf{a})g^{\dagger}(\mathbf{a}) = 1, \tag{D.25}$$

which, combined with Eq.(D.24), yields the result

$$G(\mathbf{R}) = g(-\mathbf{a})G(\mathbf{R}+\mathbf{a}+\mathbf{b})g(-\mathbf{b}), \quad |\mathbf{R}+\mathbf{a}+\mathbf{b}| > |\mathbf{a}+\mathbf{b}|. \tag{D.26}$$

We also note the relations

$$\begin{aligned} G(\mathbf{R})_{L0} &= \sqrt{4\pi}H_L(\mathbf{R}) \\ G(\mathbf{R})_{0L} &= \sqrt{4\pi}H_L(-\mathbf{R}) \end{aligned} \tag{D.27}$$

which follow from Eq.(D.10).

It is useful to note that $G(\mathbf{R}+\mathbf{a})$ can be obtained from $G(\mathbf{R})$ even in the case in which $a > R$. This can be accomplished through an expansion around, rather than through, the pole of the Green function by means of an inherently convergent process. We consider the vector $\mathbf{a}$ as a sum of vectors $\mathbf{a} = \sum_{\alpha=1}^{N} \mathbf{a}_\alpha$, chosen so that the inequalities

$$\left| \mathbf{R} + \sum_{\alpha=1}^{n} \mathbf{a}_\alpha \right| > |\mathbf{a}_{n+1}| \tag{D.28}$$

are satisfied for all $n \le N$. We can then write,

$$G(\mathbf{R}+\mathbf{a}) \equiv G\left(\mathbf{R} + \sum_{\alpha=1}^{N} \mathbf{a}_\alpha\right)$$

$$= G\left(\mathbf{R} + \sum_{\alpha=1}^{N-1} \mathbf{a}_\alpha\right) g(\mathbf{a}_N)$$

$$= \left[G\left(\mathbf{R} + \sum_{\alpha=1}^{N-2} \mathbf{a}_\alpha\right) g(\mathbf{a}_{N-1})\right] g(\mathbf{a}_N)$$

$$= \cdot \quad \cdot \quad \cdot$$

$$= \{\{[G(\mathbf{R})g(\mathbf{a}_1)]\, g(\mathbf{a}_2)\} \cdot \quad \cdot \quad \cdot \}\, g(\mathbf{a}_N), \tag{D.29}$$

where the braces indicate the order of operations. We designate this process of expanding $G(\mathbf{R})$ with the symbol $\odot$, and we have

$$G(\mathbf{r} + \mathbf{a}) = G(\mathbf{R}) \odot g(\mathbf{a}), \tag{D.30}$$

for *all* vectors $\mathbf{R}$ and $\mathbf{a}$. (Note that this process may not always be possible in strictly one-dimensional cases, but the need to invoke it arises only in two and three dimensions where it is always possible to expand around the pole of the Green function.)

References

[1] M. Danos and L. C. Maximon, J. Math. Phys. **6**, 766 (1965).

Appendix E
The Two-Dimensional Square Cell

In this appendix we provide an illustrative example of the formalism presented in Section 3.5 by examining explicitly the case of a two-dimensional square cell. In particular, we show by means of direct calculation that the basis functions obtained through a solution of the Schrödinger equation in the presence and the absence of a potential in the moon region of a cell coincide in the interior of the cell. Before presenting the results of the numerical calculations, we quote a few relevant formulae from two-dimensional scattering theory.

The Green function for the two-dimensional Helmholtz equation can be obtained from the identities given in Appendix C and is given by the expressions[1],

$$G(\mathbf{r},\mathbf{r}') = \begin{cases} \frac{-i}{4} H_0(\kappa|\mathbf{r}-\mathbf{r}'|), & \text{if } E \geq 0; \\ \frac{-1}{2\pi} K_0(\kappa|\mathbf{r}-\mathbf{r}'|), & \text{if } E < 0. \end{cases} \tag{E.1}$$

Here $\kappa = \sqrt{|E|}$ and where H_0 is a Bessel function of the third kind and K_0 is a modified Bessel function [1]. The Green function can be expanded in cylindrical harmonics as

$$G(\mathbf{r},\mathbf{r}') = \frac{-1}{2\pi} \sum_{\ell=-\infty}^{\infty} I_\ell(\kappa r_<) K_\ell(\kappa r_>) \mathrm{e}^{\mathrm{i}\ell(\phi-\phi')} \tag{E.2}$$

[1]Our reversion in this appendix alone to the usual notation for the cylinder functions, as opposed to the notation $\mathcal{H}$ and $\mathcal{J}$ of Appendix C, should not cause confusion.

for $E < 0$. The Green function for $E \geq 0$ may be obtained from the relations

$$I_\ell(z) = i^{-\ell} J_\ell(iz) \tag{E.3}$$

$$K_\ell(z) = \frac{\pi}{2} i^{\ell+1} H_\ell(iz). \tag{E.4}$$

Using the Green function expansion in the equations that determine $\psi_L(\mathbf{r})$ yields

$$\begin{aligned}\psi_\ell(r)\mathrm{e}^{\mathrm{i}\ell\phi} &= \sum_{l'=-\infty}^{\infty} [I_{\ell'}(\kappa r)\mathrm{e}^{\mathrm{i}\ell'\phi} C_{\ell'\ell}(r) - \frac{1}{2\pi} K_{\ell'}(\kappa r)\mathrm{e}^{\mathrm{i}\ell'\phi} S_{\ell'\ell}(r)] \\ &= \sum_{\ell'=-\infty}^{\infty} \mathrm{e}^{\mathrm{i}\ell'\phi}\psi_{\ell'\ell}(r),\end{aligned} \tag{E.5}$$

where

$$C_{\ell'\ell}(r) = \delta_{\ell'\ell} + \frac{1}{2\pi}\int_0^r K_{\ell'}(\kappa r')\mathrm{e}^{-\mathrm{i}\ell'\phi'} V(r',\phi')\psi_\ell(r')\mathrm{e}^{\mathrm{i}\ell\phi'} r'\mathrm{d}r'\,\mathrm{d}\phi', \tag{E.6}$$

and

$$S_{\ell'\ell} = \int_0^r I_{\ell'}(\kappa r')\mathrm{e}^{-\mathrm{i}\ell'\phi'} V(r',\phi')\psi_\ell(r')\mathrm{e}^{\mathrm{i}\ell\phi'} r'\mathrm{d}r'\,\mathrm{d}\phi'. \tag{E.7}$$

These formulae can be made to look more symmetrical by combining the factor $-\frac{1}{2\pi}$ with the function $S_{\ell'\ell}$

$$\psi_\ell(r)\mathrm{e}^{\mathrm{i}\ell\phi} = \sum_{l'=-\infty}^{\infty} [I_{\ell'}(\kappa r)\mathrm{e}^{\mathrm{i}\ell'\phi} C_{\ell'\ell}(r) + K_{\ell'}(\kappa r)\mathrm{e}^{\mathrm{i}\ell'\phi} S_{\ell'\ell}(r)], \tag{E.8}$$

$$C_{\ell'\ell}(r) = \delta_{\ell'\ell} + \frac{1}{2\pi}\int_0^r K_{\ell'}(\kappa r')\mathrm{e}^{-\mathrm{i}\ell'\phi'} V(r',\phi')\psi_\ell(r')\mathrm{e}^{\mathrm{i}\ell\phi'} r'\mathrm{d}r'\,\mathrm{d}\phi', \tag{E.9}$$

and

$$S_{\ell'\ell} = -\frac{1}{2\pi}\int_0^r I_{\ell'}(\kappa r')\mathrm{e}^{-\mathrm{i}\ell'\phi'} V(r',\phi')\psi_\ell(r')\mathrm{e}^{\mathrm{i}\ell\phi'} r'\mathrm{d}r'\,\mathrm{d}\phi'. \tag{E.10}$$

Defining $V_\ell(r)$ by

$$V_\ell(r) = \frac{1}{2\pi}\int_0^{2\pi} \mathrm{d}\phi V(r,\phi)\mathrm{e}^{\mathrm{i}\ell\phi} \tag{E.11}$$

yields

$$C_{\ell'\ell}(r) = \delta_{\ell'\ell} + \frac{1}{2\pi}\int_0^r K_{\ell'}(\kappa r')V_{\ell-\ell'}(r')\psi_\ell(r')r'\,\mathrm{d}r' \tag{E.12}$$

and

$$S_{\ell'\ell} = -\frac{1}{2\pi}\int_0^r I_{\ell'}(\kappa r')V_{\ell-\ell'}(r')\psi_\ell(r')r'\mathrm{d}r'\,\mathrm{d}\phi'. \tag{E.13}$$

If V is symmetrical about the x-axis so that $V(r,-\phi) = V(r,\phi)$, then $V_\ell(r) = V_{-\ell}(r)$ and the relations

$$C_{\ell'\ell} = C_{-\ell'-\ell} \tag{E.14}$$

and

$$S_{\ell'\ell} = S_{-\ell'-\ell} \tag{E.15}$$

hold. In this case the states will be either odd or even with respect to $\phi \to -\phi$. For even states (symmetric about the x-axis), we have

$$\begin{aligned}\psi_\ell(\mathbf{r}) &= \sum_{l'=-\infty}^{\infty} [I_{\ell'}(\kappa r)\cos(\ell'\phi)C_{\ell'\ell}(r) + K_{\ell'}(\kappa r)\cos(\ell'\phi)S_{\ell'\ell}(r)] \\ &= \sum_{\ell'=0}^{\infty} \epsilon_{\ell'}\cos(\ell'\phi)\{I_{\ell'}(\kappa r)[C_{\ell'\ell}(r) + C_{-\ell'\ell}(r)] \\ &\quad + K_{\ell'}(\kappa r)[S_{\ell'\ell}(r) + S_{-\ell'\ell}(r)]\}\end{aligned} \tag{E.16}$$

where $\epsilon_{\ell'} = 1$ if $\ell' \neq 0$ and $\epsilon_0 = 0.5$.

We can define $\tilde{C}_{\ell'\ell}(r) = C_{\ell'\ell}(r) + C_{-\ell'\ell}(r)$ and similarly $\tilde{S}_{\ell'\ell}(r) = S_{\ell'\ell}(r) + S_{-\ell'\ell}(r)$ which allows us to write the equations for calculating basis functions for symmetric states in the form

$$\tilde{C}_{\ell'\ell}(r) = \delta_{\ell\ell'} + \sum_{\ell''=0}^{\infty} \epsilon_{\ell''} \int_0^r r'\mathrm{d}r' K_{\ell'}(\kappa r')[V_{\ell''-\ell'} + V_{\ell''+\ell'}]\psi_{\ell''\ell}(r'), \tag{E.17}$$

$$\tilde{S}_{\ell'\ell}(r) = -\sum_{\ell''=0}^{\infty} \epsilon_{\ell''} \int_0^r r'\mathrm{d}r' I_{\ell'}(\kappa r')[V_{\ell''-\ell'} + V_{\ell''+\ell'}]\psi_{\ell''\ell}(r'), \tag{E.18}$$

and

$$\psi_\ell(\mathbf{r}) = \sum_{\ell'=0}^{\infty} \epsilon_{\ell'}\cos(\ell'\phi)\{I_{\ell'}(\kappa r)\tilde{C}_{\ell'\ell}(r) + K_{\ell'}(\kappa r)\tilde{S}_{\ell'\ell}(r)\}. \tag{E.19}$$

E.1 Numerical Results (*)

We now provide a numerical illustration of the formal results established in Section 3.6, namely, that the solution of the Schrödinger equation for a potential inside a sphere coincides over the domain of a cell contained in the sphere with that obtained when the potential in the cell is acting alone. In other words, the two solutions coincide in the interior of the cell regardless of the potential in the moon region. The numerical results presented here were obtained for the case of a two-dimensional square cell, using the explicit expressions quoted above.

In the present numerical studies, we consider the solution, $\psi_L^\Omega(\mathbf{r})$, of the Schrödinger equation associated with a two-dimensional cell in the shape of a square and having a constant potential, $V_0 = -5.0$. This solution can be found in either of two ways. The first method of solution involves the direct application of the formalism presented above, namely, the solution of the coupled-channel Eqs.(E.17)–(E.19). This is a viable procedure, but its successful implementation depends on the convergence of the internal sums in the expression for the basis function, Eq.(E.19). In that expression, one must carry the summation on the right-hand side of the equation to values of l' which in general must exceed the value of the basis function index, l, in order to yield converged results. The magnitude of this internal index depends in general on the value of the outer index, l.

Now, it is important to note the interplay of these two indicies and how it affects the convergence of the basis function. In solving for the basis function $\psi_L^{\Omega}(\mathbf{r})$, corresponding to a given value of l, one can determine the phase functions $C_{ll'}$ and $S_{ll'}$ through Eqs.(E.17) and (E.18) to arbitrarily high values of l'. As the basis function is not known exactly and is itself being determined through the coupled channel equations, the values of the phase functions become less and less reliable with increasing l'. This is because the exact evaluation of the phase functions requires the exact value of the basis function, which in turn is given only if the phase functions are exactly known. Thus, one may obtain fairly stable, apparently convergent results as a function of the internal summation index, but as that index approaches its cut-off value (where the phase functions are not very accurately known) the basis function may begin to show signs of instability.

This behavior is illustrated in Fig. E.1 in connection with the two-dimensional square cell for the basis function corresponding to $l = 0$, and with a cut-off value of the internal sum $l' = 64$. As shown here, the basis function, $\psi_L^{\Omega}(\mathbf{r})$, appears to converge with increasing l', but begins to diverge as l' approaches 64, its cut-off value. Thus, the coupled channel equations yield solutions that behave *asymptotically* with increasing L. The divergences can, however, be pushed further and further away from the region of interest by solving these equations for higher and higher values of the internal summation index. They disappear completely when the internal sums are carried to full convergence, that is, to infinite order. Because of this, the solutions can be said to converge rigorously as L increases, with asymptotic behavior setting in because of the truncations of the various sums that are often required in numerical calculations. This last effect is illustrated in Fig. E.2. In this case the functions $\underline{C}$ and $\underline{S}$ were determined "exactly" up to $L = 64$ through the use of Eqs.(3.143) and (3.144). The basis functions that occur in these equations are those corresponding to a circle with a constant potential and can be readily evaluated

$$\psi_L^S(\mathbf{r}) = (\beta/\alpha)^{\ell} J_{\ell}(\alpha r) e^{i\ell\phi}, \tag{E.20}$$

where $\alpha = \sqrt{E}$ and $\beta = \sqrt{E - V} = \sqrt{E + V_0}$, and ϕ denotes the angle made by the vector $\mathbf{r}$ with the positive x-axis. Thus, the use of these expressions amounts to carrying the internal summations to convergence in

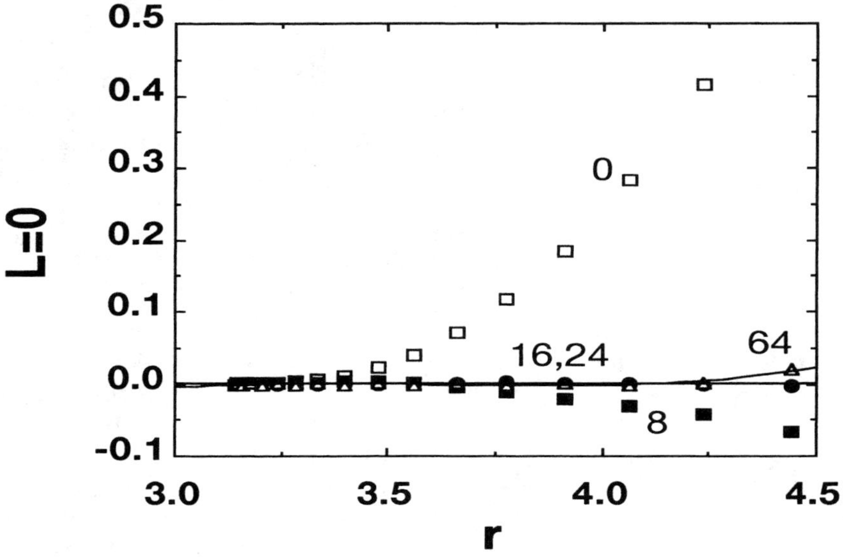

FIGURE E.1. Convergence of the cell wave function with increasing external summation index for a fixed cut-off value of the internal summation at $L = 64$.

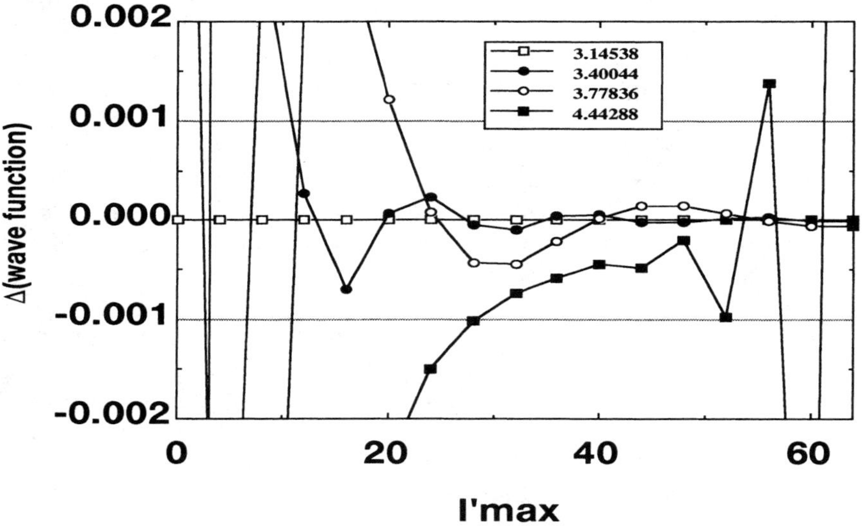

FIGURE E.2. Convergence of the cell basis function with increasing external summation index when the internal summations are carried to full convergence. Here, $l = 0$ and $l' = 64$.

Eq.(E.12). Increasing the internal summation index leads to properly convergent behavior by the basis function up to $L = 64$, as is indicated in Fig. E.2. These results indicate that the calculational procedure corresponding to the direct solution of the coupled channel equations is in principle correct, but must be used with care in numerical applications. In other words, these equations are satisfied by the exact values of the phase functions and of the basis functions, but yield only asymptotically convergent values when the internal summations are truncated at a finite value of the corresponding index. At the same time, when these equations are applied to realistic potentials, whose strength is often concentrated inside the MT sphere, reliable results may be obtained even when all expansion variables, even those corresponding to internal summations, are truncated at fairly small values.

The second approach that can be followed in solving for the cell basis functions is based on the formal results of Section 3.6. We saw there that a cell basis function is independent of the potential in the moon region of the cell and that it coincides inside the cell with the basis function for a sphere (circle) which contains the potential in the moon region. This affords one the possibility of solving first for the basis function of the sphere and then determining the cell phase functions through an additional integration over the moon region; see Eqs.(3.143) and (3.144).

For the case of the square cell with a constant potential, $-V_0$, the obvious choice for the potential in the moon region is also $-V_0$, leading to a spherically symmetric potential inside the sphere (circle) and a straight-

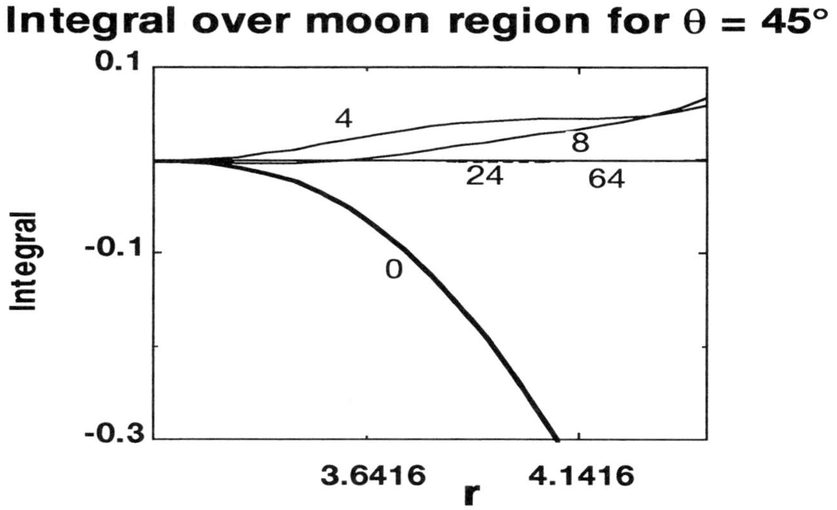

FIGURE E.3. Integrals over the moon region for a square cell with a constant potential along a ray from the MT radius to the corner of the cell.

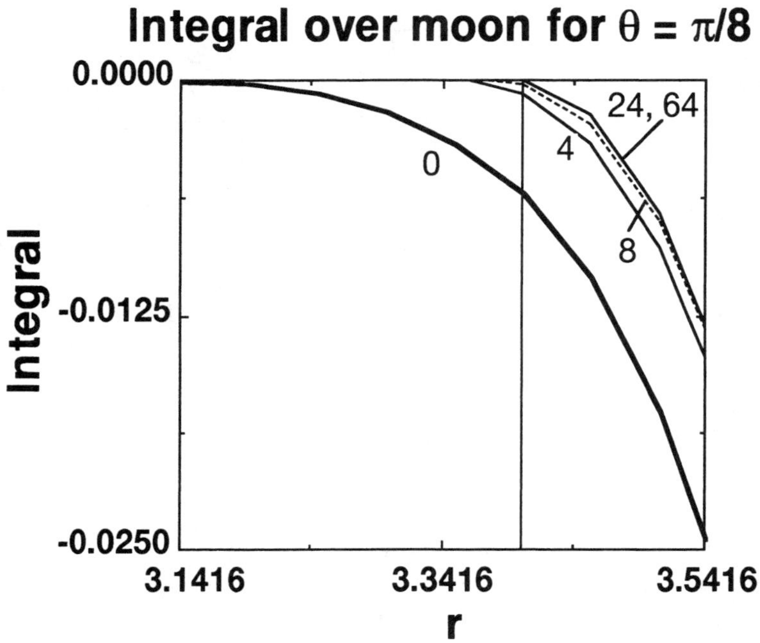

FIGURE E.4. Integrals over the moon region for a square cell with a constant potential along a ray from the MT radius to the edge of the cell at an angle of $\pi/8$.

forward determination of the sphere basis functions, Eq.(E.20). In realistic cases the choice of the potential to be used in the moon region is not as obvious. Thus, in general one cannot expect to eliminate completely the divergent behavior associated with the finite truncation of internal sums, although this behavior can be brought largely under control by a judicious choice of the potential in the moon regions. In the present case, Eq.(E.20) yields exact expressions for the basis functions of the sphere and thus for the cell basis functions inside the domain of the cell where the two coincide. The fact that these two sets of basis functions do indeed coincide over the domain of the cell is illustrated in Figs.E.3 and E.4, for the case of the square cell. It follows from Eqs.(3.125) and (3.126) and from the relations in Eqs.(3.143) and (3.144) that inside the domain of a cell, Ω, the cell basis function differs from that of the sphere by an integral over the moon region,

$$\psi_\ell^\Omega(\mathbf{r}) = (\alpha/\beta)^\ell J_\ell(\beta r)\mathrm{e}^{\mathrm{i}\ell\phi}$$
$$- (\alpha/\beta)^\ell \sum_{\ell'} \mathrm{e}^{\mathrm{i}\ell\phi'} V_0 \int_M \frac{\mathrm{d}^3 r}{2\pi} \Big[K_{\ell'}(\alpha r')\mathrm{e}^{-\mathrm{i}\ell'\phi'} J_\ell(\beta r')\mathrm{e}^{\mathrm{i}\ell\phi'} I_{\ell'}(\alpha r)$$
$$- I_{\ell'}(\alpha r')\mathrm{e}^{-\mathrm{i}\ell'\phi'} J_\ell(\beta r')\mathrm{e}^{\mathrm{i}\ell\phi'} K_{\ell'}(\alpha r)\Big], \tag{E.21}$$

where the various functions occurring in this expression have been introduced in the last subsection. Note that the second term inside the brackets in this expression involves the wrong expansion of the Green function so that if the sum over l' were carried out before the integral were evaluated the sum would diverge (the argument, r', in the regular solution, $I_{l'}(\alpha r')$, lies in the moon region and can become larger than the length, r, of the argument of the irregular function, $K_{l'}(\alpha r)$, which lies inside the cell.) However, as we agued above, and as we saw in Section 3.7, performing the integral first leads to converged results.

Now, if the functions $\psi_l^{\Omega}(\mathbf{r})$ and $\psi_l^{S}(\mathbf{r})$ are to coincide when $\mathbf{r}$ lies in the cell then the second term on the right-hand side of Eq.(E.21), the integral over the moon region, must vanish. That this is indeed the case is demonstrated in Figs.E.3 and E.4. These figures show the value of that term as a function of the internal angular-momentum summation index, indicated along the curves, and as a function of the length, r, along two different directions: In Fig. E.3, r varies along a line from the MT radius to the corner of the cell, while in Fig. E.4 the direction of $\mathbf{r}$ makes an angle of $\pi/8$ with the positive x-axis. In both cases, the integral, and hence the difference between the cell and sphere wave functions approaches zero with increasing l'. Thus the converged values of these two basis functions coincide inside the cell. We also note that when $\mathbf{r}$ lies in the moon region, rather than inside the cell, the integral over the moon region no longer vanishes as the two basis functions are no longer identical there.

These results demonstrate the characteristic feature of conditionally convergent sums and integrals–namely, that the analytic behavior of such sums depends on the order of summation. Thus, the sum over L involving the moon region in Eq.(3.136) would diverge for some values of $\mathbf{r}'$ if it were performed prior to the integral. But, when it follows the integral, the sum converges. In fact, the integral itself vanishes, as was illustrated above in connection with the two-dimensional square cell. Clearly, as Eqs.(3.134) and (3.135) are connected only conditionally, the sums and integral in them are not interchangeable, and the convergence behavior of one expression cannot be used to infer that of the other. Quite generally, expressions that rely on conditional derivations must be examined for convergence on a case by case basis.

References

[1] Jerome Spanier and Keith B. Oldham *An Atlas of Functions* (Hemisphere Publishing Co., New York, 1987).

Appendix F
Formal Scattering Theory

F.1 General Comments

In this appendix, we provide a summary of the mathematical formalism that leads to the time-independent Lippmann–Schwinger equation. We will justify rigorously some of the results quoted in Chapter 4, and clarify the role of the boundary conditions (initial conditions) of scattering theory. In the process, we indicate explicitly how a time-independent picture is obtained from the time-dependent one. Our discussion follows closely that of Schmid and Ziegelman [1]. Additional details can be found in the work of Bohm [2].

Two-particle scattering theory, of interest to us here, is concerned with the motion of two particles under the influence of an interaction potential. The Hamiltonian operator describing this motion has the form

$$H = \frac{k_1^2}{2m_1} + \frac{k_2^2}{2m_2} + V, \tag{F.1}$$

where $\mathbf{k}_1$ and $\mathbf{k}_2$ are the momenta, and m_1 and m_2 the masses of the two particles. In classical as well as in quantum mechanics, it is convenient to treat separately the motion of the center of mass of the system from the relative motion of the two particles. Defining a center-of-mass momentum, $\mathbf{k}$, and a relative momentum, $\mathbf{p}$,

$$\mathbf{k} = \mathbf{k}_1 + \mathbf{k}_2 \tag{F.2}$$

and

$$\mathbf{p} = \frac{m_2\mathbf{k}_1 - m_1\mathbf{k}_2}{m_1 + m_2} \tag{F.3}$$

along with the reduced mass

$$m = \frac{m_1 m_2}{m_1 + m_2} \tag{F.4}$$

and relative coordinate, $\mathbf{r} = \mathbf{r}_1 - \mathbf{r}_2$, where $\mathbf{r}_1$ and $\mathbf{r}_2$ are the coordinates of the two particles in the laboratory, we can express the Hamiltonian in the form,

$$H = \frac{p^2}{2m} + V \equiv H_0 + V. \tag{F.5}$$

The Hamiltonian H_0 corresponds to the motion of a particle of mass m in free space, that is, with the interaction, V, switched off. In the form of Eq.(F.5), the Hamiltonian operator can also be used to describe the motion of a particle incident upon a stationary center of force (e.g., an electron scattered by a much more massive nucleus), and is thus appropriate for a discussion of the electron states in solids.

F.2 Initial Conditions and the Møller Operators

Within a time-dependent picture, the scattering of two particles, or of a single particle by a stationary field, is described by the time-dependent Schrödinger equation, Eq.(2.1) (with $\hbar = 1$),

$$\mathrm{i}\frac{\partial}{\partial t}\Psi_\alpha^+(t) = H\Psi_\alpha^+(t). \tag{F.6}$$

As in Chapter 4, the index α denotes all observables necessary to define completely the initial state.

The formal solution of Eq.(F.6) is given by

$$\Psi_\alpha^+(t) = \mathrm{e}^{-\mathrm{i}Ht}\Psi_\alpha^+ \tag{F.7}$$

where Ψ_α^+ is a time-independent wave-packet.

As was discussed in Chapter 4, the initial (or boundary) condition governing the motion of the states $\Psi_\alpha^+(t)$ can be expressed verbally as follows: The scattering state, $\Psi_\alpha^+(t)$, develops from a state which is free in the infinite past and whose properties are fully characterized by α.

The initial condition stated above can be expressed in mathematical terms along the following lines: At $t = 0$ a reference wave-packet, Φ_α, is introduced, that is, an experimentally prepared state, whose time evolution is described by the Hamiltonian, H_0, of free space,

$$\Phi_\alpha(t) = \mathrm{e}^{-\mathrm{i}H_0 t}\Phi_\alpha. \tag{F.8}$$

We wish to impose the condition that Ψ_α^+ and Φ_α coincide in the limit of the infinite past. However, a pointwise agreement between these two functions would not be sensible as wave packets disperse in the course of time. Consequently, we demand instead that the norm of the difference of these two states as $t \to -\infty$ vanishes,

$$\lim_{t\to-\infty} ||\mathrm{e}^{-\mathrm{i}Ht}\Psi_\alpha^+ - \mathrm{e}^{-\mathrm{i}H_0t}\Phi_\alpha|| = 0. \tag{F.9}$$

This is a meaningful condition as the norm implies an integration over all space. The limit of the norm is called a *strong limit* and will subsequently be denoted by "S-limit". Equation (F.9) removes one important difficulty which was implicit in the verbal expression of the initial condition. That description becomes ill-defined due to the fact that a wave packet which is confined to the laboratory at a finite time, say at $t = 0$, becomes rather diffuse in the infinite past (unless its production itself is included in the formalism). The use of a reference wave packet, Φ_α, which is also confined to the laboratory at the same time as Ψ_α^+, and the S-limit in Eq.(F.9) allows us to circumvent this difficulty. It should be noted that the limit in Eq.(F.9) does not exist for every potential, but its existence has been proven for a large class of potentials which decrease with distance faster than the Coulomb potential. This class is broad enough to include most (if not all) the potential functions one is likely to encounter in determining the electronic structure of solids.

Equation (F.9) can also be written in the form,

$$\Psi_\alpha^+ = s - \lim_{t\to-\infty} \mathrm{e}^{\mathrm{i}Ht}\mathrm{e}^{-\mathrm{i}H_0t}\Phi_\alpha \equiv \Omega^+\Phi_\alpha, \tag{F.10}$$

where the S-limit implies the vanishing of the norm of the difference of the two sides of this equation.

In addition to Ψ_α^+, states Ψ_α^- are also introduced which are such that they coincide with a free sates in the infinite future. These, rather non-physical states, play an important role in formal scattering theory.

F.3 The Møller Wave Operators

Equation (F.10) contains the definition of the Møller wave operator, Ω^+, which has the property that when applied to a free state, Φ_α, produces the scattering state, Ψ_α^+,

$$\Psi_\alpha^+ = \Omega^+\Phi_\alpha. \tag{F.11}$$

In general, we can define the operators,

$$\Omega^\pm = s - \lim_{t\to\mp\infty} \mathrm{e}^{\mathrm{i}Ht}\mathrm{e}^{-\mathrm{i}H_0t}. \tag{F.12}$$

The operator Ω^- produces the scattering state Ψ_α^- which goes over to the free state Φ_α in the limit $t \to +\infty$. Note that because the Hamiltonian operators H and H_0 do not commute, the two exponentials in Eq.(F.12) cannot be combined into a single one, $e^{iHt}e^{-iH_0t} \neq e^{i(H-H_0)t}$. It will be convenient in our subsequent discussion to use an alternative interpretation of the Møller operators. This representation follows upon replacement of the time limit by an *Euler limit*,

$$\begin{aligned}\Omega^{\pm} &= s - \lim_{t\to\mp\infty} e^{iHt}e^{-iH_0t} \\ &= s - \lim_{\epsilon\to 0} \pm\epsilon \int_{\mp\infty}^{0} dt e^{\pm\epsilon t} e^{iHt} e^{-iH_0t},\end{aligned} \tag{F.13}$$

which can be shown to be valid [2] under the assumption that the strong limit on the left exists.

To understand the procedure of an Euler limit, we apply it to an ordinary function, $f(t)$, which goes to f_∞ as $t \to -\infty$. We easily see that

$$\int_{-\infty}^{0} \epsilon e^{\epsilon t} dt = 1, \tag{F.14}$$

and that finite time intervals do not contribute to the integral

$$\lim_{\epsilon\to 0} \int_{-\infty}^{0} \epsilon e^{\epsilon t} f(t) dt. \tag{F.15}$$

Upon setting

$$\epsilon t = x \tag{F.16}$$

we have

$$\int_{-\infty}^{0} \epsilon e^{\epsilon t} f(t) dt = \int_{-\infty}^{0} e^{\epsilon x} f\left(\frac{x}{\epsilon}\right) dx, \tag{F.17}$$

so that as ϵ goes to zero, $f(\frac{x}{\epsilon})$ goes to f_∞ and can be put in front of the integral.

It should be pointed out that the introduction of an Euler limit causes time to disappear from our two-particle scattering theory. This will become clearer in the next section where we derive a time-independent equation that incorporates exactly the boundary (initial) condition for scattering discussed above. We now return to further discussion of the Møller operators.

¿From the definition, Eq.(F.12), it follows that for any finite time τ, we have the relation

$$\begin{aligned}s - \lim_{t\to\mp\infty} e^{iH(t+\tau)}e^{-iH_0(t+\tau)} &= \Omega^{\pm} \\ &= e^{iH\tau}\Omega^{\pm}e^{-iH_0\tau}.\end{aligned} \tag{F.18}$$

Through differentiation, we obtain

$$0 = \frac{d\Omega^{\pm}}{d\tau}$$

$$= e^{iH\tau}(H\Omega^{\pm} - \Omega^{\pm}H_0)e^{-iH_0\tau}, \tag{F.19}$$

which leads to the *intertwining* relation,

$$H\Omega^{\pm} = \Omega^{\pm}H_0. \tag{F.20}$$

The last equation can be used to prove an important result of scattering theory, namely, that ψ^+ states with a well-defined value of the momentum (energy) evolve out of free states with the same energy. The proof requires application of a Møller operator on a wave packet of sharp momentum which consequently must fill all space and, therefore, is not normalizable (square integrable). That this procedure is meaningful for a large class of potentials has been shown by Faddeev [3]. One finds that the domain of the Møller operator generally consists of the Hilbert space of free states which are square integrable and hence normalizable. By performing a limiting procedure toward sharp states on the wave packets, the domain can be extended to include states of sharp momentum, e.g., plane waves.

We now apply the Møller operators, $\Omega^{\pm}$, to a momentum eigenstate,$|\mathbf{p}\rangle$, instead of a wave packet, Φ_α, thus obtaining scattering states, $|\mathbf{p}\rangle^{\pm}$. These are the states that include an "incident " plane wave and an "outgoimg" spherical wave, as discussed in Chapter 3. That the states $|\mathbf{p}\rangle^{\pm}$ correspond to the same sharp energy, $E = \frac{p^2}{2m}$, follows from the intertwining relation, Eq.(F.20),

$$\begin{aligned} H|\mathbf{p}\rangle^{\pm} &= H\Omega^{\pm}|\mathbf{p}\rangle = \Omega^{\pm}H_0|\mathbf{p}\rangle \\ &= \Omega^{\pm}\frac{p^2}{2m}|\mathbf{p}\rangle = \frac{p^2}{2m}|\mathbf{p}\rangle^{\pm}. \end{aligned} \tag{F.21}$$

In general, the Møller operators map the proper (+) and improper (−) free Hilbert spaces onto the corresponding scattering states of H. The bound states of H are not reached by this process and, therefore, the Møller operators are not unitary, but only isometric,

$$\Omega^{\pm\dagger}\Omega^{\pm} = 1. \tag{F.22}$$

The last two expressions reveal an important property of the normalization of the scattered states. In fact, they show that these states have the same normalization as the free-particle states from which they evolve,

$$\langle\psi^{\pm}|\psi^{\pm}\rangle = \langle\mathbf{p}|\Omega^{\pm\dagger}\Omega^{\pm}|\mathbf{p}\rangle = \langle\mathbf{p}|\mathbf{p}\rangle. \tag{F.23}$$

When the Møller operators are applied to the bound states, Ψ_n, of H they give zero, as follows from the relation,

$$\begin{aligned} \langle\Omega^{\pm\dagger}\Psi_n|\mathbf{p}\rangle &= \langle\Psi_n|\Omega^{\pm}|\mathbf{p}\rangle \\ &= \langle\Psi_n|\mathbf{p}\rangle^{\pm} = 0. \end{aligned} \tag{F.24}$$

The last step in the equation above follows because the bound and scattering states are eigenstates of H with different energies. Because the expression on the far left equals zero, and the free states, $|\mathbf{p}\rangle$, form a complete set, we have

$$\Omega^{\pm}|\Psi_n\rangle = 0. \tag{F.25}$$

It follows that the Møller operators are unitary only when the spectrum of H does not include bound states. It follows easily from the definition of the Møller operators, for example, Eq.(F.11), and the fact that they are isometric that the solutions of the Lippmann–Schwinger equations have the same normalization as the free states from which they evolve

$$\langle\Psi_\alpha^{\pm}|\Psi_\alpha^{\pm}\rangle = \langle\Phi_\alpha|(\Omega^{\pm\dagger}\Omega^{\pm}|)\Phi_\alpha^{\pm}\rangle = \langle\Phi_\alpha|\Phi_\alpha\rangle. \tag{F.26}$$

This statement expresses the conservation of probability (number of particles) in an event of elastic scattering. It follows from the normalization of the basis set, $\{\Phi_\alpha\}$, that the states $\{\Psi_\alpha^{+}\}$ satisfy the relation

$$\langle\Phi_\alpha|\Phi_\beta\rangle = \langle\Psi_\alpha^{+}|\Psi_\beta^{+}\rangle = \delta_{\alpha\beta} \tag{F.27}$$

and thus form an orthonormal basis set (at least in a δ- function sense.) A relation similar to that in the last equation holds also for the states $\{\Psi_\alpha^{-}\}$.

F.4 The Lippmann–Schwinger Equation

The Møller operators can be used to derive a time-independent equation for the scattering states that properly incorporates the condition at the infinite past (initial conditions). Suppressing the explicit indication of the S-limit, we use Eq.(F.13) to write,

$$\begin{aligned}
|\mathbf{p}\rangle^{\pm} &= \lim_{\epsilon\to 0} \pm\epsilon \int_{\mp\infty}^{0} \mathrm{d}t \mathrm{e}^{\pm\epsilon t}\mathrm{e}^{\mathrm{i}Ht}\mathrm{e}^{-\mathrm{i}H_0 t}|\mathbf{p}\rangle \\
&= \lim_{\epsilon\to 0} \pm\epsilon \int_{\mp\infty}^{0} \mathrm{d}t \mathrm{e}^{\pm\epsilon t}\mathrm{e}^{\mathrm{i}Ht}\mathrm{e}^{-\mathrm{i}Et}|\mathbf{p}\rangle \\
&= \lim_{\epsilon\to 0} \pm\epsilon \int_{\mp\infty}^{0} \mathrm{d}t \mathrm{e}^{(H-E\mp\epsilon)t}|\mathbf{p}\rangle \\
&= \lim_{\epsilon\to 0} \pm\mathrm{i}\epsilon(E\pm\epsilon-H)^{-1}|\mathbf{p}\rangle,
\end{aligned} \tag{F.28}$$

where the exponentials after the first line can be combined since ϵ and E are scalars. Now, defining the Green function operator, or resolvent,

$$G(z) = (z-H)^{-1}, \tag{F.29}$$

where z is a complex number, we obtain the relation,

$$|\mathbf{p}\rangle^{\pm} = \lim_{\epsilon\to 0} \pm\epsilon G(E\pm\epsilon)|\mathbf{p}\rangle. \tag{F.30}$$

This equation between the scattering states, $|\mathbf{p}\rangle^{\pm}$, and the Green function operator for the system marks *the transition from the time-dependent picture of scattering to the time-independent one.* We are now ready to derive a time-independent integral equation for the scattering states.

To this end, we note that the Green function satisfies the following two identities,

$$G(z') - G(z) = (z - z')G(z')G(z) \\ = (z - z')G(z)G(z'), \tag{F.31}$$

and the *Dyson equation,*

$$G(z) = G_0(z) + G_0VG(z). \tag{F.32}$$

The last equation follows upon the definition of the Green function for free motion

$$G_0(z) = (z - H_0)^{-1}, \tag{F.33}$$

and the obvious relation

$$V = H - H_0 = {G_0}^{-1}(z) - G^{-1}(z). \tag{F.34}$$

We now insert the second identity, Eq.(F.31), into Eq.(F.30) and obtain the relation,

$$|\mathbf{p}\rangle^{\pm} = \lim_{\epsilon\to 0} \pm\epsilon \left[G_0(E\pm\epsilon) + G_0(E\pm\epsilon)VG(E\pm\epsilon)\right]|\mathbf{p}\rangle. \tag{F.35}$$

It is easy to see that when $V = 0$ $(H = H_0)$, we have

$$\pm\, \epsilon G_0(E\pm\epsilon)|\mathbf{p}\rangle = \pm\epsilon(E\pm\epsilon - H_0)^{-1}|\mathbf{p}\rangle = |\mathbf{p}\rangle, \tag{F.36}$$

so that Eq.(F.35) takes the form

$$|\mathbf{p}\rangle^{\pm} = |\mathbf{p}\rangle + G_0(E\pm 0)V|\mathbf{p}\rangle^{\pm}, \tag{F.37}$$

where the notation $E \pm \mathrm{i}0$ indicates the limit $\epsilon \to 0$. This is the time-independent Lippmann–Schwinger equation for the scattering states, $|\mathbf{p}\rangle^{\pm}$, which properly reflects the time-dependent initial conditions of scattering theory. In the coordinate representation, the state $|\mathbf{p}\rangle$ describes a plane wave and, with the free-particle Green function given by Eq.(B.3), the scattering states far away from the field of force are seen to consist of a plane wave and an outgoing spherical wave. The time-independent Lippmann–Schwinger equation can become the starting point for the development of multiple scattering theory, an approach taken in this work.

References

[1] Erich W. Schmid and Horst Ziegelmann, *The Quantum-Mechanical Three-Body Problem,* Viewveg Tracts in Pure and Applied Physics (Pergamon Press, Oxford, 1974), Chapter 2.

[2] A. Bohm, *Quantum Mechanics* (Springer-Verlag, New York, 1979).
[2] J. M. Jauch, Helv. Phys. Acta **31**, 127 (1958).
[3] L. D. Faddeev, *Mathematical Aspects of the Three-Body Problem in the Quantum Scattering Theory*, Israel Program for Scientific Translations, Jerusalem, (1965), Chapters 1 and 10.

Appendix G
Irregular Solutions to the Schrödinger Equation

Just as there are two types of solutions to the Helmholtz equation, that is, the regular solutions $J_L(\mathbf{r})$ and the irregular solutions, $N_L(\mathbf{r})$ or $H_L(\mathbf{r})$, which can be combined to generate its Green function, both regular and irregular solutions to the Schrödinger equation are needed to generate its Green function.

Here we derive a procedure for calculating the irregular functions for a nonspherical potential that parallels that presented in Chapter 3 for the regular functions. The regular functions, $\phi_L(\mathbf{r})$, were obtained by integrating outward from the origin as indicated by Equations (3.75) and (3.76). The irregular functions, denoted here by $F_L(\mathbf{r})$, can be calculated by integrating inward. In defining one form of the regular solution we specified that it be equal to $J_L(\mathbf{r})$ at the origin. For the irregular functions, we find it convenient to specify their values on a sphere that circumscribes the cellular potential. Let us specify that the irregular functions are equal to $H_L(\mathbf{r})$ on the circumscribing sphere. Because the irregular functions satisfy the Schrödinger equation, or equivalently, the Lippmann–Schwinger equation, we can write

$$F_L(\mathbf{r}) = \sum_{L'} \bar{D}_{LL'} H_{L'}(\mathbf{r}) + \int_0^R \mathrm{d}\mathbf{r}' G_0(\mathbf{r}, \mathbf{r}') v(\mathbf{r}') F_L(\mathbf{r}'), \tag{G.1}$$

where $G_0(\mathbf{r}, \mathbf{r}')$ is the Green function for the Helmholtz equation, R is the radius of the circumscribing sphere, and $\bar{D}_{LL'}$ is a constant matrix.

Expanding G_0, we have

$$F_L(\mathbf{r}) = \sum_{L'} \bar{D}_{LL'} H_{L'}(\mathbf{r}) - i\sqrt{E} \sum_{L'} \left[J_{L'}(\mathbf{r}) \int_r^R \mathrm{d}\mathbf{r}' H_{L'}(\mathbf{r}') v(\mathbf{r}') F_L(\mathbf{r}') \right.$$
$$\left. + H_{L'}(\mathbf{r}) \int_0^r \mathrm{d}\mathbf{r}' J_{L'}(\mathbf{r}') v(\mathbf{r}') F_L(\mathbf{r}') \right]. \tag{G.2}$$

If we require that $F_L(\mathbf{r}) = H_L(\mathbf{r})$ for $\mathbf{r}$ on the surface of the circumscribing sphere, we have

$$\bar{D}_{LL'} = \delta_{LL'} + i\sqrt{E} \int_0^R \mathrm{d}\mathbf{r}' F_L(\mathbf{r}') v(\mathbf{r}') J_{L'}(\mathbf{r}') \tag{G.3}$$

and

$$F_L(\mathbf{r}) = H_L(\mathbf{r}) - i\sqrt{E} \sum_{L'} \left[\int_r^R \mathrm{d}\mathbf{r}' F_L(\mathbf{r}') v(\mathbf{r}') H_{L'}(\mathbf{r}') J_{L'}(\mathbf{r}) \right.$$
$$\left. + \int_r^R \mathrm{d}\mathbf{r}' F_L(\mathbf{r}') v(\mathbf{r}') J_{L'}(\mathbf{r}') H_{L'}(\mathbf{r}) \right]. \tag{G.4}$$

This can be written as

$$F_L(\mathbf{r}) = \sum_{L'} d_{LL'}(r) H_{L'}(\mathbf{r}) - q_{LL'}(r) J_{L'}(\mathbf{r}) \tag{G.5}$$

where

$$d_{LL'}(r) = \delta_{LL'} + i\sqrt{E} \int_r^R \mathrm{d}\mathbf{r}' F_L(\mathbf{r}') v(\mathbf{r}') J_{L'}(\mathbf{r}') \tag{G.6}$$

and

$$q_{LL'}(r) = -i\sqrt{E} \int_r^R \mathrm{d}\mathbf{r}' F_L(\mathbf{r}') v(\mathbf{r}') H_{L'}(\mathbf{r}') \tag{G.7}$$

The functions $d_{LL'}(r)$ and $q_{LL'}$ can be obtained by solving a set of coupled differential equations analogous to Eqs.(3.75) and (3.76),

$$\frac{\mathrm{d}d_{LL'}(r)}{\mathrm{d}r} = -i\sqrt{E} \int \mathrm{d}\hat{r}' F_L(\mathbf{r}') v(\mathbf{r}') J_{L'}(\mathbf{r}') \tag{G.8}$$

and

$$\frac{\mathrm{d}q_{LL'}(r)}{\mathrm{d}r} = i\sqrt{E} \int \mathrm{d}\hat{r}' F_L(\mathbf{r}') v(\mathbf{r}') H_{L'}(\mathbf{r}') \tag{G.9}$$

with boundary conditions at $r = R$,

$$d_{LL'}(R) = \delta_{LL'}, \tag{G.10}$$

$$q_{LL'}(R) = 0. \tag{G.11}$$

In order to be useful in generating the Green function the irregular solution must satisfy

$$\int_S d\mathbf{S}\cdot\sum_{L'}[\phi_L(\mathbf{r})\nabla F_{L'}(\mathbf{r}).-\nabla\phi_L(\mathbf{r})F_{L'}(\mathbf{r})] = \begin{cases} \delta_{LL'}, & \mathbf{r}=0\in S; \\ 0, & \text{otherwise.} \end{cases} \quad \text{(G.12)}$$

This result follows from the definition of the Green function $[\nabla^2 + E - V(\mathbf{r})]G(\mathbf{r},\mathbf{r}') = \delta(\mathbf{r}-\mathbf{r}')$ which can be used to show that

$$\int_\Omega d\mathbf{r}\phi_L(\mathbf{r})[\nabla^2 + E - V(\mathbf{r})]G(\mathbf{r},\mathbf{r}') - [\nabla^2 + E - V(\mathbf{r})]\phi_L(\mathbf{r})G(\mathbf{r},\mathbf{r}') = \begin{cases} \phi_L(\mathbf{r}'), & \mathbf{r}'\in\Omega; \\ 0, & \text{otherwise,} \end{cases} \quad \text{(G.13)}$$

if $\mathbf{r}'$ is within the volume Ω. Otherwise the integral vanishes. Use of Green's theorem to convert the integral over Ω into a surface integral over the surface S that bounds Ω and use of the expansion of the Green function in terms of regular and irregular functions leads to

$$\sum_{L'}\int_S d\mathbf{S}\cdot[\phi_L(\mathbf{r})\nabla F_{L'}(\mathbf{r}) - \nabla\phi_L(\mathbf{r})F_{L'}(\mathbf{r})]\phi_L(\mathbf{r}') = \begin{cases} \phi_L(\mathbf{r}'), & \mathbf{r}\in S; \\ 0 & \text{otherwise,} \end{cases} \quad \text{(G.14)}$$

from which Eq.(G.12) follows. This condition on the Wronskian of the irregular and regular solutions will be satisfied for the choice of solutions such that $F_L(\mathbf{r}) = H_L(\mathbf{r})$ and $\phi_L(\mathbf{r}) = J_L(\mathbf{r}) + \sum_{L'} t_{LL'}H_{L'}(\mathbf{r})$ for values of $|\mathbf{r}|$ which exceed the radius of the sphere circumscribing the cell. It should be noted that any amount of the regular solution can be added to the irregular solution and it will still satisfy the Schrödinger equation as well as Eq.(G.12).

Appendix H
Displacement of Irregular Solutions

The representability theorem, proved in Chapter 4 in connection with the regular solutions of the Schrödinger equation, can be extended to the irregular solutions as well. The theorem allows one to expand the wave function (or basis functions) inside a given cell in terms of basis functions associated with adjacent cells. For example consider two cells, 1 and 2, as depicted schematically in Fig. H.1, and set the potential inside both equal to zero. The solution of the Schrödinger equation inside either cell corresponding to a given L is $J_L(\mathbf{r})$. The wave function for the system (say a plane wave) can be written as a linear combination of such solutions. Now, the solution $J_L(\mathbf{r}_1)$ associated with cell 1, as indicated in the figure, can be expanded in terms of solutions $J_L(\mathbf{r}_2)$ centered about the geometric origin of cell 2,

$$J_L(\mathbf{r}_1) = \sum_{L'} g_{LL'}(\mathbf{R}_{12}) J_{L'}(\mathbf{r}_2). \tag{H.1}$$

The expansion in this equation holds for arbitrary values of the vectors $\mathbf{r}_1$, $\mathbf{r}_2$, and $\mathbf{R}_{12}$.

Now, let us consider the expansion of an irregular solution, say $H_L(\mathbf{r}_1)$. For values of $\mathbf{r}_2$ confined inside the sphere inscribed in cell 2.[1] In this case, we have

$$H_L(\mathbf{r}_1) = \sum_{L'} G_{LL'}(\mathbf{R}_{12}) J_{L'}(\mathbf{r}_2). \tag{H.2}$$

---

[1]Actually, the expansion holds for vectors $\mathbf{r}_2$ that lie inside a sphere centered at the center of cell 2 and has radius smaller than $|\mathbf{R}_{12}|$.

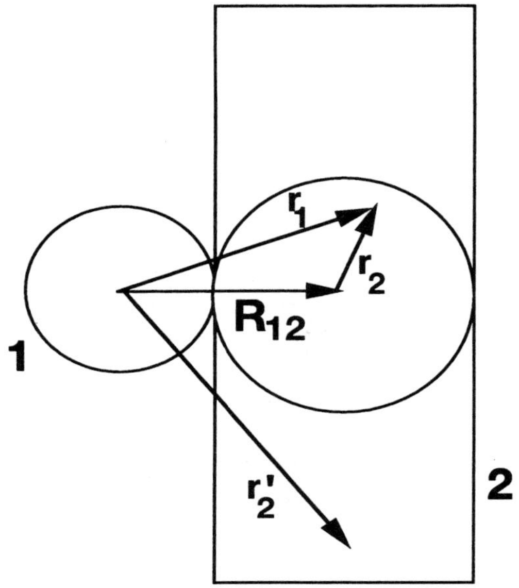

FIGURE H.1. Two adjacent cells in free space.

However, for vectors $\mathbf{r}_2$ confined inside cell 2 but whose length exceeds that of $\mathbf{R}_{12}$, such as vector $\mathbf{r}_2'$ in the figure, the last expansion diverges.

This divergence can be made to disappear if the representability theorem is viewed within the context of conditionally convergent summation procedures. Instead of expanding in the manner indicated in the last expression, we write

$$H_L(\mathbf{r}_1) = \sum_{L'} g_{LL'}(\mathbf{b}) \sum_{L''} \Big\{ G_{L'L''}(\mathbf{R}_{12} + \mathbf{b}) J_{L''}(\mathbf{r}_2) \Big\}. \tag{H.3}$$

The choice of the vector $\mathbf{b}$ in the expansion above has been discussed at length in the body of the book (Chapter 9). This vector is chosen so that the double sum in the last expression converges for all vectors in cell 2 if carried out in the order indicated by the brackets.

Now, let the internal sum in Eq.(H.3), that over L'', be taken to infinity, thus always exceeding any finite value of the outer sum, the sum over L'. Then, we can write

$$H_L(\mathbf{r}_1) = \lim_{L' \to \infty} \sum_{L'} a_{LL'} J_{L'}(\mathbf{r}_2), \tag{H.4}$$

where

$$a_{LL''} = \sum_{L''} g_{LL'}(\mathbf{b}) \Big\{ G_{L'L''}(\mathbf{R}_{12} + \mathbf{b}) \Big\}, \tag{H.5}$$

where in practical terms, L' never exceeds the value of L'' and may need to be constrained to values considerably smaller than L''. Equation (H.4) has precisely the form of the representability theorem; it allows the expansion of the solutions in a given cell to be written in terms of the *regular* solutions associated with an adjacent cell. When interpreted within the formal construct of a convergent process, the representability theorem applies both to regular and irregular solutions.

Appendix I
The Inverse t-Matrix at Large L

The inverse t-matrix defined in Eq.(5.95) may be simplified if ℓ is sufficiently large. In order to demonstrate this simplification it is convenient to write Eq.(5.95)as

$$m_\ell = -\left(\frac{\gamma_\ell^{\tilde{h}} - \gamma_\ell^R}{\gamma_\ell^{\tilde{j}} - \gamma_\ell^R}\right)\frac{\tilde{h}_\ell}{\tilde{j}_\ell}, \tag{I.1}$$

where the various logarithmic derivatives are defined by

$$\gamma_\ell^{\tilde{h}} = \frac{z^0 \tilde{h}'_\ell(z^0)}{\tilde{h}_\ell(z^0)} = \frac{\ell - z^0 \tilde{h}_{\ell+1}}{\tilde{h}_\ell,} \tag{I.2}$$

$$\gamma_\ell^{\tilde{j}} = \frac{z^0 \tilde{j}'_\ell(z^0)}{\tilde{j}_\ell(z^0)} = \frac{\ell + z^0 \tilde{j}_{\ell+1}}{\tilde{j}_\ell,} \tag{I.3}$$

and

$$\gamma_\ell^R = r_n^0 \left[\frac{\mathrm{d}R_\ell(r,E)/\mathrm{d}r}{R_\ell(r,E)}\right]_{r=r0}. \tag{I.4}$$

For square-well potentials the logarithmic derivative of the radial wave function is given by

$$\gamma_\ell^R = \frac{z_s j'_\ell(z_s)}{j_\ell(z_s)} = \frac{\ell - z_s j_{\ell+1}(z_s)}{j_\ell(z_s)} \tag{I.5}$$

where $z_s = \sqrt{E+V} r^0$.

In the limit of large ℓ these logarithmic derivatives can be approximated by

$$\gamma_\ell^{\tilde{h}} \to -(\ell+1); \qquad \ell >> \kappa r^0, \tag{I.6}$$

$$\gamma_\ell^{\tilde{j}} \to \frac{\ell - E r^{0 2}}{2\ell+3}; \qquad \ell >> \kappa r^0, \tag{I.7}$$

and for square-well potentials

$$\gamma_\ell^R \to \frac{\ell - (E+V) r^{0 2}}{2\ell+3}; \qquad \ell >> \sqrt{E+V} r^0. \tag{I.8}$$

The approximate expression for m_ℓ for square-well potentials given in the text, Eqs. (5.126) and (5.127), follows from substituting Eqs.(I.6)–(I.8) in Eq.(I.1).

Appendix J
The Structure Constants at Large L

The structure constants $G^m_{\ell,\ell'}$ satisfy the following expansion property,

$$G^m_{\ell,\ell'}(R-r) = \sum_{\ell''} f^m_{\ell,\ell''}(r) G^m_{\ell'',\ell'}(R), \tag{J.1}$$

where both vectors $\mathbf{R}$ and $\mathbf{r}$ are along the z-axis and the matrix $f^m_{\ell,\ell'}(r)$ is defined by

$$f^m_{\ell,\ell'}(r) = (-1)^{\ell-\ell'}\sqrt{(2\ell+1)(2\ell'+1)} \sum_{\ell''=|\ell-\ell'|}^{\ell+\ell'} (2\ell''+1) d^m(\ell,\ell',\ell'')\tilde{j}_{\ell''}(\kappa r). \tag{J.2}$$

Now we use $\tilde{j}_\ell(\kappa r) \to \frac{(\kappa r)^\ell}{(2\ell+1)!!}, \quad r \to 0$, and

$$d^m(\ell,\ell-1,1) = \frac{\sqrt{(\ell-m)(\ell+m)}}{(2\ell-1)(2\ell+1)} \tag{J.3}$$

to obtain

$$\left.\frac{\mathrm{d}f^m_{\ell,\ell'}(r)}{\mathrm{d}r}\right|_{r=0} = -\kappa\left[\sqrt{\frac{(\ell-m)(\ell+m)}{(2\ell-1)(2\ell+1)}}\delta_{\ell,\ell'+1} + \sqrt{\frac{(\ell'-m)(\ell'+m)}{(2\ell'-1)(2\ell'+1)}}\delta_{\ell+1,\ell'}\right]. \tag{J.4}$$

We take the derivative of Eq.(J.1),

$$\mathrm{d}G^m_{\ell,\ell'}(R)\mathrm{d}R = -\mathrm{d}G^m_{\ell,\ell'}(R-r)\mathrm{d}r\Big|_{r=0}$$

$$= - \sum_{\ell''=|\ell-\ell'|}^{\ell+\ell'} \mathrm{d}f^m_{\ell,\ell''}(r)\mathrm{d}r\Big|_{r=0} G^m_{\ell'',\ell'}(R)$$

$$= \kappa\sqrt{(\ell-m)(\ell+m)(2\ell-1)(2\ell+1)}G^m_{\ell-1,\ell'}(R)$$

$$+ \kappa\sqrt{\frac{(\ell-m+1)(\ell+m+1)}{(2\ell+1)(2\ell+3)}}G^m_{\ell+1,\ell'}(R) \tag{J.5}$$

to obtain a recursion relation for $G^m_{\ell,\ell'}$,

$$G^m_{\ell+1,\ell'}(R) = \sqrt{\frac{(2\ell+1)(2\ell+3)}{(\ell-m+1)(\ell+m+1)}}\frac{\mathrm{d}G^m_{\ell,\ell'}(R)}{\mathrm{d}(\kappa R)}$$

$$-\sqrt{\frac{(2\ell+3)(\ell-m)(\ell+m)}{(2\ell-1)(\ell-m+1)(\ell+m+1)}}G^m_{\ell-1,\ell'}(R). \tag{J.6}$$

We have for $\ell = m$,

$$G^m_{m,\ell'}(R) = (-1)^m\sqrt{(2m+1)(2\ell'+1)}$$

$$\times \sum_{\ell''=\ell'-m}^{\ell'+m} (2\ell''+1)d^m(m,\ell',\ell'')\tilde{h}_{\ell''}(\kappa R), \tag{J.7}$$

from which all $G^m_{\ell,\ell'}$ can be obtained. In the large ℓ limit, with the azimuthal quantum number m fixed, the dominant term in the above equation comes from $\ell'' = \ell' + m$,

$$d^m(m,\ell',\ell'+m) = \frac{(-1)^m(2\ell'-1)!!}{(2\ell'+2m-1)!!}\sqrt{\frac{(2\ell'+1)(2m+1)!!(\ell'+m)!}{(2\ell'+2m+1)(2m)!!(\ell'-m)!}}, \tag{J.8}$$

and, therefore, we obtain the approximate expression

$$G^m_{m,\ell'}(R) \approx \sqrt{\frac{(2\ell'+1)!!(2\ell'-1)!!(2m+1)!!}{(\ell'+m)!(\ell'-m)!(2m)!!}}\frac{1}{(\kappa R)^{\ell'+m+1}}. \tag{J.9}$$

In the limit $\ell >> m$, the recursion relation for $G^m_{\ell,\ell'}$ yields

$$G^m_{\ell,\ell'}(R) \approx \sqrt{\frac{(2\ell-1)(2\ell+1)}{(\ell-m)(\ell+m)}}\frac{dG^m_{\ell-1,\ell'}(R)}{d(\kappa R)}$$

$$\approx \frac{(2\ell-1)!!}{(2m-1)!!}\sqrt{\frac{(2\ell+1)(2m)!}{(2m+1)(\ell-m)!(\ell+m)!}}\frac{d^{\ell-m}G^m_{m,\ell'}(R)}{d(\kappa R)^{\ell-m}}. \tag{J.10}$$

Substituting in Eq.(J.9) and using the Stirling formula for the factorials, one immediately obtains Eq.(5.128) for $G^m_{\ell,\ell'}$.

Appendix K
Conversion of Volume Integrals

The technique to convert the volume integral of the radial wave function for a muffin-tin potential into a surface integral can be generalized to non–muffin-tin cases. The non–muffin-tin formula is needed here because of the need to integrate the wave function outside both square wells. We start from the Schrödinger equation at energy E,

$$\left\{\nabla^2 + E - V(\mathbf{r})\right\}\psi(\mathbf{r}, E) = 0 \tag{K.1}$$

and at energy $E + \delta E$,

$$\nabla^2\psi^*(\mathbf{r}, E+\delta E) + \{E + \delta E - V(\mathbf{r})\}\psi^*(\mathbf{r}, E+\delta E) = 0. \tag{K.2}$$

We now multiply the first equation by $\psi^*(\mathbf{r}, E + \delta E)$ and subtract the second equation multiplied by $\psi(\mathbf{r}, E)$, to obtain,

$$\psi(\mathbf{r}, E)\psi^*(\mathbf{r}, E{+}\delta E)\delta E = \psi(\mathbf{r}, E)\nabla^2\psi^*(\mathbf{r}, E{+}\delta E) - \psi^*(\mathbf{r}, E{+}\delta E)\nabla^2\psi(\mathbf{r}, E). \tag{K.3}$$

After taking the limit $\delta E \to 0$, we integrate the above equation over the volume of the cell and use Green's theorem, and finally obtain,

$$\int \mathrm{d}^3\mathbf{r}|\psi(\mathbf{r})|^2 = \oint \mathrm{d}^2S\hat{n}\cdot\left[\frac{\partial\psi(\mathbf{r})}{\partial E}\nabla \;-\; \nabla\frac{\partial\psi(\mathbf{r})}{\partial E}\right]\psi(\mathbf{r}) \equiv \left[\frac{\partial\psi(\mathbf{r})}{\partial E}, \psi(\mathbf{r})\right]_S . \tag{K.4}$$

One should be careful when using this formula, especially for bound states. Because the energy derivative of the Schrödinger equation can be taken only when no boundary conditions are specified, the boundary of the integration must be where the boundary conditions are matched to avoid including discontinuities of the wave functions into the integral.

Appendix L
Energy Derivatives

In this appendix we derive an expression that relates $\dot{G}^{ij} = dG^{ij}/dE$ to Wronskian-like surface integrals involving J, H, $\dot{J}$, and $\dot{H}$. Here G^{ij} is a real-space KKR structure constant, or, more precisely, it is the expansion coefficient of an irregular solid harmonic centered at site i in terms of regular solid harmonics centered at site j,

$$H_L^i = H_L(\mathbf{r}_i) = \sum_{L'} G_{LL'}(\mathbf{R}_{ij}) J_{L'}(\mathbf{r}_j). \tag{L.1}$$

In the above equation $\mathbf{r}_i = \mathbf{R}_{ij} + \mathbf{r}_j$ and the expansion is convergent if $R_{ij} > r_j$.

Using a dot to represent differentiation with respect to E and a notation that suppresses the angular-momentum indices, we have

$$\dot{H}^i = \dot{G}^{ij} J^j + G^{ij} \dot{J}^j. \tag{L.2}$$

Forming a Wronskian surface integral with H^j over cell j and using the fact that $[J^i, H^j] = \delta_{ij}$ we have

$$-\dot{G}^{ij} = G^{ij}[\dot{J}^j, H^j]_j - [\dot{H}^i, H^j]_j. \tag{L.3}$$

The second term on the right-hand side of the above equation is an integral that encloses cell j. It can equally well be written as an integral over the surfaces of all cells other than cell j with the assumption that the integral over the surface at infinity vanishes. At negative energy this is certainly valid since both H^i and H^j vanish exponentially. We shall assume that it is generally valid since H is regular at infinity. Thus

$$- [\dot{H}^i, H^j]_j = \sum_{k \neq j} [\dot{H}^i, J^k]_k G^{kj}, \tag{L.4}$$

which gives us

$$- \dot{G}^{ij} = G^{ij}[\dot{J}^j, H^j]_j + [\dot{H}^i, J^i]_i G^{ij} + \sum_{k \neq i,j} [\dot{H}^i, J^k]_k G^{kj}. \tag{L.5}$$

Finally we use Eq.(L.2) again and obtain

$$- \dot{G}^{ij} = G^{ij}[\dot{J}^j, H^j]_j + [\dot{H}^i, J^i]_i G^{ij} + \sum_{k \neq i,j} G^{ik}[\dot{J}^k, J^k]_k G^{kj}. \tag{L.6}$$

Appendix M
Convergence of the Secular Matrix

Here we summarize our previous discussion on the convergence of the secular equation. We have seen that under certain conditions on the lengths of vectors one can derive a secular equation that has the MT form, for example, $1 - [Gt] = 0$. This form, however, may not be the most convenient to implement in a computational procedure for evaluating the scattering path operator, for example, $\tau = t[1 - [Gt]]^{-1}$ or the spectrum of eigenvalues. If, for example, an expansion in terms of the t-matrix,

$$\tau_{ij} = t_i\delta_{ij} + t_i[G_{ij}t_j](1 - \delta_{ij}) + t_i \sum_k [G_{ik}t_k][G_{kj}t_j] + \cdots \tag{M.1}$$

is used to compute the scattering path operator, divergences in angular-momentum summations may result. It may be useful, therefore, to give expressions that are known to converge if sums are performed in the proper order. Because a collection of scatterers satisfying the restrictions of the MT geometry can be described by the usual MT form of the secular equation (with appropriate convergence induced in the nonspherical case), we discuss two cases that violate the MT conditions.

First, we consider the case in which intracell vectors, $\mathbf{r}$, are smaller than intercell vectors, $|\mathbf{R}| > \mathbf{r}$. In this case the convergence of Eq. (M.1) indicates that the secular matrix can be written in the form $\tau = M^{-1} = t[1-[Gt]]^{-1}$. The bracket around the product Gt indicates that the internal summation in this product must be carried to convergence *before* the inverse is computed. Alternatively, the external angular momentum indices in M must not exceed those of the internal summation in Gt and must be kept sufficiently smaller so as not to change the order in which sums are performed.

Now, the convergence of $\tau_{LL'}$ in terms of L and L' must be studied with the internal sum in Gt carried out to higher values than L and L' to guarantee convergence. Under these conditions, the secular equation takes the form

$$\det[1 - [Gt]] = 0. \tag{M.2}$$

Again, the internal summations in Gt must exceed the values of L entering the calculations of the determinant by a sufficient margin to guarantee convergence.

Second, we consider the case of convex cells in which some intracell vectors may exceed the distance between cells, $r > R$. In this case, the secular matrix can be written in the form

$$\tau = t\left[1 - g(\mathbf{a})[G(\mathbf{R} + \mathbf{a})t]\right]^{-1}. \tag{M.3}$$

Again, the values of L in the sums enclosed in brackets, for example, $[Gt]$, must exceed thise in other summations or external indices to guarantee convergence. The secular equation can be written in the form,

$$\det\left[1 - g(\mathbf{a})[G(\mathbf{R} + \mathbf{a})t]\right] = 0. \tag{M.4}$$

It is to be noted that all of these expressions involve the complete separation of the potential from the structure of a scattering assembly. Even though the last two expressions may be computationally more cumbersome than the corresponding expressions in the MT case, they show clearly that, at least formally, MST retains its basic and useful feature of separability even in the case of space-filling, nonspherical potentials.

Appendix N
Summary of MST

In this appendix we summarize the basic equations of MST.

N.1 General Framework

The time-independent Schrödinger equation (in atomic units),

$$\left[-\nabla^2 + V(\mathbf{r})\right]\psi(\mathbf{r}) = E\psi(\mathbf{r}), \tag{N.1}$$

can be written as the integral Lippmann–Schwinger equation

$$\psi(\mathbf{r}) = \chi(\mathbf{r}) + \int G_0(\mathbf{r}-\mathbf{r}')V(\mathbf{r}')\psi(\mathbf{r}')\mathrm{d}^3r', \tag{N.2}$$

where $\chi(\mathbf{r})$ is a solution of the homogeneous equation,

$$(\nabla^2 + E)\chi(\mathbf{r}) = 0, \tag{N.3}$$

and $G_0(\mathbf{r}-\mathbf{r}')$ is the free-particle propagator

$$G_0(\mathbf{r}-\mathbf{r}') = -\frac{1}{4\pi}\frac{\mathrm{e}^{\mathrm{i}k|\mathbf{r}-\mathbf{r}'|}}{|\mathbf{r}-\mathbf{r}'|}, \tag{N.4}$$

with $k = \sqrt{E}$. In the angular-momentum representation, we have the expansion

$$G_0(\mathbf{r}-\mathbf{r}') = \sum_L J_L(\mathbf{r}_<)H_L(\mathbf{r}_>), \tag{N.5}$$

where $J_L(\mathbf{r}) = j_\ell(kr)Y_L(\hat{r})$ and $H_L(\mathbf{r}) = h_\ell^+(kr)Y_L(\hat{r})$, with $j_\ell(kr)$ and $h_\ell^+(kr)$ being the spherical Bessel function and the Hankel function of the first kind, respectively, and $Y_L(\hat{r})$ is a spherical harmonic of order L. In Eq.(N.5), $\mathbf{r}_>$ ($\mathbf{r}_<$) denotes the larger (smaller) of the vectors $\mathbf{r}$ and $\mathbf{r}'$. The functions $J_L(\mathbf{r})$ and $H_L(\mathbf{r})$ are the regular and irregular solutions, respectively, of the free-particle Schrödinger equation. The Neumann function, $N_L(\mathbf{r}) = n_\ell(kr)Y_L(\hat{r})$, can also be used in place of the Hankel function.

N.2 Single Potential

The solution of Eq.(N.1) for a single, spatially bounded potential can be written in the form,

$$\psi(\mathbf{r}) = \sum_L A_L \psi_L(\mathbf{r}), \tag{N.6}$$

where

$$\psi_L(\mathbf{r}) = \sum_{L'} \left[c_{LL'}(r) J_L(\mathbf{r}) - s_{LL'}(r) H_L(\mathbf{r}) \right], \tag{N.7}$$

with the phase function matrices, $\underline{c}(r)$ and $\underline{s}(r)$, being determined by the equations,

$$\begin{aligned} \frac{\mathrm{d}c_{L'L}(r)}{\mathrm{d}r} &= \int_r \mathrm{d}\Omega H_{L'}(\mathbf{r}')V(\mathbf{r}')\psi_L(\mathbf{r}') \\ \frac{\mathrm{d}s_{L'L}(r)}{\mathrm{d}r} &= \int_r \mathrm{d}\Omega J_{L'}(\mathbf{r}')V(\mathbf{r}')\psi_L(\mathbf{r}'). \end{aligned} \tag{N.8}$$

Here, $|\quad\rangle$ denotes a vector whose components are indexed by L, as discussed in Appendix D. Outside a sphere bounding the potential, the phase functions assume their asymptotic constant values, $\underline{C}(k)$ and $\underline{S}(k)$, where the dependence on energy is explicitly noted. In this asymptotic region, the wave function can be written in the form,

$$|\psi(\mathbf{r})\rangle = \underline{C}(k)|J_L(\mathbf{r})\rangle - \underline{S}(k)|H_L(\mathbf{r})\rangle. \tag{N.9}$$

Other forms for $|\psi(\mathbf{r})\rangle$ include

$$|\psi(\mathbf{r})\rangle = \underline{C}'(k)|J_L(\mathbf{r})\rangle - \underline{S}'(k)|N_L(\mathbf{r})\rangle, \tag{N.10}$$

and

$$|\psi(\mathbf{r})\rangle = J_L(\mathbf{r})\rangle + \underline{t}(k)|H_L(\mathbf{r})\rangle, \tag{N.11}$$

where

$$\begin{aligned} \underline{t} &= \left[\underline{C}'(k)\right]^{-1} \underline{S}'(k) \\ &= -\left[\underline{S}^{-1}(k)\underline{C}(k) + \mathrm{i}\right]^{-1}. \end{aligned} \tag{N.12}$$

The quantity $\underline{t}$ is commonly referred to as the cell t-matrix. This quantity is one of the basic building blocks of MST. The t-matrix can also be determined as the solution of the equation (with a dot denoting a derivative with respect to r)

$$\dot{\underline{t}}(r) = \int_r \mathrm{d}\Omega |Z_L(\mathbf{r})\rangle V(\mathbf{r}) \langle Z_L(\mathbf{r})|, \tag{N.13}$$

where $|Z_L(\mathbf{r})\rangle$ is the regular solution of the Schrödinger equation which on the surface of the circumscribing sphere joins smoothly to the quantity,

$$|Z_L(\mathbf{r})\rangle \to (\underline{t})^{-1}|J_L(\mathbf{r})\rangle - \mathrm{i}k|H_L(\mathbf{r})\rangle. \tag{N.14}$$

The Green function for a single scattering potential can be written in the form

$$G(\mathbf{r},\mathbf{r}') = G_1(\mathbf{r},\mathbf{r}') + \langle Z_L(\mathbf{r})|\underline{t}|Z_L(\mathbf{r}')\rangle. \tag{N.15}$$

Here, $G_1(\mathbf{r},\mathbf{r}')$ is given by the expression,

$$G_1(\mathbf{r},\mathbf{r}') = \langle Z_L(\mathbf{r})|S_L(\mathbf{r}')\rangle, \tag{N.16}$$

where $S_L(\mathbf{r})$ is that irregular solution of the Schrödinger equation which on the surface of the bounding sphere joins smoothly to $|J_L(\mathbf{r})\rangle$. In Eq.(N.15), we have set $r' > r$. The particular forms that these results assume in the case of central potentials are given in Chapter 3.

N.3 Multiple Scattering Theory

The potential $V(\mathbf{r})$ is taken to be the sum of nonoverlapping but otherwise arbitrarily shaped, spatially bounded potentials confined in cells Ω_n,

$$V(\mathbf{r}) = \sum_n v_n(\mathbf{r} - \mathbf{R}_n), \tag{N.17}$$

where $\mathbf{R}_n$ denotes the center or origin of cell Ω_n. According to the representability theorem, the wave function in cell Ω_n is now written as a linear combination,

$$\psi^n(\mathbf{r}) = \sum_L A_L^n \psi_L^n(\mathbf{r}), \tag{N.18}$$

where $\psi_L^n(\mathbf{r})$ are the single-cell functions defined in Eq.(N.7). The coefficients, A_L^n, are found as the solutions of the set of homogeneous equations,

$$\sum_m \underline{M}^{nm}|A^m\rangle = 0, \tag{N.19}$$

where $\underline{M}^{nm}$ is a matrix in angular-momentum space

$$M_{LL'}^{nm} = m_{LL'}^n \delta_{nm} - G_{LL'}(\mathbf{R}_{nm})(1 - \delta_{nm}), \tag{N.20}$$

where $\underline{m}^n = \underline{t}^{n-1}$ is the inverse t-matrix for cell Ω_n, and $G_{LL'}(\mathbf{R}_{nm})$ are the real-space structure constants associated with the intercell vectors, $\mathbf{R}_{nm}$. For a definition of the structure constants, see Appendix D. Equation (N.19) has a nontrivial solution for the coefficients $|A^n\rangle$ when the determinant of the matrix $\underline{M}^{nm}$ vanishes,

$$\det|\underline{M}| = 0. \tag{N.21}$$

This is one form of the *secular equation* of MST, whose eigenvalues determine the electronic structure of the system described by the potential $V(\mathbf{r})$.

For a translationally invariant solid, the secular equation can be written in the form

$$|\underline{m} - \underline{G}(\mathbf{k})| = 0, \tag{N.22}$$

where $\underline{G}(\mathbf{k})$ is the lattice Fourier transform of $\underline{G}(\mathbf{R}_{nm})$. The $\underline{G}(\mathbf{k})$ are commonly referred to as the Korringa–Kohn–Rostoker (KKR) structure constants. A number of alternative forms of the secular equation are derived and discussed in the body of the book.

Finally, the Green function for a system of scattering potentials takes the form

$$G(\mathbf{r}, \mathbf{r}') = G_1(\mathbf{r}, \mathbf{r}') + \langle Z_L(\mathbf{r})|\underline{\tau}^{nm}|Z_L(\mathbf{r}')\rangle \tag{N.23}$$

where $\mathbf{r}$ and $\mathbf{r}'$ are taken to lie in cells Ω_n and Ω_m, respectively, and $\underline{\tau}^{nm}$ is the nmth matrix element of the inverse of the matrix $\underline{M}^{nm}$. The quantity $\underline{\tau}^{nm}$ is the so-called *scattering path operator*, and formally satisfies the *equation of motion*,

$$\underline{\tau}^{nm} = \underline{t}^n \left[\delta_{nm} + \sum_{k \neq n} \underline{G}(\mathbf{R}_{nk})\underline{\tau}^{km}\right]. \tag{N.24}$$

The equation of motion provides a convenient starting point for the application of MST to a great number of physical problems.

Index